PAPER MACHINE CLOTHING

Sabit Adanur, B.S., M.S., Ph.D.

Associate Professor, Department of Textile Engineering
Auburn University, Alabama, U.S.A.

CRC Press

Taylor & Francis Group
Boca Raton London New York

CRC Press is an imprint of the
Taylor & Francis Group, an **informa** business

First published 1997 by Technomic Publishing Company, Inc.

Published 2019 by CRC Press
Taylor & Francis Group
6000 Broken Sound Parkway NW, Suite 300
Boca Raton, FL 33487-2742

© 1997 by Taylor & Francis Group, LLC
CRC Press is an imprint of Taylor & Francis Group, an Informa business

First issued in paperback 2019

No claim to original U.S. Government works

ISBN 13: 978-0-367-44816-5 (pbk)
ISBN 13: 978-1-56676-544-2 (hbk)

**Visit the Taylor & Francis Web site at
http://www.taylorandfrancis.com**

**and the CRC Press Web site at
http://www.crcpress.com**

Library of Congress Catalog Card No. 97-60981

Table of Contents

Foreword

Colonial papermakers would truly be amazed at the technological achievements that the paper industry has attained. Asten is fortunate to have had a major role in ensuring the success of those achievements.

Starting more than 120 years ago with simple weave structures manufactured from metal, wool, and cotton, Asten has evolved its paper machine clothing into sophisticated products made from synthetic fibers with highly engineered properties for distinct paper machine applications. These new products and technologies, combined with modern and advanced manufacturing techniques, have offered tremendous opportunities for sophisticated paper machine clothing products to meet the demanding needs of today's paper machines.

Asten has played an instrumental role in the technological shift from the art of papermaking to the science of papermaking. As papermaking processes have become more sophisticated and demanding, Asten has continually demonstrated its leadership by providing the highest quality, innovative products and services available anywhere in the paper machine clothing industry. Today our dynamic organization is providing a full range of paper machine clothing in every major global market.

The management of Asten is proud to present the first edition of *Paper Machine Cloth-ing*. We are delighted and fortunate that the book is authored by a former Asten associate, Dr. Sabit Adanur. Also, we are grateful to all Asten associates who contributed significantly to the publication of this book.

Dr. Adanur holds a Ph.D. in Fiber Polymer Science and an M.S. degree in Textile Engineering and Science from North Carolina State University, as well as a B.S. degree in Mechanical Engineering from Istanbul Technical University. Currently a professor in the Textile Engineering Department at Auburn University, Dr. Adanur spent three years at Asten as manager of Research and Development.

This first edition of *Paper Machine Clothing* is designed as a comprehensive resource for managers, engineers, professionals and production specialists who work directly or indirectly with the papermaking process. The book contains the latest technology available in the profession.

Asten is proud to be part of an alliance with the industry it serves to help introduce these new paper technologies that have kept our industry on the leading edge of a global and competitive paper industry.

We hope you will find this book educational and informative. Furthermore, we truly hope that it will be used to further advance the science of papermaking.

WILLIAM A. FINN
Asten, Inc.
President and CEO

Introduction

The first two basic necessities of *civilization* are textiles and paper. Assuming that a person's stomach is full (food is a *must* for survival not a *necessity* for civilization), his next need is a piece of cloth to cover his body to look more civilized to the eye and protect himself from the adverse effects of the environment. The second necessity to be further civilized is a piece of paper to write on or to read from to communicate and to extend his knowledge as well as to record history. There has been a direct correlation between a nation's standard of living and the amount of textiles and paper used.

Although textiles and paper are in general different products, they have some interesting commonalities between them. They both use fibers in their final products. Both of their manufacturing processes are a mixture of science and art, two of the most ancient arts, indeed. Neither textile nor paper manufacturing processes are positively controlled. That is, in either case it is hardly possible to control the individual fiber which is the smallest meaningful building block in the structure. Moreover, there are some products that may be classified as both nonwoven textiles and paper, depending on which industry you are in. These products link the two industries together. There are of course some differences between a textile product and paper. Textiles in general use longer fibers than paper. Paper uses mostly wood and other plant fibers; textiles use both natural and man-made fibers.

Nonwoven paper and textile products mostly use synthetic fibers. In a textile product, frictional and mechanical forces are the major forces that hold the textile fibers together giving the structural integrity to the product. In paper, relatively short fibers are chemically bonded together with hydrogen bonds.

Another link between the textile industry and paper industry is paper machine clothing. The paper machine clothing industry involves both textile and paper aspects which makes it fascinating. Paper machine clothing is an integral part of the papermaking process. Paper and related products such as books, newspapers, magazines, tissue, towels, bags and currency, etc., cannot be made without a textile fabric with today's technology. Therefore, textile fabrics are essential to the well being of the paper industry. Textile fabrics are also essential in the manufacture of other textile products such as nonwoven textiles.

Although this book is mainly about paper machine clothing, it also has some concise information about papermaking, paper structure and properties. Therefore, this book may serve as a bridge between the textile industry and the paper industry. Paper topics related to paper machine clothing are included for convenience to the reader. Highly sophisticated structures and properties of paper machine clothing cannot be fully comprehended or appreciated without some knowledge of pulp and paper technology, structure and properties. A further similarity between textiles and paper is seen in the terminology that is used in both industries. Some terms are used in both industries with the same or completely different meanings.

Chapter 1 is a brief introduction to pulp and paper technology. Major steps in pulp and paper manufacturing and technology are summarized. A brief history of papermaking is given.

Chapter 2 is about formation. The sheet formation mechanism is described. Design, manufacturing, testing, application and service of forming fabrics are explained in detail. Pressing is covered in Chapter 3. Theory of water removal is explained. Design, manufacturing, testing and service of press fabrics are given in detail. Drying is the subject of Chapter 4. The role of dryer fabrics in paper drying is explained. Dryer fabric design, manufacturing, testing and applications are covered.

Chapter 5 explains the paper machine auditing for forming, press and dryer sections. It gives insight to the papermaker about the type of analysis that can be done on a paper machine. Paper structure, properties and testing are covered in Chapter 6, whose main purpose is to give a concise summary to the reader about this wide area. This is helpful in design and manufacturing of paper machine clothing.

In today's highly competitive environment, no manufacturing is meaningful without good product quality and any book related to paper machine clothing and papermaking would be incomplete without a chapter on quality. Therefore, Chapter 7 is devoted to Total Quality Management (TQM) and Statistical Process Control (SPC) which are the tools used to improve quality. ISO 9000 and 14000 standards are also explained. Another equally important area, computer applications, is included in Chapter 8. A terminology of the state-of-the-art computer technology is given. Application of computers is briefly discussed. An attempt is made in Chapter 9 to review the trends for the future of papermaking and paper machine clothing.

Units present a challenge in writing a book in this field. Although desired, use of metric units throughout the book is neither practical nor convenient to the reader. Therefore, both metric and standard units are used in the book,

whichever made more "sense" in the current industry practice. A unit conversion table is given in Appendix O to convert units from one system to the other.

This book could not have been a reality without help. I thank God for everything that made it possible. I would like to thank Mr. William A. Finn, who made his company's human and technical resources freely available for the successful outcome of the book. Mr. Daniel D. Cappell provided the necessary technical information for the book and offered editorial comments. Special thanks are extended to Prof. Dr. William K. Walsh, Textile Engineering Department Head, and Prof. Dr. William F. Walker, Dean of the College of Engineering, for their leadership and support. Last but not least, this book would not have been possible without my wife Nebiye's continuous help and support.

I would like to acknowledge many people, institutions, universities and companies who have contributed to the book by providing information, pictures, graphs and data. Each chapter has been reviewed and edited by several professionals for technical content which is greatly appreciated. I would like to thank the following individuals for their help in preparation of this book: Charlie Abraham, Dave Antos, Elwood Beach, Bill Boyce, Thomas Butler, Payton Crosby, Frank Cunnane, Mark Davis, Tom Durkin, Ted Fry, Mike Harvey, Adolphe Hermens, Sam Herring, Dieter Kuckart, Bob Ledet, Ken McCumsey, Bob McIntire, Karen Moore, James Nicholaou, Prof. Dr. Gerald Ring, Ralph Sieberth, Marcel Siquet, Bill Summer, Paul Sutherland, W. Thommen, Glen Townley, Rex Treece, Lennart Vihma and Dietmar Wirtz. My sincere apologies to those whose names I might have failed to mention for their valuable time and contributions.

Since the amount of paper consumed is an indication of a nation's standard of living, the future of paper industry and paper machine clothing is bright. My hope is that this book will be useful for industry and academic professionals as well as students in shaping that future.

SABIT ADANUR

Overview of Pulp and Paper Technology

The raw material for paper is fiber. Pulping is the process of separating the fibers, suitable for papermaking, from the non-fibrous material of wood or other fibrous sources. Papermaking is the process of consolidating individual fibers in the pulp into an integrated sheet structure. Since fibers can not be used in pulp form, paper manufacturing is a natural continuation of pulp manufacturing. In other words, pulp is an intermediate product for the paper which is the final product. Figure 1.1 shows the main steps in pulp and paper manufacturing. Pulp and paper manufacturing can be done at the same location or different locations. Figure 1.2 (page 7) shows a pulp and paper mill. A complete cycle of the pulp and paper technology is shown in Figure 1.3 (page 9).

Figure 1.4 (page 10) shows the schematics of the steps in pulp and paper manufacturing using the chemical pulping method. Although the manufacture of pulp and paper is not the main topic of this book, a brief summary of these processes will be given in the following sections of this chapter for reader convenience. Since paper machine clothing is directly used during papermaking, paper machines and papermaking process are covered in more detail throughout the book. There are excellent books regarding the pulp and paper manufacturing. The reader is referred to those sources for more in-depth information on these subjects.

Although there are many types of fibers that are suitable for papermaking, wood fibers are the most widely used (Section 6.1, Chapter 6). Wood is a renewable source which is a big advantage for paper industry compared to other industries whose resources may not be renewable. Every year, billions of trees are planted to ensure a continuing source of wood (Figure 1.5 on page 11). In fact, the commercial forest growth exceeds the harvest and natural tree mortality in the United States. For example, in 1994, some 1.5 billion seedlings were planted in the United States. As a result, there are 20% more trees in the United States today than just 20 years ago. The following discussion is given for wood pulp and paper manufacturing.

1.1 Pulp Manufacturing

Pulp is defined as "fibrous material produced either chemically or mechanically (or by some combination of chemical and mechanical means) from wood or other cellulosic raw material" [2].

The wood cell has a nonliving cell wall made of cellulose fibers, hemicellulose and lignin which give strength and support to the cell wall. Cellulose is a carbohydrate, i.e., it is made of carbon, hydrogen and oxygen. It does not dissolve in water and its surface is hydrophilic. Cellulose imparts high tensile stiffness and strength to the structure. The amount of cellulose in the native wood fiber is less than 50%. Lignin holds the cellulose

FIGURE 1.1. Main steps in pulp and paper manufacturing.

fibers together in the cell wall. Therefore, lignin must be removed to separate the individual cellulose fibers which eventually become paper. Lignin is difficult to remove. Some of the hemicellulose, which has some properties of cellulose and lignin, is also removed during the pulp manufacturing process.

Pulp is manufactured in pulp mills where trees are turned into pulp by separating wood into individual fibers. This is done in several steps [1,3,4]:

Step 1: Trees are cut into logs and showered with water to remove sand and soil.

Step 2: The bark and rotted parts in logs are removed in the barker since they do not make high quality paper. There are various types of barkers. A common type of barker is a cylindrical drum with open ends (Figure 1.4). Logs enter from one end, rub against each other inside the drum until bark is removed with friction and shear, and leave the barker on the other end. Bark falls down through the openings on the bottom of the barker. It is collected and used for other purposes such as fuel generation and mulch.

Step 3: The next step depends on the method for separation of individual fibers, i.e., the pulping method. Separation of fibers can be done in many ways including mechanical, chemical and semichemical methods. Table 1.1 gives the classification of major pulping methods. Some of the more common pulping methods are briefly described here for the reader.

Stone ground wood (SGW) and refiner mechanical pulping (RMP) are mechanical pulping methods. In the stone grinding method, barked logs are pressed sideways against a rotating grinding stone (Figure 1.6 on page 12). The grinding stone separates the fibers by mechanical friction. The heat generated due to friction between the wood and grinding stone helps to soften the lignin and separates

the wood fibers from each other. Mechanical pulping does not usually remove much lignin that causes hydrophobic fiber surfaces. Water is used to prevent the stone from burning the fibers and to remove the pulp collected on the stone. The pulp made with this method is called stone ground wood pulp. In this method, the wood is changed physically only and no chemical is used. This method may produce considerable fiber damage. Stone ground wood pulp is used in newspapers and other applications that do not require long life and is usually strengthened by blending with a percentage of chemical pulp. Even so, the paper is weak and turns yellow easily. The pulp made with this method is cheaper since the process is simple and 90% of the tree is used.

In the RMP process, wood chips are shredded and ground between the rotating discs of a refiner. New RMP processes employ thermal and/or chemical presoftening of the chips.

Other mechanical pulping methods include another form of energy and/or chemical to ease the pulping process. Chemi-mechanical pulping (CMP) combines chemical and mechanical methods. The chips are partially softened with chemicals before the final mechanical method of pulping. In thermo-mechanical pulping, heat is utilized instead of chemicals to improve the simple mechanical pulping. Temperature should be carefully controlled to prevent melting of lignin. Melted lignin will reduce good bonding in the sheet by covering up the exposed cellulose.

For the chemical pulping method, the barked logs are cut and sent to the chipper on a conveyor belt. The logs are cut into small pieces (chips) with the shear action of steel knives (Figure 1.4). Then the chips are sorted out in screening trays according to their sizes. The ideal size chips are approximately 20–25 mm square and 2–3 mm thick. Over-sized chips are rechipped and under-sized chips are used for fuel generation.

In the chemical pulping method, the chips are cooked with water and chemicals in a large pressure cooker called a digester which removes lignin and other impurities from the wood (Figure 1.4). In this process, the wood structure is chemically changed. More lignin is removed than in mechanical pulping, however cellulose degradation and fibrillation are increased. Chemical pulping produces collapsed, flexible and hydrophilic fibers. Crystallinity is increased and microfibril orientation is decreased. Paper made with chemical pulp is generally whiter and stronger than purely mechanical pulps. The alkaline kraft process and acidic sulfite process are the two major methods of chemical pulping.

In the kraft process, a widely used chemical pulping method, a solution of sodium sulfide (Na_2S) and sodium hydroxide (NaOH, caustic soda), which is called white liquor, is used. The chips are cooked under pressure around 170°C in white liquor for 1–2 hours after which most of the lignin and approximately half of the hemicellulose are removed from the chips. The white liquor turns black due to impurities. After cooking, the softened pulp is sent to the blow tank by opening the valve under the digester (Figure 1.4). The softened chips are smashed against the walls of the blow tank under high velocity and disintegrate into individual fibers. The heat generated in this process is re-used. In the kraft process, approximately half of the wood is turned into fibers and the rest becomes part of the black liquor. The cooking chemicals

TABLE 1.1. Classification of Major Pulping Methods [1,3,4].

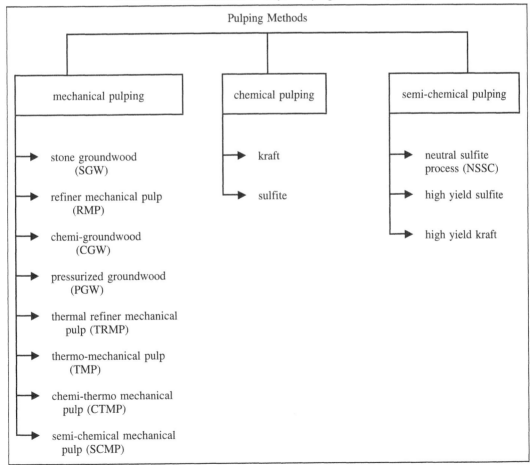

in the black liquor are recycled. The lignins, etc., are burned.

In the sulfite process, a mixture of sulfurous acid and bisulfite ion is used to remove lignin. The sulfite process produces bright pulp that is easy to bleach and refine, but with less strength than kraft pulp.

Step 4. Washing: Pulp is washed after the blow tank to remove the impurities. The cellulosic fibers are separated from the black liquor and washed with water. Figure 1.4 shows schematic of a rotary vacuum washer used for this purpose. The water is filtered through the fabric on the drum surface and collected inside the drum using vacuum. The fibers that are collected on the drum surface are deposited in a tank using a blade. Water is added into the tank for easy pumping.

Step 5. Bleaching: In some papermaking processes where a white paper is to be produced, bleaching is done to whiten the pulp which may still have a tan color because of the lignin remaining after the digester. For example, the pulp for grocery bags is not bleached; that is why they have the color of the pulp. In the bleaching process, chemical bleaches such as sodium hypochlorite are added to the pulp to remove the color by removing the remnant lignin molecules (Figure 1.4). Bleaching is a form of chemical pulping which should be done carefully under optimum conditions to prevent side effects. For example, excessive bleaching may reduce fiber strength.

Step 6. Refining (beating): The pulp is refined in a beater or refiner by fibrillating the fibers (Figure 1.4). The purpose is to brush and raise fibrils from fiber surfaces for better bonding to each other during sheet formation resulting in stronger paper. Pulp beaters process batches of pulp while refiners process pulp continuously. Equipment types include Hollander beater, Jones-Bertram beater, Chaflin refiner, Jordan refiner and disk refiners. Disk refiners provide continuous process and have become the most widely used equipment. Disk refiners can be single or double disk.

Refining is a mechanical process which causes physical change in the fibers. In a typical refining process, the fibers are sheared, crushed, twisted, stretched, cut and broken between two bar (or knife) surfaces. Shortening the fibers breaks the covalent bonds. The fiber surfaces are abraded and their primary walls are removed (Figure 6.1, Chapter 6). Fiber is fibrillated both internally and externally and its layers are delaminated. Fibrillation breaks the hydrogen bonds allowing water to be incorporated easily into the structure of the fiber. Fibers that adsorb water in this way become "hydrated" which is simply the state of a molecule being associated with water. Hydration is different than "swelling" because swelling takes place via rate dependent diffusion of solvent into a molecular structure. Fibrillation is critical in papermaking because paper strength increases by bonding of fibrils from fiber to fiber. Internal fibrillation is also called bruising or macerating the fiber. Controlled shortening of fiber length is also an objective of the beating process, but fiber shortening and external fibrillation also produce debris. Debris is cellulose particles or fiber segments that are too short to be classified as fibers or fibrous structures to bear loads. Beating has other effects as well. It affects pulp slurry viscosity. Increasing the exposed surface by beating adsorbs cationic materials such as alumina, acids, dyes, etc. Retention of precipitates and other fines also increases.

A schematic of a double disk refiner is shown in Figure 1.4. The disks have surfaces with design patterns made up of bars and grooves. Clearance between disks is critical and must be accurately controlled. One disk is stationary and the other disk rotates giving the shearing action to the pulp against the faces of the disks. This action collapses the fibers, which were roughly hollow cylindrical or semi-rectangular shapes, and roughens their surfaces (Figure 1.7 on page 12).

As shown in Figure 1.8 (page 12), refining increases tensile strength and bursting strength in the paper. However, after a brief increase at higher freeness, tearing strength declines continuously with increased refining.

Step 7. Additives: Several additives such as fillers (e.g., talc or clay), sizing agents (e.g., rosin, wax, starch, glue) and dyes are added to the furnish in stock preparation to obtain different properties. Fillers are added to improve printing properties, smoothness, brightness and opacity. Sizing agents make the paper stronger and more water resistant. Sizing can also be done on the paper machine, by surface application to the sheet.

Step 8: Finally, more water is added to the pulp to form a slurry that is ready to be pumped to the forming fabric through the headbox.

Although the properties of the original wood fiber are partially destroyed during the pulping process, the original characteristics are at least in part retained in the fiber [5]. Therefore, the properties of the resultant paper to a great extent depend on the fiber that is used. However, special properties such as improved strength, softness, moisture absorption, moisture resistance, etc., can be imparted to the paper by special treatment during the manufacturing process of the paper [1].

Important properties of pulp for good papermaking are ease of refining, dewatering ability, wet web strength, and dry strength. The pulp should be easily formed into paper on a paper machine which is called runnability.

1.2 Paper Manufacturing

Paper is a network of fibers. The principle of papermaking is relatively simple. The basic necessary ingredients of sheet forming are fiber (wood or other), water and a drainage medium. The mixture of fibers and water (called slurry), is filtered through a porous fabric. Water is drained through the fabric and wood fibers, fillers and fines are collected on the surface of the fabric; thus the sheet is formed (Figure 1.9 on page 13). After formation, the sheet is pressed and later dried to remove excess water. Although the main structure of paper is fibrous, nonfibrous components such as clays and other mineral fillers, dyes, polymers and bonding agents are added to improve formation, sheet structure and optical properties such as brightness and opacity.

Papermaking is a process that can be manually done in a lab or at home with the basic

necessary equipment. However, current papermaking technology is a highly sophisticated and continuous process which requires state-of-the-art machinery and equipment. Various chemicals, finishing and coating agents are used to help with the formation and improve the sheet properties and characteristics. There is a constant search to improve texture, print quality, control of ink bleeding, optical and other properties of papers. Nevertheless, like most textile manufacturing processes, papermaking is still a combination of art and science. The main reason for this is that the exact control of individual wood fibers during formation is hardly possible. Besides, as a natural product, there is inherent variation in wood fibers.

1.2.1 *History of Papermaking*

Papermaking is an ancient art. Approximately 4000 BC, Egyptians developed the first paper for writing purposes from a plant called papyrus. In fact, the word *paper* may have come from the word "papyrus." The stems of papyrus were thinly sliced and layers of slices were beaten into hard thin sheets and coated with glue-like material. However, the individual fibers were not separated which is different from modern papermaking principle [1].

Table 1.2 (page 23) shows a chronology of papermaking. The actual art of making paper was invented by Ts'ai Lun in China in the year 105 AD. He beat the water and mulberry tree bark mixture into a pulp. Using a screen, he collected the fibers and dried them in the sun which resulted in the first sheet of paper made from wood. With the invention of printing press by Johann Gutenberg of Germany in 1450, the demand for paper increased. In the mid 18th century, Rene de Reaumur, a French scientist, observed the pulp made by paper wasps by chewing bits of wood and mixing it with saliva which they use to make nests. Therefore, he suggested that the same procedure could be used to make paper. Despite this discovery, rag pulp remained the principle raw ingredient of paper for a long time.

Modern paper machines as we know them were invented by Nicholas Louis Robert, an

Englishman in 1798. He designed a hand-cranked machine that produced paper in a continuous roll which was a big improvement over the hand-dipping method. The Fourdrinier brothers of London bought the patent, improved and developed a machine that is named after them and still in use today. The machines, even the earliest Chinese method, all used wire meshes to form the paper on. In 1879, C. F. Dahl, a Swede, invented the kraft process by cooking the wood chips with sodium hydroxide and sodium sulfide under pressure which is a widely used process resulting in stronger paper.

Since the introduction of the first paper machine, there has been continuous increases in machine speeds and widths. Improvements and innovation in paper machine design and paper machine clothing played a key role in this progressive development. In early paper machines, woolen felts in the press and cotton felts in the dryer section were used along with the metal wires in the forming section. However, those types of fabrics were a limiting factor for higher machine speeds and new paper machines. The development of synthetic forming fabrics allowed the machine builders to design and develop twin-wire forming units which improved the sheet quality.

Today, paper is formed on polyester and nylon fabrics compatible with increased machine width and speed. In the press section, the development of all synthetic needled press fabrics allowed heavy press loadings which increased the efficiency of water removal. The increased stability of synthetic press felts met the demands of faster and wider machines. Similar improvements were made in the dryer section in speed and width with the introduction of polymeric dryer fabrics. Drying rate was increased along with the fabric life.

1.2.2 Papermaking Process

The five basic steps of the sheet forming process on a paper machine are:

- fiber dispersion and distribution onto a forming fabric

- drainage of water
- consolidation of the sheet
- compaction of the sheet
- drying

Coatings can be applied to the sheet during or after these processes. A typical paper machine has three distinct sections as shown in Figure 1.10 (page 13): forming, pressing and drying. The fabrics that are used in these sections are called forming fabrics, press fabrics (felts) and dryer fabrics (felts), respectively. Each fabric type performs specific functions. Therefore, each fabric type must be designed, engineered and manufactured differently depending on the paper grade, paper machine position and size.

Modern papermaking is a fast and massive process. For example, up to 100,000 liters of paper pulp are evenly spread within one minute onto a 7 m wide fourdrinier paper machine. Finished paper is wound on the rolls at the end of the paper machine only a few seconds after the high water content material has reached the forming fabric [7].

Forming

The fabrics that are used in the forming section are called forming fabrics. The term "forming" comes from the formation of paper. Pulp slurry, which is a mixture of approximately 99% water and 1% fiber, is pumped through the headbox evenly across the forming fabric. The forming fabric travels along the forming section of the machine like a porous conveyor belt. As the fabric moves, water is filtered through the forming fabric and wood fibers are retained on the fabric surface, thus forming the sheet (i.e., a wet paper web at this stage) as shown in Figure 1.9. At the end of the forming section, the sheet has a consistency of approximately 20% fiber and 80% water. Figure 1.11 (page 13) shows the typical sheet and water ratio through the fourdrinier machine. The forming section of the paper machine is also called the "wet-end."

FIGURE 1.2. A pulp and paper mill (courtesy of Florida Coast Paper Company).

Paper Machine

Stock Preparation, Paper Formation and Drying: Fibers are tailored and blended for each end-use requirement, formed into a mat on the paper machine at high speeds, and dried.

Pulping Stage: Raw materials pass through the digester, grinders, refiners or repulper to release the fibers for subsequent screening and cleaning. Some fibers require bleaching.

Digester

Repulping

Forest Residue/Wood Waste

Recovered Paper

Pulpwood

Fibrous Raw Materials: Recovered materials— wood wastes and recovered paper—provide 54 percent of the industry's fiber; pulpwood supplies the rest.

Machine Control Console

Quality Control: Through the use of on-line sensors and other state-of-the-art equipment, quality is constantly monitored.

Finishing

Finishing and Converting: After finishing, paper and paperboard rolls are cut, trimmed and packaged into individual products for shipment and subsequent converting.

Sheeting

Recovery for Reuse

Source Separating Paper

FIGURE 1.3. Complete cycle of the pulp (chemical) and paper technology (courtesy of American Forest and Paper Association).

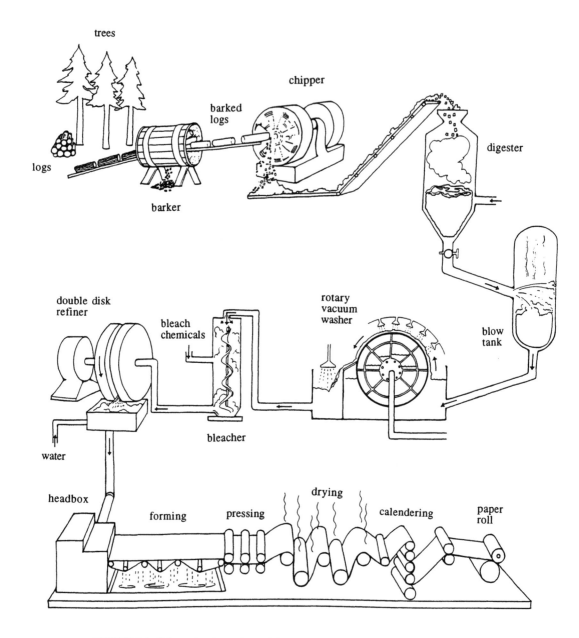

FIGURE 1.4. Schematic of wood pulp (chemical) and paper manufacturing (TAPPI, [1]).

10

FIGURE 1.5. Trees are renewable sources for the pulp and paper industry (courtesy of American Forest and Paper Association).

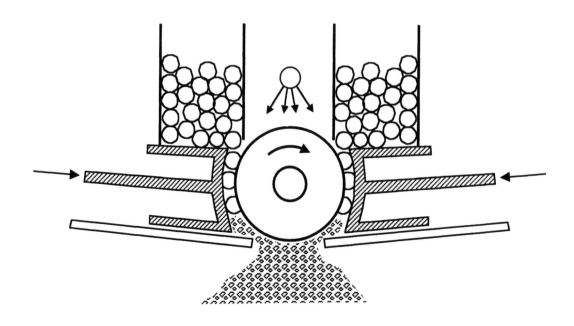

FIGURE 1.6. Schematic of mechanical grinder.

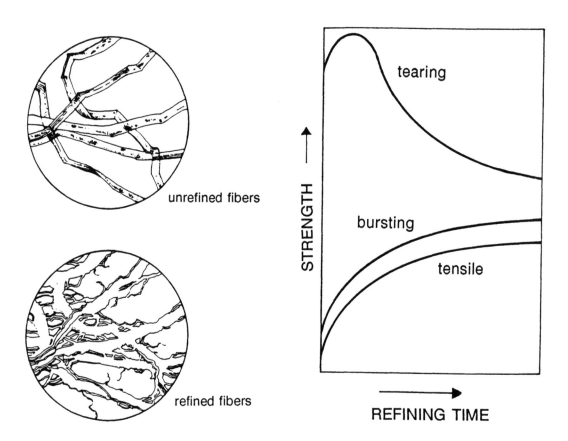

FIGURE 1.7. Fibers before and after refining (TAPPI, [1]).

FIGURE 1.8. Effect of refining on paper strength (TAPPI, [1]).

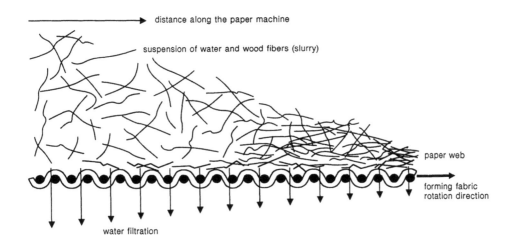

distance along the paper machine

suspension of water and wood fibers (slurry)

paper web

forming fabric
rotation direction

water filtration

FIGURE 1.9. Schematic of sheet formation (courtesy of Wellington Sears Company).

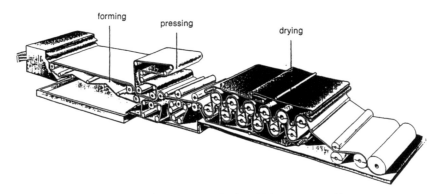

forming

pressing

drying

FIGURE 1.10. Schematic of a fourdrinier paper machine.

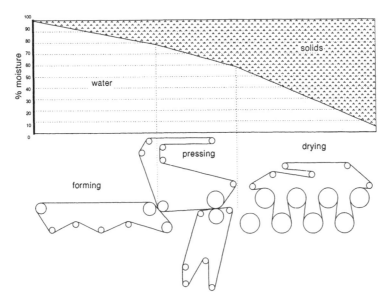

% moisture

solids

water

pressing

drying

forming

FIGURE 1.11. Sheet and water ratio through the fourdrinier machine.

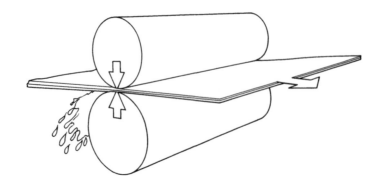

FIGURE 1.12. Schematic of pressing (courtesy of Beloit).

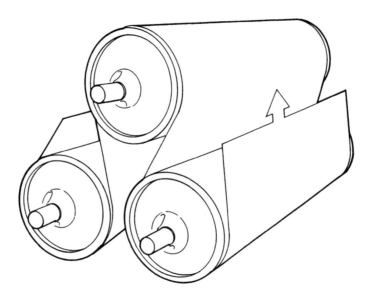

FIGURE 1.13. Schematic of drying (courtesy of Beloit).

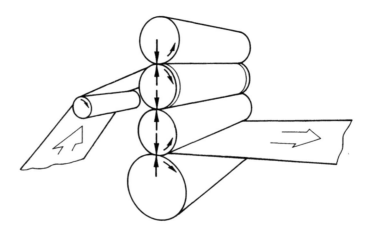

FIGURE 1.14. Calendering (courtesy of Beloit).

FIGURE 1.15. Hybrid fourdrinier paper machine (photo courtesy of Valmet).

15

FIGURE 1.16. Twin-wire paper machine (photo courtesy of Beloit).

FIGURE 1.17. Tissue products (photo courtesy of American Forest and Paper Association).

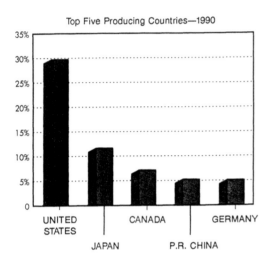

Top Five Producing Countries—1990

FIGURE 1.18. Share of world paper and paperboard capacity (sources: American Forest and Paper Association and Food and Agriculture Organization of the United Nations).

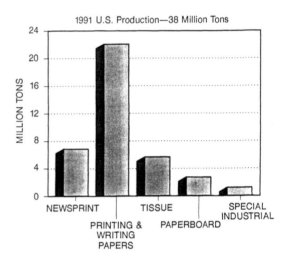

1991 U.S. Production—38 Million Tons

FIGURE 1.20. U.S. production of paper other than packaging (American Forest and Paper Association).

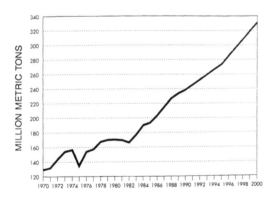

FIGURE 1.19. World demand for paper and paperboard (sources: American Forest and Paper Association, Pulp and Paper International, Food and Agriculture Organization of the United Nations).

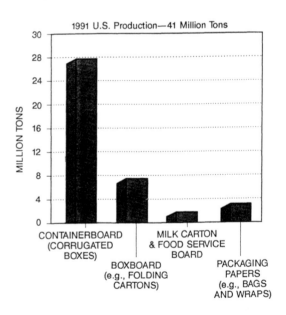

1991 U.S. Production—41 Million Tons

FIGURE 1.21. U.S. paper and paperboard packaging (American Forest and Paper Association).

FIGURE 1.22. Paper and paperboard packaging (photo courtesy of Thilmany).

19

FIGURE 1.23. Paper is used to protect perishable food products (photo courtesy of American Forest and Paper Association).

FIGURE 1.24. Corrugated containers (photo courtesy of American Forest and Paper Association).

U.S. PAPER RECOVERY RATE

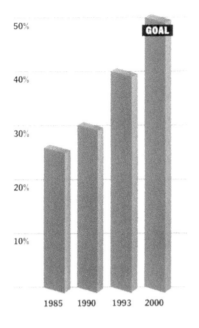

50%

40%

30%

20%

10%

GOAL

1985   1990   1993   2000

FIGURE 1.25. U.S. paper and paperboard recovery rate: ratio of paper and paperboard recovered to total amount of paper and paperboard consumed in a year (American Forest and Paper Association).

Recycled Paper Symbol

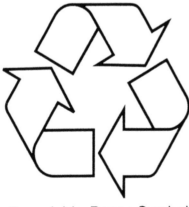

Recyclable Paper Symbol

FIGURE 1.26. Recycling symbols (American Forest and Paper Association).

TABLE 1.2. A Chronology of Papermaking [1,3,4,6].

BC	4000	Papyrus for writing purposes developed in Egypt
AD	105	Ts'ai Lun credited for invention of paper in China
	600	Mayans of Central America made paper from bark
	751	Arabs learned papermaking after battle of Semerkand
	800	Paper reportedly made by Arabs in Baghdad
	868	Wang Chieh printed first book in China
	900	Paper made in Egypt
	1085	Crusaders took over paper mill located in Toledo, Spain
	1102	Paper produced in Sicilia
	1268	Papermaking started in Italy where watermark was introduced
	1348	Paper mill established in France
	1450	Invention of printing press by Gutenberg to print the Bible; this invention increased demand for paper
	1576	First paper mill established in Russia
	1609	First newspaper with a regular publication date in Germany
	1637	Sung Ying-Hsing noted the use of waste paper for making paper in Northern China
	1690	Rittenhouse establishes a paper mill in Germantown, Pennsylvania which was the first mill in North America
	1719	Reamur, a French naturalist, suggested using wood as a fiber source
	1774	Chlorine was used with lime to bleach paper
	1798	Nicholas Louis Robert invented the first paper machine
	1800	Matthias Koops received patent for process using straw, wood and deinked waste paper which was the initial form of recovery and recycle system
	1803	Bryan Donkin developed the first successful paper machine for Fourdrinier brothers
	1805	Joseph Bramah invented cylinder machine which was patented by John Dickenson in 1809
	1807	English patent issued to Fourdrinier brothers for improved paper machine
	1817	Thomas Gilpin installed the first paper machine, which was a cylinder type, on Brandywine River near Wilmington, Delaware; Dickinson invented size press for paper machine
	1821	T. B. Compton, England, received patent for drying paper on the machine. Cloth was used for protection and better appearance
	1823	Crompton used steam dryers in England
	1826	John Marshall invented dandy roll; vacuum pumps on the fourdrinier machine
	1827	First fourdrinier paper machine installed in America at Saugerties, New York
	1844	Pulp is made by grinding
	1850	Ebart invented a flame retardant paper
	1852	Turner developed endless wire for paper machine
	1863	Patent issued to J. F. Jones of Rochester, USA for multicylinder machine with seven or more vats
	1866	First groundwood pulp mill in the United States established at Curtisville, Massachusetts
	1875	Multiply paper machine with two separate fourdrinier sections
	1879	C. F. Dahl invented the kraft process
	1909	First kraft pulp mill in the United States established at Roanoke Rapids, North Carolina
	1953	Daniel Webster of Consolidated Paper invented modern twin-wire former
	1965	Blade and roll formers introduced
	1970	Multiple combined fourdriniers
	1980	Multiple fourdrinier with top station or top wire
	1985	Roll/blade former
	1990	Retrofit blade former with adjustable drainage

Compared to pressing or drying, forming is the most critical stage in papermaking. Forming fabrics affect the final sheet properties more than press or dryer fabrics. If the sheet structure is not formed properly in the forming section, it would hardly be possible to correct it later on the paper machine [8]. Therefore, forming fabric properties must be carefully engineered. Although it depends on the type of formation, the average life of a forming fabric is three months. Forming and forming fabrics are the subject of Chapter 2.

Pressing

After the sheet is formed, it is transferred to the press section (Section 2.7). The function of pressing is to continue the water removal process that started in the forming section, consolidate the sheet, give texture to the sheet surface, support and transport the sheet. During pressing, the sheet web is compressed in the nip formed by two press rolls, one or both of which may be felted. Water is moved from the sheet into the felt(s) and may be expressed at the nip, or carried away for removal by felt vacuum boxes. Figure 1.12 (page 14) shows this process in a plain press nip. There is a considerable amount of pressure at the point of contact. Increased compression increases water removal [8–11]. Since most of the paper web is still water when entering the press section, the paper surface finish is influenced by the press fabric design. Although it depends on sheet grade and paper machine, typical sheet consistency at the end of press section is 40% fiber and 60% water. At the end of the press section, the sheet is transferred to the dryer section. The average running life of a press felt is about 30–90 days. Pressing and press fabrics are the subject of Chapter 3.

Drying

After pressing, most of the residual water in the sheet is removed by evaporation (mass transfer) in the dryer section by using steam (heat transfer). The fabrics in this section are called dryer fabrics or dryer felts. The paper sheet travels around heated cylinders where most of the remaining water is evaporated. The consistency after this section is on the order of 95% fiber and 5% water which is the typical consistency of regular paper. The dryer end of the paper machine is also referred to as "dry-end."

The drying zone can be considered as a "black box" into which wet paper, steam and air enter. The dried paper, the moist and heat-laden air, and the condensate leave the box [12]. Typically, large volume of drying cylinders evaporate more than 17,000 litres of water per hour from the paper. Figure 1.13 (page 14) shows the arrangement in the drying zone. Dryer fabric presses the sheet tight against the heated cylinder. Most of the energy used to make the sheet is consumed in the dryer section and the dryer fabric has a significant effect on drying efficiency. Dryer fabrics typically last 6–15 months. Drying and dryer fabrics are the subject of Chapter 4.

Coating and Calendering

Coating of the paper surface may be done to impart different surface properties. In calendering, paper is passed through the nip(s) of two or more rolls (Figure 1.14 on page 14). Calendering increases the smoothness and uniformity of the paper and makes it thinner. After calendering, the paper is wound onto a large roll which may be cut into smaller rolls in a winder machine.

Converting Process

After paper is manufactured, it usually goes through a converting process before it is sold to the consumer. The converting operation is critical for the function or performance of the paper. Quality deficiencies in the paper are usually revealed in the converting plant or in the printing press [5].

1.3   Paper Machines

Although the principle of papermaking has not changed since the introduction of fourdrinier machine in 1798, there have been major improvements and changes in paper machines. As a result, paper machines became larger, faster and wider with more efficiency and higher productivity. A detailed description of the paper machine types is out of the scope of this book. Therefore, brief descriptions of forming, press and dryer section configurations of paper machines are given in Chapters 2, 3 and 4, respectively. Figures 1.15 and 1.16 (pages 15 and 16) show modern paper machines.

Papermaking is a capital intensive business. A modern paper machine can be as long as a football field [the longest one in the world is 275 m (902 ft) long] and as wide as 13 m (512 inch). The height of the machine can be equal to the height of a 3–4 story building. The speed of a modern high-speed tissue machine is close to 2134 m/min (7000 ft/min) which corresponds to 2.1 km of paper produced every minute (1.3 miles/min). A paper machine costs several hundred million dollars to build which makes the cost of downtime very high. Therefore, efficient operation of the machine is extremely important which demands high quality fabrics with long life at high speeds and harsh environmental conditions such as high temperature, tension and humidity.

Depending on the size of the paper machine, the cost of one complete set of clothing on a sizable machine can be approximately a million dollars. The life of a typical fabric can be in the range of 0 days to 1+ year. If a fabric is damaged beyond the repair during installation or if it is not functioning properly from the moment of installation, then it must be taken off immediately.

1.4 Paper Grades

There are various grades of paper depending on the pulping method or end use. Table 1.3 shows paper grades according to the American Forest and Paper Association, AF&PA (formerly American Paper Institute, API).

TABLE 1.3. Paper Grades (AF&PA).

Paper Grade	Area (sq ft) per Unit Basis Weight (1 lb)
Newsprint	3000
Tissue	2880 or 3000
Linerboard	1000
Corrugating medium	1000
Kraft papers	3000
Uncoated groundwood	3300
Coated groundwood	3300
Uncoated free sheet	3000 or 3300
Coated free sheet	3300
Pulp	1000
Bleached paperboard	1000 or 3000
Recycled paperboard	1000

1.4.1 *Uncoated Free Sheet*

Chemical pulps (sulfate, sulfite, soda, cotton linters or vegetable fiber) are used to make uncoated free sheet papers, with occasional additions of up to 10% mechanical fiber or bleached chemi-thermomechanical pulp (BCTMP) and recycled fibers. Free sheet is a paper free of mechanical wood pulp. Table 1.4 shows uncoated free sheet properties. Many of the major producers have converted some or all of their machines to alkaline-based paper.

Uncoated free-sheet papers are used for office and business printing (copiers, computer printers), business forms and envelopes, publishing grades, commercial printing and writing (stationery).

The surface of a printing paper must accept and retain the ink and present it to the reader in an optimum manner. For good print quality, the sheet surface must be smooth with a uniform pore structure. The fibers on the surface must also be securely bonded to the sheet structure.

1.4.2 *Coated Free Sheet*

Coated free-sheet papers contain less than 10% mechanical (groundwood) pulp with no groundwood in Superpremium, No. 1 and No. 2 grades, and some or none in Nos. 3 and 4, and even some in No. 5. Table 1.5 shows general properties of coated free sheet.

TABLE 1.4. Uncoated Free-Sheet Properties.

Furnish:	100% Bleached kraft, usually 50+% hardwood
pH:	7.0+ Alkaline
Filler:	Calcium carbonate
Opacity:	Obtained by calcium carbonate; critical requirement ash 10–20% range
Basis weight:	38–80 lbs
Formation:	Critical on bond sheet, also important are MD/CD tensile, bulk, uniform product
Smoothness:	Surface smoothness critical for end use
Printability:	Will be printed directly on sheet
	a) Xerox paper
	b) Forms

Coated two-side (C2S) papers represent about 88% of coated free-sheet grades. C2S papers are used for magazines (e.g., *National Geographic*), annual reports, some catalogs, advertising brochures and inserts, and other commercial printing. Coated one-side paper (C1S) is used for labels, with some used in books, commercial printing, business forms, envelopes and converted products other than labels.

Nos. 1 and 2 generally range from 60 lb to 100 lb (27.22 kg to 45.36 kg) basis weights, and Nos. 3 and 4 grades from 40 lb to 90 lb. Most coated paper (about 75 to 85%) has a glossy surface, the rest a combination of dull, matte and embossed. Gloss finishes on the Gardner glossmeter are 45 or more for glossy, down to 25 for dull and 20 or below for matte.

1.4.3 *Uncoated Groundwood*

Uncoated groundwood paper grades span the spectrum from near-newsprint to almost coated paper quality. Uncoated groundwood papers include rotogravure newsprint, machine finished (MF) offset, directory and some closely related catalog papers, forms (mostly blended with chemi-thermomechanical pulp), various quality levels of supercalendered (SC), and the newer soft calendered grades. Essentially most uncoated groundwood papers have higher brightness levels and smoother surfaces than newsprint. Supercalendering and soft calendering add a glossy surface often competing with lightweight coated (LWC) groundwood papers. Brightness levels range from 64 to 74 (or higher) Elrephro, compared to 56 to 62 for standard newsprint and 68 to 84 for LWC. Basis weights range from 20 to 45 lb/3300 sq ft (9 to 20 kg/307 m^2). Primary uses include preprinted newspaper inserts, and direct mail flyers, catalogs (e.g., JC Penney), lower-cost business forms, Sunday newspaper magazines, some newspapers and telephone directories. Table 1.6 shows uncoated groundwood sheet properties.

TABLE 1.5. Properties of Coated Free Sheet.

Furnish:	100% Bleached kraft
	from: 20% softwood, 80% hardwood to 80% softwood, 20% hardwood
pH:	4.8–8.0. Acid to alkaline depending on coating
Filler:	10–15% Sheet is "base loaded" at 10–15% ash usually from coated broke, also add purchased fillers—filler clay or calcium carbonate
Opacity:	Critical; opacity may be obtained from coating
Basis weight:	30–150 lbs. Very heavy weights referred to as enamels
Formation:	Subject gets involved in the heavy basis weights due to coat weight and percent fillers
Printability:	Often sheetfed—roto
	Some high-speed volume printers
	Job lot printers

TABLE 1.6. Uncoated Groundwood Sheet Properties.

Furnish:	70+% Groundwood
Filler:	Clay or calcium carbonate
pH:	4.8–8.0 Depending on whether filler is clay or calcium carbonate
Opacity:	Critical due to basis weight
Basis weight:	30–32 lb
Formation:	Critical; poor formation will be a major problem when the sheet is supercalendered
Smoothness:	Generally seen as a supercalender problem; smoothness greatly affected by filler
Printability:	Printed by large regional printers who use this as filler between weekly issues

1.4.4 *Coated Groundwood*

Coated groundwood (mechanical) papers are used primarily in commercial printing, magazines (e.g., *Time, Newsweek*), catalogs, Sunday newspaper supplements, newspaper advertising inserts (coupon pages), directories and books. Table 1.7 shows properties of coated groundwood.

Coated groundwood papers contain 10% or more mechanical (groundwood) pulp. Coated groundwood includes the most popular designation, lightweight coated (LWC), though basis weights for coated groundwood papers go into medium and heavy weights, with LWC topping out at 40 lb or 45 lb, depending upon the producer. Coated groundwood papers are almost universally coated on two sides (C2S). Coated groundwood also includes some newer, less expensive grades, such as machine-finished coated.

Coated groundwood papers are commonly made in the 32 to 50 lb per 3300 sq ft range, although basis weights can run as low as 26 lb and as high as 70 lb. Coating weight is approximately 30% of total sheet weight for LWC papers. Basis weights for coated groundwood paper continue to decline in an effort by papermakers to help publishers, direct marketers, and commercial printers save money on postage weight-related costs. News magazines commonly use 32 lb, and catalogs have often gone down to 24 lb.

1.4.5 *Tissue*

The 6 million ton U.S. sanitary hygiene papers business is divided into two major segments: the consumer market and the commercial/industrial (C&I) market. Consumer products represent about 65% of sanitary papers volume and are sold to the public through supermarkets and other retail outlets. Also called the "away from home" market, C&I products represent the remaining volume, sold for uses in factories, schools, offices, restaurants, hotels and hospitals. Figure 1.17 (page 17) shows various tissue products.

1.4.6 *Bleached Paperboard*

Bleached paperboard, also known as solid bleached sulfate (SBS) board, is a premium paperboard grade that is produced from a furnish

TABLE 1.7. Coated Groundwood Sheet Properties.

Furnish:	45–50% Bleached groundwood 50–55% Bleached northern kraft
pH:	4.5–4.8 Acid
Filler:	Base sheet is at 10–15% ash usually from coated broke
Opacity:	Critical; lighter weight sheets make this a key property
Basis weight:	38–48 lb, Coated weight
Formation:	Headbox freeness lower compared to free sheet because of furnish; sheet forms differently
Printability:	Large printers, high speed roto presses

containing at least 80% virgin-bleached wood pulp. Most SBS is clay-coated to improve its printing surface and may also be polyethylene-coated for wet-strength food packaging. Basis weights vary from 40 to 100 lb/1000 sq ft (18.1 to 45.3 kg/93 m^2), and calipers typically range from 12 pt to 28 pt for packaging grades. Basis weights and calipers for bleached printing bristols are slightly lower.

Bleached paperboard is primarily used for folding cartons, milk cartons, and other packaging products that require superior folding, scoring, and printing characteristics. SBS is also used for disposable cups and plates, food containers, and preprint linerboard for high graphic corrugated boxes and displays. Some mills that produce SBS packaging board also manufacture lightweight bleached bristols used by commercial printers for paperback book covers, telephone directory covers, greeting cards, postcards, baseball cards, merchant displays, etc.

1.4.7 Linerboard

Major end use of linerboard is in domestic corrugated container production. Linerboard is made in a wide range of basis weights. Standard weight in the United States is 42 lb/1000 sq ft, but other important weights are 26 lb, 33 lb, 69 lb and 90 lb. High performance linerboard provides comparable strength at lower basis weights.

Recycled linerboard production based on 100% wastepaper has been growing rapidly in the United States and accounts for an increasing percentage of industry capacity. Unbleached kraft linerboard statistics also include mottled and white top linerboard and tube, drum and miscellaneous grades.

1.4.8 Recycled Paperboard

Recycled paperboard is made from recovered wastepaper. It is a multi-ply material which can be manufactured in various weights and calipers using various types of wastepaper. Recycled paperboard may be clay coated, which improves its printability. The most important end uses are folding cartons, corru-

gated containers, rigid setup boxes, paper tubes, cans and drums, gypsum wallboard, solid fiber partitions, book covers and binders, and insulation board.

1.4.9 Corrugating Medium

Corrugating medium, the middle fluting material used in corrugated containers, is made from both semichemical pulp and recycled fiber. Semichemical medium is not supposed to contain more than 25% recycled fiber content, but an increasing amount is probably exceeding this industry definition. The share of total medium capacity held by semichemical board has declined to about 67% in 1994 from 78% in 1980, while recycled has grown to 33%. Basis weights for corrugating medium are 22, 26, 33, 36 and 40 lb/1000 sq ft. The standard sheet weight is 26 lb, which accounts for about 62% of production, down from more than 80% in the mid-1970s. Heavier weights have been growing in popularity with the trend toward boxes offering increased compression strength.

1.4.10 Kraft Paper

Grocery bags and sacks are the largest market for bleached and unbleached kraft packaging papers, accounting for about 50% of total shipments, followed by multiwall shipping sacks, 31%; other bag and sack, 4%; and wrapping and converting papers, 15%. Basis weights for most bag and sack papers range from 30 lb to 80 lb/3000 sq ft.

1.4.11 Market Pulp

Chemical paper grade market pulp is sold in the open market and excludes captive pulp used on site or shipped to affiliated mills in the same country. Exports are generally included, except for shipments from Canadian mills to affiliated U.S. mills. Pulp is divided into grades based on whether it is made from softwood or hardwood fiber, produced by the sulfate (kraft) or sulfite pulping process, and the amount of bleaching.

Northern bleached softwood kraft (NBSK)

is considered the premium pulp grade because of its long fiber length and strength. U.S. pulp statistics include about 2 million tons of fluff pulp production used in diapers and other absorbent hygiene products. Dissolving and mechanical pulp are excluded from the statistics. Canada has about 1.6 million tons of mechanical pulp capacity, mainly bleached chemithermomechanical pulp (BCTMP).

1.4.12 *Newsprint*

Newsprint is uncoated paper used for the printing of newspapers. Not less than 65% of the total fiber content of newsprint consists of wood fibers obtained by a mechanical process, with other standards addressing smoothness, weight and ash content. Virtually all U.S. and Canadian newsprint capacity is on a nominal 30 lb basis weight standard, and a typical sheet falls into the 58 to 60 brightness range.

1.4.13 *Nonwovens*

A nonwoven product is a manufactured sheet, web or batt of directionally or randomly oriented fibers, made by bonding or entangling fibers through mechanical, thermal or chemical means. The fibers may be of natural or man-made origin. Nonwoven products form a bridge between paper and textile products. Nonwoven manufacturing processes are explained in Chapter 2.

The application areas of nonwoven products are ever increasing. Some examples are medical products, diapers, interlinings, shoe components, wipes, floor coverings, mattress covers, disposable clothing, filters, insulation, soil stabilization, luggage and automotive components.

1.4.14 *Specialty Papers*

This category is for grades that do not neatly fit the previously mentioned grades. Specialty papers are often used as a carrier for polyester, foil or other applications such as computer paper, thermo imaging papers, etc. The specialty paper market is characterized by smaller machines. Physical properties vary greatly. Plastics are a major competitor for these grades.

1.5 End-Use of Paper

The United States is the world's largest manufacturer of pulp, paper and paperboard. The production capacity of the United States is approximately equal to the combined capacity of the next four largest producing nations (Figure 1.18 on page 18). Although the United States has only 5% of the world's population, with 12% of the world's paper and paperboard mills, it has 30% of the world's paper and paperboard production and with 16% of the world's pulp mills, it has 35% of the world's wood pulp production [13]. The U.S. pulp and paper industry is one of the nation's top ten manufacturing industries. There are approximately 550 facilities in the United States making paper, paperboard and building products. Since 1980, the U.S. pulp and paper industry invested more than $110 billion to increase productivity and capacity. Over $40 billion of this investment is done since 1990. In 1992, the U.S. pulp and paper industry produced over 82 million tons of paper and paperboard and 10 million tons of market pulp. Figure 1.19 (page 18) shows the world demand for paper and paperboard.

Paper is critical to life today. Americans use more than 90 million short tons of paper and paperboard every year. Paper can be used in three different ways:

(1) As a structural material for load-bearing and other applications; e.g., grocery bags, paper boxes, construction materials, etc.

(2) As a printing and communication material, e.g., books, newspapers, notebooks, etc.

(3) As a material to absorb liquids (e.g., tissue, paper towel) or resist moisture (e.g., paper plates and cups).

Sometimes, paper has to perform more than one function at the same time. Table 1.8 lists various types of papers according to end use.

TABLE 1.8. Major Paper Types.

1. Newsprint	Tablet	Corrugated medium
Catalog	Uncoated ground	Linerboard
Directors	Writing	Unbleach carton
Lightweight news	Xerox	Bleached brown
Newsprint	Glassine	5. Board Stock
RotoGrav	Food wrap	Cylinder stock
Special news	Glassine	Bag medium
2. Fine Paper	Greaseproof	Box board
Fine Speciality	Label backing	Chip board
Carbonizing	Bristol, Cover & Tag	Core/tube stock
Cigarette	Book cover	Cylinder board
Crepe	Bristol	Gypsum liner
Foil back	File folder	Part stock
MG wrap	Index stock	Special cylinder
Parchment	Post card	Test linerboard
Photo	Poster	Bleached Board
Spec fine	Tag stock	Bleached board
Wax spec	Tab card	Coated board
Fine Printing	3. Tissue and Absorbent	Cup stock
Banks/bonds	Filter	Food board
Bible/Quran	Napkin	Milk carton
Coated fine	Sanitary	Uncoated board
Computer	Saturating	6. Nonpaper
Envelope	Towel	Glass mat
Fine print	Tissue/towel	HTL
Special groundwood	Tissue wrap	Roofing
Lightweight coated	Wadding	Nonwovens
Mimeo bond	4. Brown and Kraft	Filtration
Offset	Brown	Food processing
Publication	Bag and wrap	Textiles
Supercalender	Core stock	

There are an estimated 12,000 different paper products [14].

Figure 1.20 (page 18) shows the U.S. production of paper other than packaging. Communication would be extremely difficult without paper. The world would be quite different without books, newspapers, fax messages, etc. Sanitary paper products provide comfort, convenience and safety. Paper towels, diapers, napkins, toilet paper and facial tissue have improved the quality of personal life, improved sanitary conditions and made household cleaning easier (Figure 1.17). Specialty papers include technical papers, saturating grades and certain laminating papers. They are used to protect humans and machines. Paper

filters are used to filter dust, pollen and other airborne particles. This is especially critical in hospitals and high precision engineering of electronic and machine parts. Paper seals and gaskets protect precision engine parts from dirt and sediment.

Figure 1.21 (page 18) shows the paper and paperboard packaging production in the United States. Paper and paperboard packaging is the most widely used packaging which is highly effective in preserving and transporting consumer goods (Figure 1.22 on page 19). Paper and paperboard packaging is used to wrap food, to protect perishables like milk, juice and frozen foods (Figure 1.23 on page 20).

Corrugated containers are ideal for efficiently moving products. Corrugated containers have replaced wooden boxes and crates. Today, 95% of all manufactured goods are shipped in corrugated boxes (Figure 1.24 on page 21).

1.5.1 *Recycling*

Paper and related products are recyclable and biodegradable. The U.S. pulp and paper industry is one of the most successful industries in recycling. More than 400 paper and related facilities use recovered paper as a raw material for papermaking. The percentage of paper and paperboard recycling is ever increasing. In 1995, more than 40% of all the paper in the United States was recycled paper (Figure 1.25 on page 22). That was 50% more than was recovered in 1988. Old corrugated containers (OCC), old newspapers, magazines, office papers, etc., provide valuable raw material to papermaker. For example, in 1995, 70% of the corrugated cardboard in the United States was recovered for recycling and reuse. Old newspapers are the second highest product recycled in the United States. In 1995, more than six out of every ten newspapers were recovered. In recent years, 60% of the tissue was recovered. Figure 1.26 (page 22) shows recycling symbols that depict recycled and recyclable products.

In 1993, the U.S. paper industry set a goal to recover—for recycling and reuse—50% of all paper used in the year 2000. There are several organizations that promote pulp and paper recycling. Sixteen leading U.S. and Canadian companies formed the 100% Recycled Paperboard Alliance (RPA-100%) to promote the benefits and increase the use of 100% recycled paperboard. Other organizations for recycling include American Paperboard Packaging Environmental Council, Paper Grocery Bag Council and Paper Recycling Advocates.

The U.S. pulp and paper industry invested heavily in environmental improvement. Today, the U.S. pulp and paper industry uses 60% less water to produce 50% more products than 20 years ago. Biochemical oxygen demand has been decreased 70% [14].

1.6 References

1 Burdette, J., Conway, L., Ernst, W., Lanier, E., and Sharpe, J., *The Manufacture of Pulp and Paper: Science and Engineering Concepts,* TAPPI Press, 1988.

2 Smook, G. A., *Handbook of Pulp and Paper Terminology,* Argus Wilde Publications, 1990.

3 Smook, G. A., *Handbook for Pulp and Paper Technologists,* Joint Textbook Committee of the Paper Industry, TAPPI (Atlanta, USA) and Canadian Pulp and Paper Association (Montreal, Canada), 1989.

4 Graham, L. L., *Introduction to Pulp and Paper Properties and Technology Course Notes,* The Paper Science Department, University of Wisconsin/Stevens Point, 1991.

5 Bristow, J. A., and Kolseth, P., *Paper Structure and Properties,* International Fiber Science and Technology Series/8, Marcel Dekker, Inc., New York, 1986.

6 Thorp, B. A., and Kocurek, M. J., Eds., *Pulp and Paper Manufacture, Vol. 7 Paper Machine Operations,* Joint Textbook Committee of the Paper Industry, TAPPI (Atlanta, USA) and Canadian Pulp and Paper Association (Montreal, Canada), 1991.

7 Hoechst Aktiengesellschaft, Trevira® Technical Service, Hoechst Celanese, April 1991.

8 Adanur, S., *Wellington Sears Handbook of Industrial Textiles,* Technomic Publishing Co., 1995.

9 Reese, R. A., "Pressing Operations," in Thorp, B. A. and Kocurek, M. J., Eds., *Pulp and Paper Manufacture, Volume 7: Paper Machine Operations,* TAPPI Press, 1991.

10 Wicks, L. D., "Press Section Water Removal Principles and Their Applications," *Southern Pulp & Paper Manufacturer,* Jan. 1978.

11 Bliesner, W. C., "Sheet Water Removal in a Press: Time to Review the Fundamentals," *Pulp&Paper,* Sept. 1978.

12 Paper Machine Felts and Fabrics, Albany International, 1976.

13 *Paper: Linking People and Nature,* The US Pulp, Paper and Paperboard Industry, American Paper Institute, Inc., 1992.

14 *Investing for Success,* American Forest and Paper Association, Washington DC, 1995.

General References

Anderson, S. L., "The Outer Limits of Paper Recovery and Recycling," *TAPPI Journal,* April 1997.

Britt, K. W., *Handbook of Pulp and Paper Technology,* Van Nostrand Reinhold, 1970.

Dalzell, D. R., "A Comparison of Paper Mill Refining Equipment," *TAPPI,* 44(4), 1961.

The Dictionary of Paper, American Paper and Pulp Association, 1965.

Hardacker, K. W., and Brezinski, J. P., "Individual Fiber Properties of Commercial Pulps," *TAPPI,* 56(4), 1973.

Kline, J. E., *Paper and Paperboard: Manufacturing and Converting Fundamentals,* Miller Freeman, 1982.

Lavigne, J. R., *An Introduction to Paper Industry Instrumentation,* Miller Freeman, 1977.

Lavigne, J. R., *Instrument Applications for the Pulp and Paper Industry,* Miller Freeman, 1979.

Lavigne, J. R., *Pulp and Paper Dictionary,* Miller Freeman, 1986.

Leider, P. J., and Nissan, A. H., "Understanding the Disk Refiner: The Mechanical Treatment of the Fibers," *TAPPI,* 60(10), 1977.

Quick Facts about America's Forest and Paper Industry, American Forest and Paper Association, 1995.

Rydholm, S. A., *Pulping Processes,* Wiley, 1965.

1.7 Review Questions

1 Define paper. What are the major differences between paper and textile products?

2 Compare kraft and sulfite pulping processes for their effects on pulp properties.

3 Why is refining necessary? How does it affect the paper properties?

4 Explain the major steps in pulp and paper manufacturing.

5 What are the possible effects of repeated recycling on wood fiber properties? Explain.

Forming

This chapter is about sheet formation, forming fabrics and the forming section of the paper machine. The general mechanism of sheet formation is explained briefly. Forming fabric design, manufacturing, properties and testing are explained in detail. Finally, forming section configurations of paper machines are summarized.

Formation of the sheet is the most critical phase of paper manufacturing. If the sheet is not formed properly in the forming section, there is hardly anything that can be done to correct it later on in the remaining sections of the paper machine. Many sheet properties such as opacity, smoothness, tensile and tear ratio and formation are directly controlled by the forming section of the machine. The manner in which the filler, fiber and water flows on the forming section are handled controls sheet properties and contributes to the way the presses and dryer section will react to the fiber mat. The way the sheet behaves in the paper converting operations, particularly in the areas of roll condition and converter runnability, is a function of the operation of the wet end of the paper machine.

2.1 Sheet Formation Mechanism

There are various forming section configurations of paper machines and although the principle of sheet formation is the same, the process of sheet formation depends on the machine configuration. Therefore, sheet formation process and equipment may differ from machine to machine. Forming section configurations are given in Section 2.5. A rather detailed analysis of sheet formation on fourdrinier type machines is also given there. In this section, a generic description of sheet formation is given. Formation as a paper property is discussed in Chapter 6.

There are five basic processes which take place on the forming section of the paper machine. Each is interrelated with the other and each is necessary for good operation of the machine [1].

(1) The first step in the operation is to take the properly prepared furnish (mixture of water, fiber, fillers, fines and additives) from the stock preparation area and dilute it to a consistency low enough to permit easy relative motion between all the particles in the furnish, particularly the fiber. Consistency is a measure of the % solids in the stock slurry. The typical consistency at the headbox is around 0.5–1.0%. The turbulence developed in the piping, pumps, cleaners and the headbox results in a very uniformly dispersed stock slurry. Typical consistencies of paper grades at headbox are given in Table 2.1.

(2) The slurry is distributed across the width of the machine by the headbox and slice. A good headbox will distribute the fiber uniformly while maintaining good fiber

TABLE 2.1. Typical Papermaking
Consistencies at Headbox [2].

Paper Grade	Headbox Consistency Range (%)
Tissue	0.2–0.5
Newsprint	0.6–0.75
Bond	0.6–0.9
Corrugated and bleached board	0.7–1.0
Linerboard	0.4–0.5

dispersion. The uniformity of the distribution directly affects the uniformity of both the machine and cross machine basis weight as well as the area to area weight uniformity of the final sheet.

(3) The uniformly, well distributed, diluted slurry must be delivered onto the forming fabric through the slice or nozzle in such a way as to promote the formation of a uniform fiber mat on the top surface of the forming fabric. At the same time the water must begin to drain away through the forming fabric, thus developing the fiber mat (Figure 1.9).

(4) As the free water is being drained and while the fiber mat is still in a wet compressible state, it must be consolidated and compacted in a controlled way so as to increase fiber-to-fiber contact and gradually close up the porous structure of the web. Compactness of the filtered fiber mat increases along the movement direction of the fabric. Fiber mat thickness also

increases. The main factors that affect filtration of the furnish are temperature, forming fabric structure and properties, consistency, level of refining, fiber properties, additives and machine drainage elements [2]. Flow rates must be controlled to give acceptable levels of fines, filler and fiber retention. Fiber-to-fiber contact and pore structure greatly affect the final sheet properties.

(5) The web at the point of removal from the fabric must be as dry as possible to attain high wet web strength. Wet web strength increases with increased dryness. From the headbox to the point of transfer to the press section, gravity and induced drainage forces work together to drain as much water from the fiber mat as possible to give maximum dryness. The forming fabric itself, due to its structure, will contribute to the dryness of the sheet by letting water drain freely while providing support for the fiber mat being built on the top of the fabric.

The formation of the sheet is the result of physical interactions during the forming process. Factors involved in sheet formation are fiber dispersion, fiber orientation and degree of packing. The reactivity that cellulose fibers have with water to produce swelling and softening is crucial to sheet formation and structure. Hydrodynamic variables of the sheet forming process are the viscosity of slurry, drainage, turbulence and shear (Figure 2.1). Oriented shear is the result of the difference between fabric speed and jet speed

DRAINAGE   ORIENTED SHEAR   TURBULENCE

FIGURE 2.1. Hydrodynamic forces in sheet formation [3].

from the headbox which is called "jet-to-wire" ratio.

A measure of drainage ability is the freeness of the pulp. Freeness is the rate at which the water drains from the slurry. The amount of debris and fines in the slurry reduces freeness. There are two types of freeness testers: Canadian Standard Freeness (CSF) Tester and Schopper-Riegler Tester. TAPPI test method T 227 is designed to measure the freeness using the CSF tester. It should be noted that the freeness measurement does not predict the drainage on the paper machine. Drainage time of pulp is measured with TAPPI method T 221.

Unlike textile structures, which are also fibrous structures like paper, there is no interlacing (woven and braided textiles) or interlooping (knitted or tufted textiles) of fibers during sheet formation. During formation of the sheet, individual wood fibers are deposited on top of each other one after another. As a result, the paper structure looks like a stack of fibers. Therefore, theoretically, the last fiber deposited should be the easiest to pull out of the sheet structure first (before compaction in press and dryer sections) without disturbing the other fibers in the structure. Figure 2.2 shows the schematic comparison of various textile and paper structures.

During formation in the forming section and later in the press section, suction and compression forces consolidate the sheet web. These forces also cause the fibers on the web surface to conform to the topography of the forming fabric which may result in a wire mark in the sheet. Due to the contact between surface fibers and fabrics and metal surfaces during manufacturing, the structure of the paper surface is different from the rest of the sheet structure. In addition, the two surfaces of the sheet may also be quite different. In general, pulp properties that give a denser sheet structure also result in a smoother surface [4].

During formation, fibers tend to orient in the direction of the motion of the forming fabric, i.e., along the paper machine direction (MD). The tensile strength of paper increases with the increase in fiber orientation. Therefore, the tensile strength in the MD direction

may be different than that of the cross machine direction (CD).

Flocculation, which is the tendency of fibers to agglomerate, results in poor formation. Sheet formation quality requires that fibers are randomly oriented in the headbox and fibers and fillers are uniformly distributed. Constant turbulence is generated in the headbox to disperse flocs. Then, dewatering should be done fast enough to prevent formation of new flocs.

It is generally accepted that traditional sheet forming methods produce a layered structure where fibers lie approximately parallel to the paper surface. However, in high consistency forming, fibers are arranged as a three-dimensional network structure with fibers oriented to some extent in the thickness direction [5]. Table 2.2 shows the comparison of low and high consistency handsheet mechanical properties. A high consistency sheet has a drastically higher Scott Bond value than that of low consistency sheet. Thickness and bending stiffness are also increased. Compression strength slightly increases in high consistency forming while tensile strength decreases.

Various paper properties such as strength, appearance, amount of directionalism and dimensional stability are affected by formation. Therefore, some of these factors can be used to evaluate formation.

While the fibers are formed into a sheet, non-fibrous materials such as fillers in the furnish should be retained in the sheet structure as well. Various fillers are used to give different properties to the paper. Table 2.3 shows the typical size range of some filler materials. Fillers should be retained on the first pass on the forming fabric. For best retention of fillers, fines should form flocs which contradict the prevention of fiber flocculation for good formation. To increase retention, retention aids are used. High retention allows for lower headbox consistency which improves formation. Increased retention increases basis weight and strength of the paper.

As the sheet is formed on the forming fabric, the fiber network starts acting like a filter in addition to the action of forming fabric. This further helps with the retention of fines

plain weave (top view)

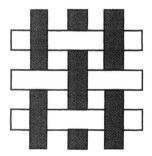

regular braid (top view)

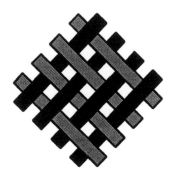

weft knit (top view)

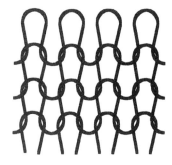

tufted (cross-section)

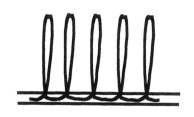

nonwoven (top view or
cross-section)

paper (cross-section)

FIGURE 2.2. Schematics of textile and paper structures.

*TABLE 2.2. Mechanical Properties of Low Consistency (LC) and
High Consistency (HC) Formed Handsheets [6].*

	Density (kg/m^3)	Bending Stiffness/ Grammage (Nm^7/kg^3)	Tensile Index (kNm/kg)	Compression Index (kNm/kg)	Scott Bond (J/m^2)
LC sheet	650	1.00	60	18	75
HC sheet	500	1.30	50	20	380
Ratio HC/LC	0.76	1.30	0.83	1.11	5.06

and fillers. The efficiency of the fiber network as a filter depends on its thickness and density, filtered particle size, stock freeness and hydrodynamic conditions such as pressure drop and fluid viscosity. The mechanisms for retention are mechanical integration, adsorption, chemical bonding and colloidal interaction. Dyes, sizing and bonding agents can be adsorbed or chemically bonded. Cellulose fines and mineral fillers are in general poorly retained. The probability of fiber interception by the forming fabric depends on several factors [3]:

- ratio of fabric grid spacing to fiber length
- initial position of fiber
- fiber friction
- drainage velocity
- fiber width and length
- fiber flexibility

2.2 Forming Fabrics

The forming fabric is the vehicle which ties together the pieces of the delivery zone in the forming section of the paper machine. The forming fabric has three distinct functions:

- It must allow water to pass through its structure.

TABLE 2.3. Typical Size of Filler Materials [2].

Filler	Size (μm)
TiO$_2$	0.2–0.5
CaCO$_3$	0.2–0.5
Clay	0.5–1.0
Talc	1.0–10.0

- It must support, retain and form the sheet.
- It must act as a conveyor belt transporting the sheet to the press section.

The top surface of the forming fabric acts as a filter cloth to create the base upon which the fibers are deposited to form the fiber mat or sheet. The geometry of the surface on which the sheet is formed contributes to sheet properties such as wire mark, linting and sheet smoothness. The better the quality of the support, the better will be the quality of the fiber mat and the higher will be the retention of fines, fillers and fibers on the supported side of the mat. In many grades of fine and printing papers, better retention will help minimize the two sidedness of the sheet surface properties which affect printing and other converting operations.

The bottom side of the fabric contributes to the conveyor belt characteristics of the fabric. Most of the life-reducing wear will occur on the bottom side because the wear-producing elements such as rolls, foils and flat box covers are in contact with the bottom surface. Machines with high drag loads require heavier duty fabrics to withstand the stretching forces and wear over the forming boards, foils, vacuum equipment and rolls. Drag and fabric wear (and life) are all related.

Since the fabric is somewhat elastic, it undergoes changes in length as it traverses the loop around the forming section of the paper machine causing wear over roll surfaces and machine elements. It is easy to see that conditions as a conveyor belt would be better if large, strong yarns could be used to resist stretch and increase the volume of yarn available to wear away. This is the opposite of the type of yarns we need to improve sheet quality by better fiber support.

The total fabric structure contributes to the ability of the fabric to act as a drainage medium. On the machine, the fabric has the capability to drain the headbox flow in a very short time, provided the flow consists only of water, as is the case during the initial start-up. It is only when the other parts of the furnish are added to the flow that the fabric appears to show significant drainage resistance. Once the fiber mat is deposited on the top of the fabric, the mat becomes the major source of drainage resistance. In the early part of the forming process, the way in which the fiber mat is built up governs the drainage. In the later part of the table, the ability to generate a pressure drop from the top of the sheet to the bottom of the fabric governs the drainage.

Mechanically a forming fabric must have:

- good wear resistance
- resistance to stretching, narrowing, skew, puckering, ridging and wrinkling
- guide-ability
- drive-ability
- resistance to high pressure shower and other damages
- the ability to be cleaned and have the sheet knocked off

A forming fabric should give the papermaker:

- good first pass retention
- good formation
- reduced power consumption
- reduced offset linting
- less two sidedness
- good strength properties
- proper surface topography to achieve desired paper properties

To achieve these tasks compromises are necessary. Forming fabrics are highly complex structures. A detailed understanding of each paper machine is needed to be sure that the right fabric is designed and used.

2.2.1 *Forming Fabric Designs*

Forming fabrics have more effect on the final sheet properties than press or drying fabrics which makes their design very critical. As a result, the design and manufacturing of forming fabrics require very special considerations and careful engineering.

There are basically three types of forming fabric designs used in the papermaking industry: single layer, two layer and three layer. Before we discuss these designs, a clarification needs to be made concerning manufacturing and installation of paper machine fabrics. A fabric can be woven in two ways as shown in Figure 2.3:

(1) Flat weaving
(2) Endless weaving

Flat woven fabrics have to be joined (seamed) to make an endless fabric for use on the paper machine. Warp direction on the weaving machine becomes machine direction (MD) on the paper machine and filling direction on the weaving machine becomes the cross machine direction (CD) on the paper machine. In endless weaving, the fabric is woven as an endless belt. The warp direction on the weaving machine becomes the cross machine direction (CD) on the paper machine and the filling direction on the weaving machine becomes the machine direction on the paper machine.

During weaving of the fabric, cross machine direction yarns are also called "shute" yarns. The term "shute," which is extensively used in the paper machine clothing industry, comes from the "shooting" of the yarn across the weaving machine with a shuttle. Another term for the filling yarn is "weft" which is commonly used in the textile industry.

Single Layer Designs

Single layer fabrics have one layer of warp yarns and one layer of weft yarns. Most single layer fabrics are woven with two to eight harnesses. Figure 2.4 shows warp profiles of some single layer fabrics.

Flat Woven

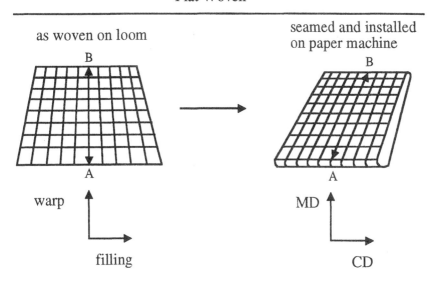

as woven on loom

seamed and installed on paper machine

Endless Fabric

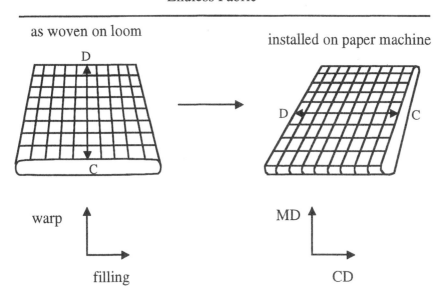

as woven on loom

installed on paper machine

FIGURE 2.3. Flat and endless fabric weaving and installation.

Plain weave design requires a minimum of two harnesses ("sheds") to weave. Paper machine clothing manufacturers use the term "shed" instead of "harness" to indicate the number of yarns in a repeat unit of the fabric.

A "shed" is the opening between the warp yarns during weaving. A harness is a mechanical device to which warp yarns are attached. Machine direction (warp) yarns pass over and under alternate cross machine direction yarns.

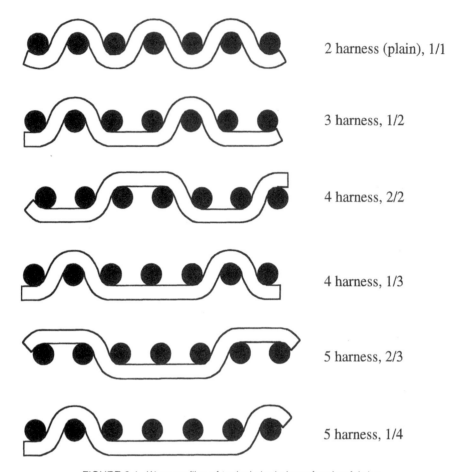

FIGURE 2.4. Warp profiles of typical single layer forming fabrics.

The meshes are usually coarse and used primarily in board, pulp and industrial cloth applications. During operation, single layer fabrics wear on both machine direction and cross machine direction yarns. Figure 2.5 shows a plain weave fabric. Plain weave gives minimum sheet marking but it is not a practical single layer fabric for normal papermaking.

Three harness (1/2 twill) designs are woven with the machine direction yarn passing over one cross machine direction yarn then under two cross machine direction yarns. The crossover is offset by one yarn each time, which forms a diagonal pattern in the fabric. This imaginary line is called twill line. This type of fabric is not utilized in many applications today.

Four harness designs can be achieved in different ways. Four harness twills are woven with the machine direction yarn passing under one cross machine direction yarn then over three cross machine direction yarns. The crossovers are offset by one, to form a diagonal pattern in the fabric similar to three harness but with longer yarn surfaces. This type of fabric is generally not used in paper manufacturing today.

Four harness broken twill is similar to the four harness twill, however, the crossovers are offset in a broken pattern. This type of pattern is sometimes referred to as a crowfoot pattern. The coarse mesh fabrics in this design are used in brown paper and are generally run with the long cross machine knuckle toward the sheet for better release. The medium mesh fabrics are used in the newsprint industry and are run with the long cross machine knuckle toward the machine

FIGURE 2.5. Photomicrograph of a plain weave fabric.

for better wear. The finer meshes in this weave are also run with the long cross machine knuckle toward the machine and are used for high quality printing, copying and writing papers.

In four harness double twill the machine direction yarn passes over two cross machine direction yarns then under two cross machine direction yarns. The crossovers are offset by one to form a twill pattern.

Five harness design also has different combinations. Figure 2.6 shows schematic of a five harness, 1/4 twill design. The warp yarn passes over 4 and under 1 weft yarn. The

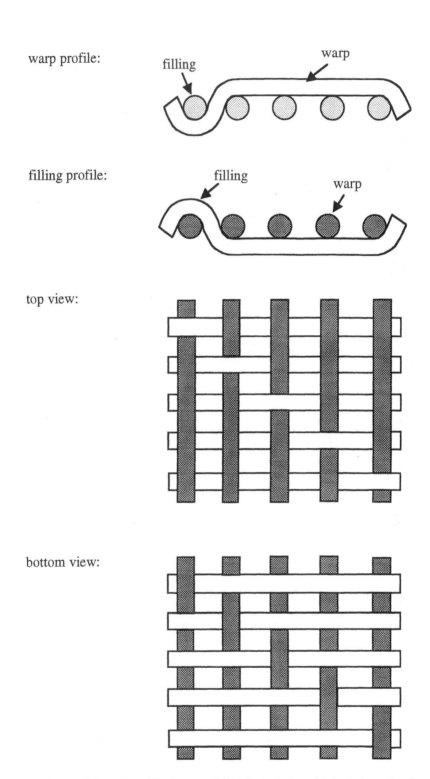

warp profile:

filling

warp

filling profile:

filling

warp

top view:

bottom view:

FIGURE 2.6. Schematics of five harness (1/4) twill single layer fabric design unit cell.

42

FIGURE 2.7. Top view of a five harness single layer broken twill fabric.

crossover points are adjacent so that they form a twill line.

Five harness broken twills are also woven with the machine direction yarn passing over four cross machine direction yarns then under one cross machine direction yarn. However, the crossovers are staggered to prevent a twill pattern as shown in Figure 2.7. These fabrics have low profile knuckling and smoother surfaces. Fabrics in this weave can be run with long machine direction knuckles to the sheet or long cross machine direction knuckles to the sheet depending on the grade of paper to be run. Suitable for a variety of paper grades, it is more easily adaptable for optimizing fiber retention than other designs.

Five harness 2/3 twills are woven with the machine direction yarn passing over three cross machine direction yarns and under two cross machine direction yarns (Figure 2.8). The crossover is offset by one yarn to form a twill pattern. This weave is used primarily in brown paper of all grades from lightweight linerboard, corrugating medium and bag to heavy weight linerboard. It offers good wear and release characteristics. These fabrics are

normally run with the longer machine direction knuckles to the sheet, but also can be run with the longer cross machine knuckles to the sheet for better release without sacrificing much life.

Two Layer Designs

In two (double) layer forming fabrics, there is one layer of warp yarns and two layers of filling yarns. There are two major variations of two layer fabrics: a) standard two layer and b) two layer extra designs.

a) Standard Two Layer Design

Figure 2.9 shows schematics of various standard two layer designs. There are several different weave patterns available in both seven and eight harness two layer weave patterns. These weaves can be used for all grades of paper as well as for industrial applications.

Four harness two layer fabrics have the wear knuckles in the machine direction. These fabrics generally employ heavyweight yarns. This weave is suitable for pulp, some brown

FIGURE 2.8. Five harness 2/3 twill.

paper and industrial applications such as fiberglass mat. Seven harness two layers offer a smooth surface and higher guiding efficiency over other double-layer weaves. This design is suitable for all grades of paper. Eight harness two layers are used on all critical finish grades.

Two layer designs offer high fiber support. The top filling diameters, coupled with the larger bottom fillings give the fabric unique properties. The result is improved formation, retention and life. Figure 2.10 shows a photomicrograph of an eight harness two layer fabric. Figure 2.11 shows warp and filling profiles of a four harness two layer fabric.

Advantages of two layer fabrics versus single layer:

(1) Greater dimensional stability and resistance to stretching
(2) Greater resistance to wear and damage
(3) Improved formation

- higher drainage capacity but at a lower rate (slower initial drainage)
- more uniform fiber distribution
- more fiber support
- more fiber retention
- less fiber deflection—less linting
- less two sidedness

Fabric stability is enhanced as a result of an increased number of MD yarns coupled with the beam-like structure developed by stacking the CD yarns. This provides for a forming surface with increased cross directional support, while at the same time maximizing the wear surface and reducing the load on the paper machine elements. This type of fabric is considerably less susceptible to deformation caused by the constantly changing tension, as it runs its course around the machine.

The unique potential of a two layer offers a long CD knuckle on both the sheet side and wear side. This offers two advantages. Larger

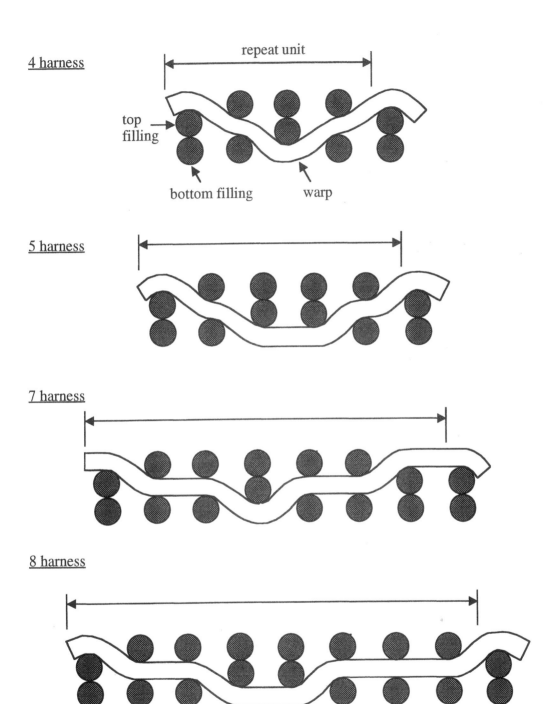

4 harness

repeat unit

top filling

bottom filling

warp

5 harness

7 harness

8 harness

FIGURE 2.9. Standard two layer fabric warp profiles.

45

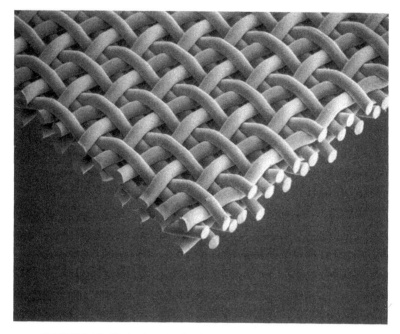

FIGURE 2.10. Photomicrograph of an eight harness two layer fabric.

yarn diameter is generally more effective in reducing wear. The long CD knuckle down shows the lowest wear rate. Two layers are generally designed with a smaller CD yarn on the sheet side for sheet quality and a larger CD yarn on the machine side for wear resistance.

The overall drainage capacity of a two layer design is greater than the comparable single layer. The two layer actually reduces the drainage rate in the initial forming zones where whitewater losses are the most dramatic. Water and solids must change direction of travel several times; reducing the velocity minimizes the amount of solids lost in the forming process. The increase in fiber support also reduces the penetration of the mat into the fabric mesh, which normally would retard drainage capacity and adversely affect fabric hygiene. Drainage capacity, which is directly proportional to the caliper of the forming fabric, is substantially increased. The stacking of the CD yarns affects the drainage capacity (void volume) of the fabric. This explains why a two layer with no projected open area is as effective as a single layer having a 20% open area or more.

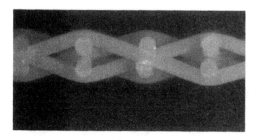
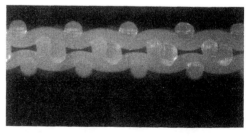

warp profile   filling profile

FIGURE 2.11. Warp and filling profiles of a four harness two layer fabric.

FIGURE 2.12. Top surface of a seven harness two layer fabric.

It is a well established fact that fibers tend to orient in the machine direction particularly at the bottom portion of the jet. Due to the hydrodynamic effect in the headbox, the fibers leaving the headbox are oriented in the jet direction before hitting the forming fabric. Therefore, the run attitude of the knuckles in the CD direction will dramatically improve the random distribution of fibers by turning the flow. The two layer design increases the number of CD yarns to improve this random distribution, while at the same time increasing the fiber support. Two layer fabric designs offer increased fiber support (Section 2.2.4) by substantially increasing the yarns per unit area and reducing the penetration of the fiber mat into the forming fabric surface. The fiber support of a two layer is approximately 20% greater than that of a single layer. Figure 2.12 shows the top surface of a seven harness two layer fabric. The two layer design offers greater weave variation than a single layer and hence better chances to enhance overall sheet appearance through desired sheet side surface characteristics.

An essential consideration in dealing with fiber support is knuckle area and height. The shape, number and location of knuckles have a dramatic effect on the fabric operating characteristics and machine performance. The two layer design allows optimization of fiber support in the forming zone. Running the long CD knuckle on the sheet side enhances formation and reduces fiber penetration. Two layer fabrics can be surfaced on the sheet side to increase the fiber support surface area (Section 2.2.3). The enhanced sheet release is an advantage of the two layer structure.

Wire marking is the result of optical and physical characteristics of the sheet caused by point to point density and surface differences of the fabric topography. Two layer designs minimize this point to point density difference by reducing the MD knuckle height above the plane of the fabric, utilizing small diameter MD yarns. This transfers a substantial amount of crimp from the MD yarns to the CD yarns, which reduces wire marking. Figure 2.13 shows crimp levels of yarns in a two layer fabric. Surfacing two layer fabrics enhances the monoplane characteristics of the fabric, thus further reducing wire mark.

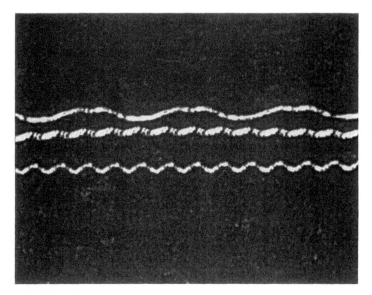

FIGURE 2.13. Crimp levels in a two layer fabric (top: warp, middle: top filling, bottom: bottom filling).

Explanation of X-Diagrams

X-diagrams are graphic representations of weave patterns. Figure 2.14 shows the X-diagram of a typical seven harness two layer design. Boxes in the vertical direction represent warp or machine direction (MD), boxes in the horizontal direction represent filling or cross machine direction (CD).

The X's show where outer warp knuckles are located on both the top and bottom sides of the fabric and the positions of the top and bottom warp knuckles in relation to one another, when observing the fabric from the top layer through to the bottom layer.

The blank boxes show where the filling is exposed in both layers. The following is shown in the top diagrams:

- The fabric is a seven harness repeat.
- In the machine direction, a warp knuckle is formed at a first pair of filling yarns in the top layer and at a fourth pair of filling yarns in the bottom layer for each warp yarn as shown by the warp profile.
- Warp knuckles are formed in a 1, 3, 5, 7, 2, 4, 6 filling sequence. This is a broken weave in a two step offset with the prominent diagonal running right.

- In the cross machine direction, the filling is exposed over six of the seven warp yarns in both layers, that is, over six, under one as shown by the filling profile.

For filling yarns of equal size, stacking is obtained by exact positioning of the cross machine direction yarns in the top layer over those of the bottom layer as shown in Figure 2.15(a). For filling yarns of nonequal size, stacking means good vertical positioning of the cross machine direction yarn whereby the top layer yarn (smaller) is 100% over the bottom layer yarn (larger) as shown in Figure 2.15(b).

Nonstacking is the nonvertical positioning of the cross machine direction yarns in a two layer fabric design [Figure 2.15(c)].

Low Density Two Layer Fabrics

The original designs of two layer fabrics were successfully run on many newsprint and fine paper applications. Most of these applications have required special showering to clean the fabrics. For other applications, such as tissue and brown paper, this same type of fabric has not proved very successful for several reasons:

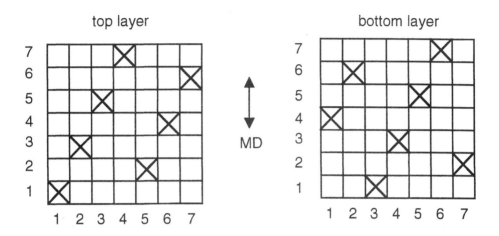

Weave Pattern
X-Diagram

FIGURE 2.14. Analysis of a seven harness two layer fabric design.

- drainage too slow resulting in poor formation
- inability to knock off the sheet or edge trim
- inability to clean heavy soil causing fabric to plug up

To overcome these problems, two layer fabrics were developed that had characteristics more similar to single layer so far as drainage and flow characteristics are concerned. These fabrics were called low density double layer (LDDL) fabrics. The advantages of low density two layer fabrics were:

- faster drainage
- easier to clean
- easier to knock off the sheet

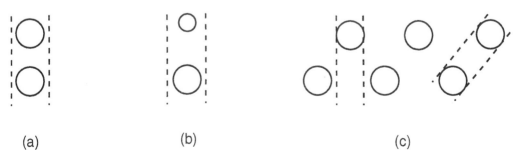

FIGURE 2.15. Stacking versus nonstacking.

- special showering not required in most cases
- improved wear over single layer
- dimensional stability
- damage resistance

The drainage index of a low density two layer design is approximately 5% greater than comparable single layer designs. Unlike traditional double layer designs, the low density two layer structure has projected open area so that sheet knock-off and fabric cleaning functions can be performed with the showering of a standard single layer.

On average, the fiber support of the low density two layer is 25% greater than that of normal single layers. This design gives substantial life increase compared to single layer fabrics, due to the uniqueness of the high plane difference between the CD and MD yarns. This difference results in the wear being transferred to the CD yarn. Thus, low density two layer design protects the load bearing yarn (MD yarn) to resist wear and increase fabric life.

b) Two Layer Extra Design

This is a two layer design consisting of two cross machine direction (filling) yarns on the sheet side for every cross machine direction yarn on the wear side (Figure 2.16). The two top filling yarns are designed to increase the fiber support index of the fabric (Figure 2.17). The bottom filling yarn is of an increased diameter to ensure long life.

There are several operating advantages of two layer extra design over conventional single and double-layer designs. Improvement in sheet formation is the main advantage. The high filling yarn count ensures fabric stability. This reduced tendency towards ridging improves life. The unique relationship of the filling yarns in the vertical plane ensures ease of cleaning. Figure 2.18 shows the warp and filling profiles of a typical two layer extra fabric.

The high fiber support, coupled with the reduction in crossover points improves sheet smoothness and reduces wire mark. These two physical properties are major considerations for the papermaker. Two layer extra forming fabrics incorporate additional CD yarns into the fiber support area or sheet side of the fabric. Each of these yarns alternates with a pair of stacked cross-direction yarns in the fabric construction. This construction results in increased fiber support and helps to eliminate fiber deflection without sacrificing drainage capability. Typically, standard two layer fabrics for fine paper and newsprint have 50–65 yarns per inch on the sheet side whereas two layer extra weave would have around 90.

Drainage characteristics are similar to two layer designs. The two layer extra drains the same as a standard two layer of the same permeability initially, but more efficiently later on due to the higher drainage index. Water and solids must change direction of downward flow several times; thereby reducing the velocity, under conditions of high hydraulic pressure. This reduces the amount of solids lost in the forming process. The two layer extra has double the drainage index of a standard two layer.

The two layer extra fabric gives improved sheet appearance and quality. Field results have shown that most cases of linting problems previously experienced with single layer and regular double layer designs are significantly reduced with two layer extra type designs. Formation problems caused by weak dimensional stability in single layers are eliminated with the two layer extra as a result of the enhanced number of MD and CD yarns per unit area. Two sidedness of the paper is also reduced to a minimum with the two layer extra fabric.

In a recent development of two layer extra designs, all top side yarns are the same size and woven into the fabric structure to maximize sheet quality and process stability. This design is especially suitable when maximum water is removed at low consistency. This design offers very high retention, formation and sheet runnability. In some two layer extra designs, finer double warp yarns are used against machine elements instead of a single large warp yarn.

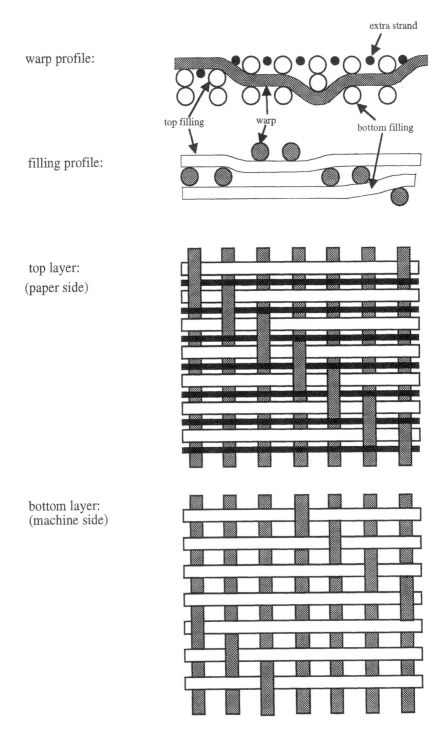

warp profile:

extra strand

top filling

warp

bottom filling

filling profile:

top layer:
(paper side)

bottom layer:
(machine side)

FIGURE 2.16. Schematic of a seven harness two layer extra fabric design.

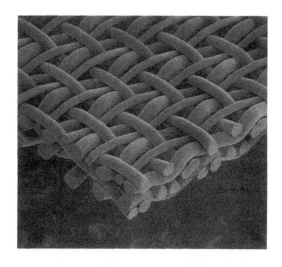

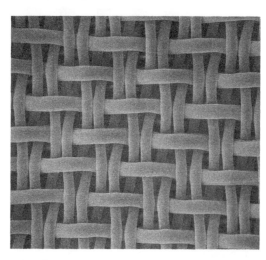

FIGURE 2.17. Photomicrographs of a typical two layer extra design; middle: top view, bottom: bottom view (magnification: 18×)

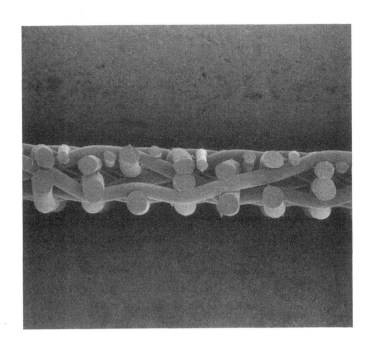

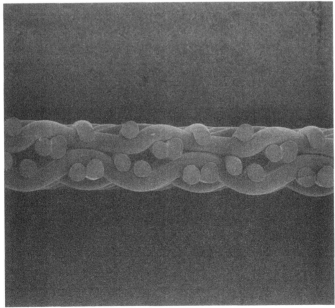

FIGURE 2.18. Warp (top) and filling profiles of a typical two layer extra fabric.

Three Layer Designs

Three layer fabrics have two separate fabrics—top and bottom—connected with a binder yarn (stitch) as shown in Figure 2.19. The fine top side of the fabric improves formation and fines retention. It also reduces wire mark, which is a major consideration in fine papers. The coarse bottom layer is specifically engineered to ensure long fabric life (Figure 2.20). The large filling and warp yarns allow decreased running loads.

The stitch yarn can be an additional yarn or part of the top or bottom layer fabrics. The additional stitch yarn is generally placed in a location where it is protected from wear by the large bottom yarns. Wear of stitch yarn can cause delamination of the two layers and failure of the forming fabric. Figure 2.21 shows X-diagram and warp and filling profiles of a typical three layer design.

Advantages of three layer designs:

- improved sheet support: a fine sheet side provides over 5000 fiber support points per square inch. On average, the fiber support of a three layer is 25% greater than that of standard double layers.
- improved formation and stability
- reduced two-sidedness
- increased flexibility in design by allowing each layer to be optimized independently
- longer life due to larger diameter wear yarn

The three layer fabric has improved sheet appearance and quality. Field results have shown that many cases of two sidedness previously experienced with double-layer designs are significantly improved with three layer designs. Formation problems caused by weak dimensional stability and the MD frame length with double layers can be eliminated with the three layer. This reduction in MD frame length (MD length is defined as the spacing between adjacent rows of CD directional yarns) has led to higher retentions due to less surface area open to embedment of the fiber from the stock jet. The three layer has a noticeable shorter MD hole, thus providing significant improvement in both fine and fiber retention.

The uniqueness of the three layer offers the opportunity to design each layer independently. The three layer fabric is designed with a fine mesh top layer to provide the ultimate sheet support. This equates to improved formation, increased fines retention and reduced two sidedness. The increase in fiber support means that the sheet releases easier, thus reducing machine draws. The rugged bottom fabric provides a durable base for increased fabric life and dimensional stability. Three layer fabrics are used for all grades of paper.

The unique potential of the three layer offers a long CD knuckle on both the sheet side and wear side. Three layer fabrics are generally designed with smaller CD yarn on the sheet side for papermaking and a larger CD yarn on the machine side for wear resistance. This offers two advantages: (a) larger yarn diameter is generally more effective in increasing life, and (b) The long CD knuckle down shows the lowest wear rate.

Finally, it should be noted that although three layer designs offer these potential advantages, in actual practice many are outperformed by two layer extra designs in retention, smoothness, wear and ring crush on brown grades. Therefore, the two layer extra design is a major competitor of three layer designs, which, in many cases, is replacing three layer applications. Cost is another factor that is on the side of the two layer extra fabrics.

Other Designs

Other forming fabric designs are also possible. In one design variation, there are three layers of weft yarn and one layer of warp yarn as shown in Figure 2.22(a). An extra fine yarn may also be used on the top layer as shown in Figure 2.22(b).

2.2.2 *Forming Fabric Design Variables*

The yarns used in warp and filling directions in forming fabrics have different properties. The rule of thumb in a forming fabric is

warp profile:

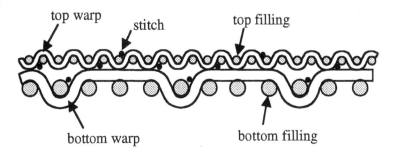

top warp   stitch   top filling

bottom warp   bottom filling

filling profile:

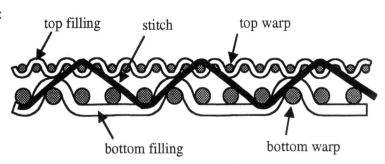

top filling   stitch   top warp

bottom filling   bottom warp

top layer:
(paper side)

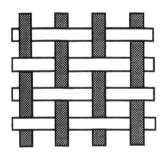

bottom layer:
(machine side)

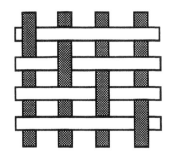

FIGURE 2.19. Schematic of three layer fabric design.

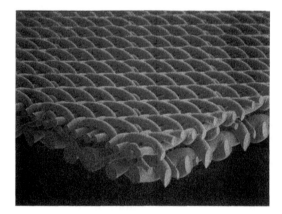

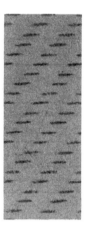

top surface impression

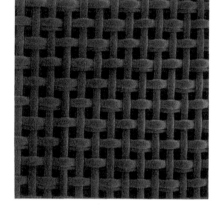

four harness top layer   plain bottom layer

FIGURE 2.20. Photomicrograph of a three layer fabric.

that CD yarn is for wear and MD yarn is for load bearing on the paper machine. Therefore, higher modulus yarns are used in MD direction to reduce stretching on the paper machine.

The fiber support and drainage characteristics of most fabric designs can be varied by changing yarn diameter, mesh count, weave and yarn predominance.

By proper selection of yarn diameter and mesh, it is possible to obtain a variety of hole sizes and open areas to meet the required machine conditions. If, for example, it has been determined that the hole size of a fabric is adequate, but a different percent open area is needed, a change in MD yarn diameter and mesh will change the percent open area [7]. Figure 2.23(a) illustrates this situation.

Similarly, if the percent open area is satisfactory but more sheet support is required, a similar change can be made to give a greater number of smaller holes with the same percent open area. In Figure 2.23(b), an additional MD yarn is brought into the picture. By proper selection of the additional number of MD yarns, the percent open area could be held constant.

As the fabrics get finer with increasing mesh count and smaller yarns, the bridging

X - diagrams

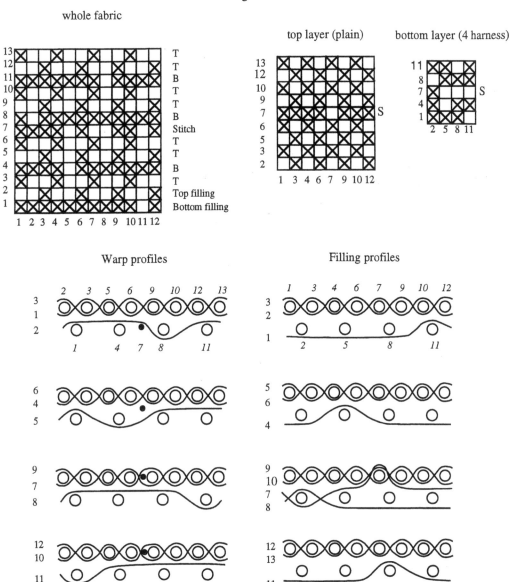

FIGURE 2.21. X-diagram and warp and filling profiles of a typical three layer design.

location moves higher in the structure of the fabric (Figure 2.63 in Section 2.3). Depth of penetration of the knuckle becomes more favorable, and at the same time the location of the bottom of the fiber-filler-fines plug moves upward.

The crowding of the MD yarns by the CD yarns increases with increasing MD yarn count. Because of finishing and operating tensions at the points of crossover contact, the MD yarn tends to ride up the side of the CD yarn knuckle. This results in a lower height of the knuckle.

The closely spaced MD yarns in the span between CD yarns and the climbing effect of the MD yarn at the knuckle combine to give

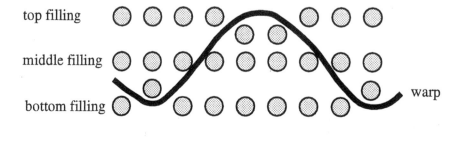

top filling

middle filling

bottom filling

warp

(a)

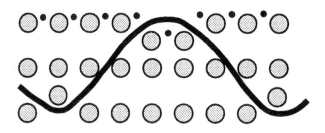

(b)

FIGURE 2.22. Triple shute weave.

a fabric with a sheet side surface that is smooth compared with the surface of a fabric with a lower MD yarn count. This smoother sheet-side surface will improve the surface smoothness and density uniformity of the final sheet.

Fabric warp profiles are used to determine whether a multilayer design is symmetric or nonsymmetric. Consideration must be given to a full warp cycle. Two layer symmetry or non-symmetry is determined by the warp path between the filling layers as referenced from the short warp knuckle. Symmetric designs have equal division or arrangement on either side of a common reference plane or division line as shown in Figure 2.24. There is a balance of opposite corresponding parts in size, shape or position.

Nonsymmetric designs have unequal division or arrangement on either side of a common reference plane or division line as shown in Figure 2.25. Opposite corresponding parts are disproportionate in size, shape or position.

2.2.3 *Manufacture of Forming Fabrics*

The design and manufacture of a forming fabric for a paper machine is a complex task. In general, forming fabrics are paper machine specific, i.e., quite often a forming fabric is custom designed for a particular position on a particular paper machine. Papermakers demand a long-lasting forming fabric which will enable them to produce high quality paper products. Forming fabrics are manufactured to meet various demanding requirements which include stability, wear resistance, proper drainage capacity, uniform fiber distribution, no wire marking and proper retention. Figure 2.26 shows the flow chart for manufacturing of forming fabrics.

Forming Fabric Materials

Until the 1950s, fabrics used for forming were made of phosphor bronze or stainless

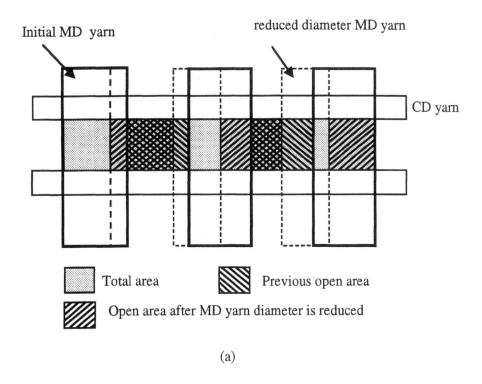

(a)

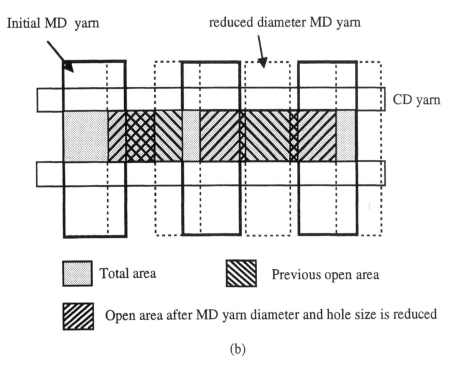

(b)

FIGURE 2.23. Changing the percent open area in a forming fabric.

7 harness

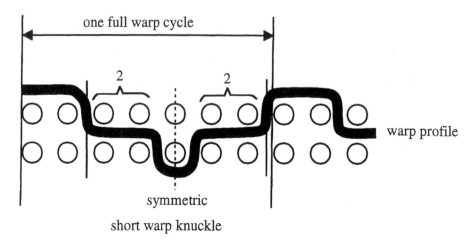

8 harness

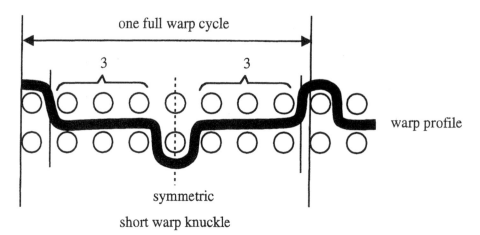

FIGURE 2.24. Examples of symmetric fabric designs.

steel, and because they were made of metal, called wires. There were very few design variations. The average fabric life was one to two weeks. Fabric stability was good, but the forming fabric was prone to damage due to creases and metal fatigue. Metal fabrics had high drainage rates and poor fiber retention and sheet release. Figure 2.27 shows a three harness, metal wire which was a common design before synthetic fabrics. The first synthetic forming fabric was used in 1958 (Table 2.4). Although fabric life was not improved

initially, sheet quality was better. By mid-1970s, 75% of all paper machine clothing had mono- or multifilament yarns. Multilayer forming fabrics were introduced in the mid-1970s. In 1976, three layer fabrics were introduced. Today almost all of the forming fabrics are made with high molecular weight and high modulus monofilament polyester. For wear resistance, alternating nylon yarns can be used in the bottom layer of multilayer fabrics (Figure 2.28). However, nylon is not dimensionally stable under wet conditions which limits

7 harness

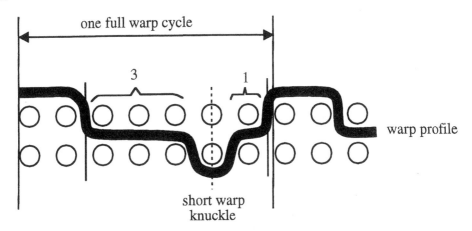

8 harness

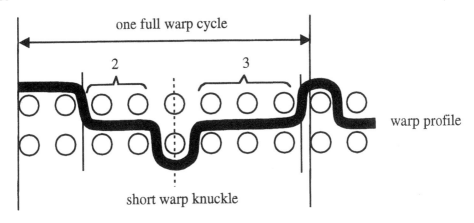

FIGURE 2.25. Examples of nonsymmetric fabric designs.

its use in forming fabrics. Nevertheless, with the increase of alkaline papermaking, more and more nylon yarns are being used along with polyester. Polyester has good stress-strain properties and it is resistant to acid, but is subject to hydrolysis under alkaline conditions. Nylon also has more memory than polyester; therefore, it takes longer to recover.

Figure 2.29 shows specific heat of polyester and nylon yarns. Specific heat is an indication of energy needed to bring the yarn up to a certain temperature. Nylon takes more heat to get it to a certain temperature than polyester. Therefore, to reach the same temperature, the heatset amount should be increased. Mass of the yarn also has an effect.

Nylon and polyester have the following comparative properties:

- For flexural modulus, nylon is softer than polyester.
- For compression resistance (and impact resistance), nylon is superior. Polyester tends to fracture under pressure. That

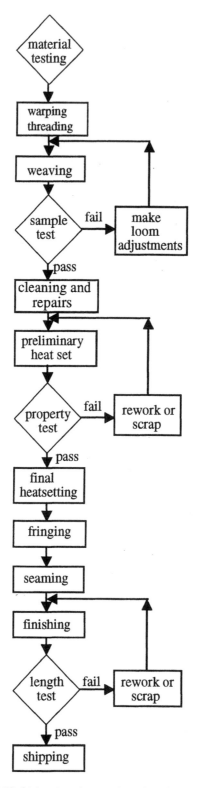

FIGURE 2.26. Major steps in manufacturing of forming fabrics.

FIGURE 2.27. A three harness, metal wire.

is why nylon is used in press fabrics (Chapter 3). These properties are inherent in the polymers.

- Heat transfer will be different for nylon and polyester.
- The specific heat of nylon is more than polyester.
- The abrasion resistance of nylon is higher than polyester.
- Polyester has higher density than nylon.

TABLE 2.4. Historical Development of Forming Fabrics.

1957	First synthetic forming fabrics patented
1958	First commercial applications of synthetic fabrics
1965	First monofilament fabrics for fine paper
1973	Monofilament fabrics for brown paper
1975	Three layer fabrics patented
1976	First three layer fabric trial on fine paper
1979	First successful double layer on newsprint
1982	Straight-thru design patented
1985	Double layer extra design patented
1986	Four layer design trial

Today, with the use of recycled fibers, contaminants in the pulp such as pitch, wax, latex, etc., may create problems during paper manufacturing. Therefore, yarns with contaminant resistance properties are in high demand. There are several ways to achieve contaminant resistance such as yarn coating, contaminant resistant additives to the polymer and fabric coating. Fabric coating also improves stability. Generally, fluorocarbon based coatings are used in fabric coating.

There are several properties of polymers that affect forming fabric manufacturing and performance. Intrinsic viscosity (I.V.), which is dimensionless, is the ratio between the time the dissolved polymer takes to travel a specific distance through a glass tube known as a capillary viscometer, in comparison to the time that a solvent alone would take to travel that distance. High I.V. resin may likely produce more slubs than low I.V. (a slub is a thick place on the yarn).

Modulus is an indication of the strength of yarn material. Modulus is the ratio of stress to strain that can be determined from the slope of the stress-strain curve. High modulus yarns

WEAR   FUNCTION

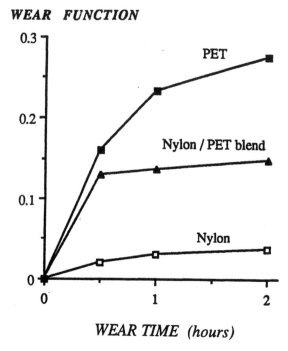

WEAR TIME (hours)

FIGURE 2.28. Comparison of nylon and polyester for wear resistance (courtesy of Textile Research Institute).

are stiff with small elongation under load. Low modulus yarns are extensible with high elongation under load.

There are several modulus values defined in the technical literature [8]: Young's modulus, initial modulus and chord modulus. ASTM (American Society for Testing and Materials) defines Young's modulus as the ratio of

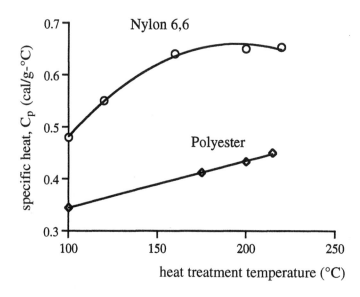

FIGURE 2.29. Specific heat of polyester and nylon (courtesy of Shakespeare Monofilament).

change in stress to change in strain within the elastic limit of the material [9]. The ratio is calculated from the stress (force per unit cross-sectional area) and strain (fraction of the original length). Initial modulus is the slope of the initial straight-line portion of the stress-strain curve. Chord modulus is the ratio of the change in stress to the change in strain between two specified points in the stress-strain curve.

Shrinkage properties of forming fabric yarns are important. High shrinkage is good to obtain low caliper fabrics. For forming fabrics, warp should have low elongation (high modulus) and filling should have low shrinkage. As drawing increases, modulus and shrinkage also increase. Drawing is the stretching of the yarn under load between the rollers (Figure 2.30). In general, increased draw results in lower abrasion resistance.

Hydrolysis resistance is another important property for forming fabric yarns. Staboxyl is an additive for hydrolysis resistance that is added to the extruder.

(1) Staboxyl I: liquid, lower molecular weight
(2) Staboxyl P: solid, higher molecular weight
(3) Staboxyl P100: solid, highest molecular weight

Nucleating agents are added to polymer to speed up its crystallization. If crystallinity is high, then the physical dimension is very stable but elongation and strength are sacrificed.

Conductive monofilaments are used in forming fabrics for nonwoven manufacturing. Polyester or nylon monofilaments can be suffused with conductive carbon to make them conductive enough to control electrostatic discharge and conduct certain low voltage currents. The outer skin of monofilament is chemically suffused with tiny conductive/carbon particles which become part of the structure of the fiber. Depending on the denier, the resistivity range is 10^3–10^6 ohm/cm. A typical 21 denier carbon suffused nylon monofilament features a resistivity level of 10^5 ohm/cm per filament [10].

Manufacture of Yarns for Forming Fabrics

Yarns for forming fabrics are produced with the melt spinning method. In melt spinning (also called extrusion), polymer chips are fed to the extruder (Figure 2.30). They are melted under heat and forced through the tiny holes, the diameter of which depends on the yarn diameter to be produced, in the spinnerette (Figure 2.31). Extruded filaments which come out in liquid form from the spinnerette are stretched and cooled with water or air and solidified. After the application of a finish, the filaments pass around a series of godets and wound onto a yarn package. The godet speeds determine the degree of orientation and speed of production. The diameter of filaments is controlled by the godet take-up speed and extrusion pump. Figure 2.32 shows a monofilament extrusion line.

Yarn Material Testing

In addition to usual production and quality testing done by the yarn manufacturer, the forming fabric manufacturers frequently test the incoming materials for various properties.

Diameter (Size)

Diameter of round or nonround monofilament yarns can be measured using a laser device or precise micrometer. The contact pressure during testing with micrometer should limit compression of the monofilament to less than 1/1000 mm.

Yarn Linear Density

It is not practical to express the linear density of yarns in weight per volume. Therefore, the units denier and tex are used to indicate the relation between weight and length of yarns.

Denier is the weight (in grams) of the length of 9000 meters of yarn. Specimens are cut to a length of exactly 1 meter and weighed together on an electronic analytical balance

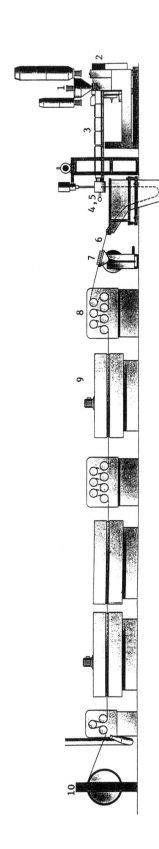

FIGURE 2.30. Schematic of melt spinning process: 1. hopper for polymer chip, 2. extruder, 3. barrel (heated), 4. spin pump, 5. spinnerette, 6. extruded filaments, 7. finish application, 8. godet, 9. oven, 10. winder (courtesy of Reifenhauser).

FIGURE 2.31. Spinnerette.

FIGURE 2.32. Monofilament extrusion line (courtesy of Asten Monotech).

TABLE 2.5. Value of K for Some Yarn Materials [11].

Material	K
Nylon 6 and Nylon 6,6	5.2
Nylon 6,10	4.92
Polyester	6.3
PEEK	5.9
HPA-40	6.13
HPA-80	6.12
HPA-14, HPA-500	6.19
Polypropylene	4.1

(accuracy of reading is 1 mg). For round monofilaments, the denier may be calculated by the formula,

$$\text{Denier} = d^2 K \qquad (2.1)$$

where d is diameter of yarn expressed in mils (1 mil = 1/1000 inch). K is a constant for a particular material whose values are given in Table 2.5 for some materials.

Yards per pound of yarn can be calculated by dividing 4,464,414 by the denier.

Example: 0.0045 inch diameter polyester.

$$\text{Denier} = (4.5)^2 \times 6.3 = 127.5$$

Yards per pound 4,464,414/127.5 = 35,015

Tex is the weight in grams of 1000 meters of yarn and is given by:

$$\text{Tex} = \text{Denier}/9 \qquad (2.2)$$

$$1 \text{ Decitex} = \text{Tex}/10 \qquad (2.3)$$

Appendix A gives the size-denier relationships for polyester and nylon monofilaments.

Tensile Properties

Tensile testing is done according to ASTM test method D 2256, Standard Test Method for Tensile Properties of Yarns by the Single Strand Method [9]. The following values are reported:

- load (kg)
- tenacity (g/denier)
- elongation at break (%)
- elongation at specific load (%)

Figure 2.33 shows the stress-strain charts for warp and filling polyester yarns.

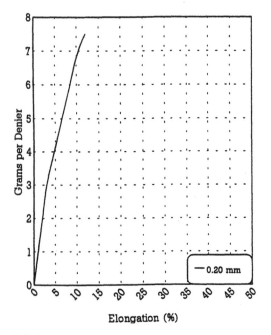 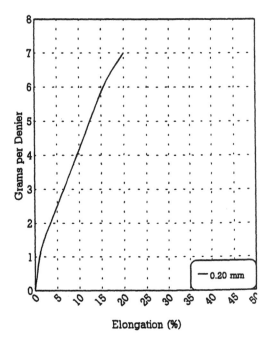

FIGURE 2.33. Stress-strain diagrams for warp (left) and filling monofilament yarns (courtesy of Shakespeare Monofilament).

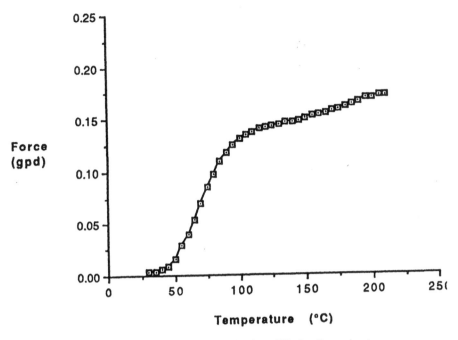

FIGURE 2.34. Force-temperature relation for a CD direction polyester yarn.

Shrinkage (%)

To measure shrinkage, specimens are cut to an exact length of 1 meter. The samples are then coiled into a ring without tension and placed in a forced air oven for 15 minutes. After removal from the oven, the specimens are cooled and their lengths are measured. Oven temperature is either 140°C (284°F) or 200°C (392°F) depending on the end use.

Shrink Force and Post Force

Shrink force and post force of the yarns are measured using a shrinkage testing device. The yarn is placed in the heat chamber of the device and the shrink force is measured with a tensiometer while the sample is in the heater to 200°C (392°F). Figure 2.34 shows the force-temperature diagram for a CD direction polyester yarn. After the removal from the heat chamber, the sample is cooled. The post force is then measured with a tensiometer.

Color

Color is measured with a colorimeter. American Association of Textile Chemists and Colorists (AATCC) has developed various test methods related to the color of textile materials [12].

Weaving

Weaving is the method of fabric formation in forming fabric manufacturing (the term *formation* in this sentence is a textile term while the term *forming* refers to paper sheet formation). Figure 2.35 shows the major components in weaving.

In weaving, two sets of yarns (warp and filling) are interlaced with right angles in established sequences. Weaving consists of five basic steps.

(1) Shedding (Figure 2.36)—the raising and lowering of the warp ends by pneumatic or hydraulically actuated heddles thus forming an opening through which the filling yarn is inserted. This opening is called "shed." The heddles are attached to and controlled by a harness.

(2) Filling insertion (Figure 2.37)—transfer of filling yarn from one side of the loom to the other. The shuttle has been the primary

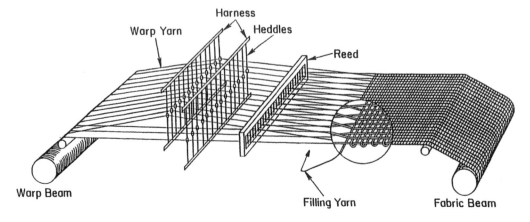

FIGURE 2.35. Basic components of weaving (courtesy of Wellington Sears Company).

device for filling insertion over the years in forming fabric manufacturing. Figure 2.38 shows major types of shuttles used for filling insertion. In pirn shuttle, a small supply of filling is held on a removable pirn which is inserted into the shuttle. When the pirn is empty, the pirn detector stops the loom, and the weaver replaces the empty pirn with a full one. From the pirn, the filling is threaded through a

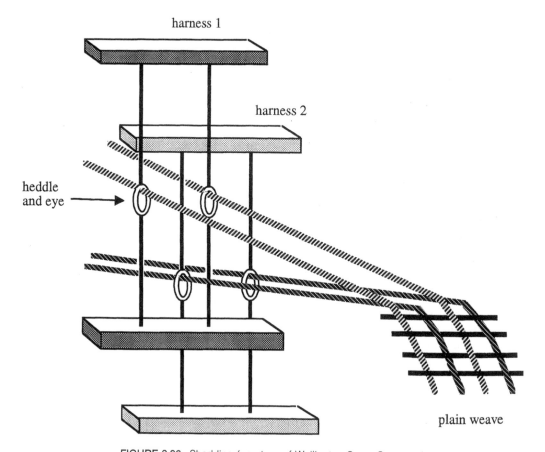

FIGURE 2.36. Shedding (courtesy of Wellington Sears Company).

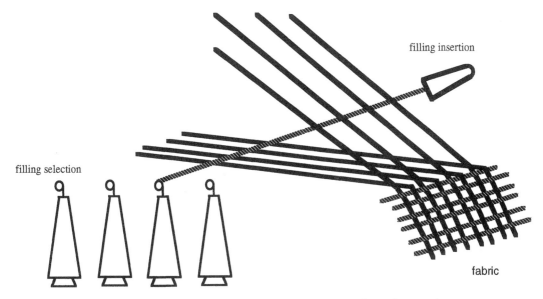

FIGURE 2.37. Filling insertion (courtesy of Wellington Sears Company).

Pirn shuttle

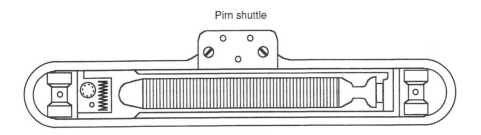

Gripper shuttle

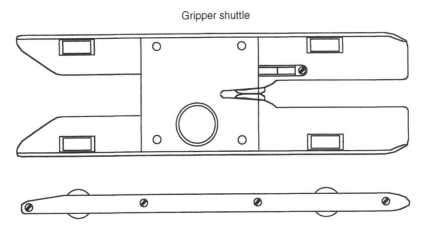

FIGURE 2.38. Shuttle types used in weaving forming fabrics.

tension device, through a filling die, around a roller, through one or two eyelets, and between two more rollers before it exits the shuttle. This type of shuttle is fired back and forth across the loom by means of spring loaded picker sticks, pneumatic or hydraulic pistons, and pays out filling from the pirn as it travels. In some looms, a metal (titanium) shuttle with four vespel wheels and four axles is used.

In the latest weaving machines for forming fabrics, projectile or rapier insertion methods are used as shown in Figures 2.39 and 2.40. In filling insertion with projectile, also called gripper, the projectile grabs the filling yarn and draws it into the shed by gliding in a rake-shaped guide. The projectile is braked in the receiving unit and conveyed to its original position under the shed. In filling insertion by flexible rapier, the pick is inserted into the shed by the right-hand gripper head, is transferred to the left gripper at mid-width and then pulled to the edge of the fabric. Modern weaving machines for forming fabrics are computer controlled. The width of these machines depends on the width of the paper machines. Today, weaving machines for forming fabrics are over 12 m (472 inches) wide for flat weaving and 30 m (1181 inches) wide for endless weaving (Figure 2.41).

(3) Beating up (Figure 2.42)—pushing the newly inserted filling yarn into the proper position in the fabric using a reed

(4) Take-up—taking up the newly made fabric onto the cloth roll

(5) Let-off—releasing warp ends from the warp beam

The first three steps are essential for the weaving process while the last two steps provide continuity of the weaving process. The let-off and the take-up work together to maintain the correct warp tension, the key factor influencing the yield of the fabric. Figure 2.43 shows the fabric construction parameters.

Warp Geometry from Guide Roll to Whip Roll

The relative position of guide roll, heddle frames and whip roll plays an important part in fabric quality. The imaginary line from the guide roll to the fabric fell is called warp line or zero line (Figure 2.44). This is especially important on multilayered fabrics, where stacking of filling yarn layers is necessary. The guide roll may be raised or lowered. Raising the guide roll decreases warp tension on up shed relative to down shed. Similarly warp tension changes can be achieved by adjusting harness travel and/or whip roll angle. The whip roll adjustment raises or lowers the cloth near face. It is primarily used to make the warp ends match the angle of race.

The distance from the reed to the fabric being woven must be maintained the same or a beat change will occur. When sheds are too open during beat-up (open shed beat-up), they may cause fluting which is a wavy appearance at the cloth fell.

Once the design, reed, loom, loom set-up, raw materials, and procedures to be used have been chosen, the only items left in weaving to "control" are:

- pick count
- yield
- fabric consistency

Pick (Shute or Filling) Count

Along with weave, mesh, void volume, caliper, holes/unit area, hole size, diagonal and open area, pick count is a very important factor in determining the air permeability of the fabric (Section 2.2.4). The air permeability, although not the only factor involved, plays a major role in determining how a fabric will drain on the paper machine. In general, the higher the air permeability, the better a fabric will drain.

Of equal importance to the target pick count is the minimization of abrupt pick count changes within a woven piece of cloth. Although a papermaker might be able to tolerate

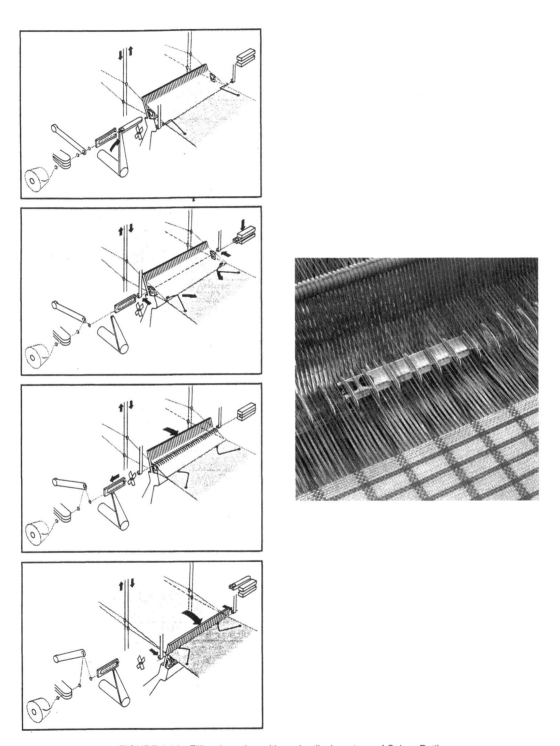

FIGURE 2.39. Filling insertion with projectile (courtesy of Sulzer Ruti).

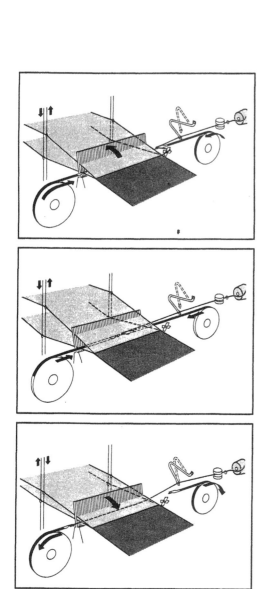

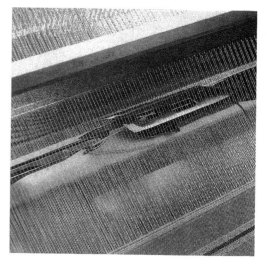

FIGURE 2.40. Filling insertion with rapier (courtesy of Sulzer Ruti).

FIGURE 2.41. State-of-the-art weaving machines for forming fabrics.

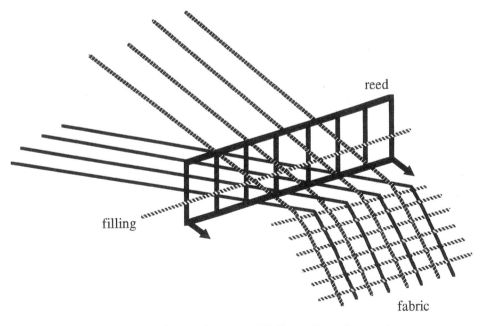

FIGURE 2.42. Beat-up (courtesy of Wellington Sears Company).

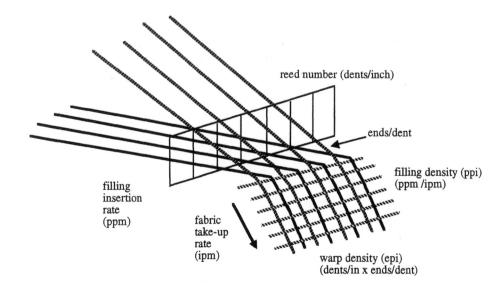

construction = epi x ppi

FIGURE 2.43. Fabric control parameters on the weaving machine (courtesy of Wellington Sears Company).

gradual and minor pick count changes, a change which occurs within a few shots (such as what occurs during start-up of a loom) may cause an abrupt change in fabric drainage on the paper machine in this very small area. These changes show up as definite weak places in the finished sheet of paper.

Contrary to popular belief, these pick count changes will remain in the fabric throughout its manufacture, i.e., they do not improve with

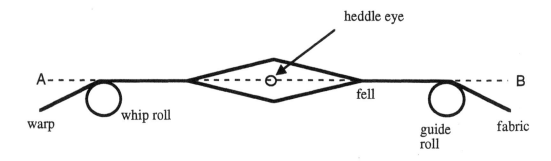

FIGURE 2.44. Warp line (zero line), AB.

preliminary heatsetting. Although the woven pick count is changed by heatsetting because the fabric is stretched, the pick count difference remains constant. Moreover, if the fabric pick count is not constant, it will make seaming more difficult.

Yield

The second attribute that can be controlled during weaving is yield. Yield is defined as the ratio of warp yarn length to the woven fabric length. Yield affects the fabric modulus. The modulus is a measure of fabric strength and stability, in that it measures how much a fabric will stretch when a given tension is applied to it.

Older paper machines were designed to run metal wires, which had a modulus of approximately six times more than the current polyester fabrics. Thus, they were built with very small adjustments for tensioning the wires or fabrics, sometimes as little take-up as a few inches on machines designed to run fabrics one hundred feet long. Today, if the fabrics are made too short, the mill cannot even get the fabric on the machine; if they are too long, the mill cannot put enough tension on the fabric to take out all the slack and provide drive friction.

The yield, along with fabric design, material type, weave, caliper, warp size, and preliminary heatset tension, determines how much a fabric will stretch in the heatsetting operation. This will affect the fabric modulus and the fabric final sizing to achieve the proper length. This in turn determines fabric stability.

As with pick count, the uniformity of the yield is also important because of the interaction between the fabric properties. Changes in yield can cause differences in fabric thickness, air permeability, stacking in two layers, pick count, fabric width and appearance, as well as in fabric modulus. Yield change during weaving causes stacking to change.

Yield can be measured with a simple method. Make a mark on the warp yarns and on the fabric at the same time. After, say, 10 cm of warp has come off the warp beam then make a mark again on the fabric and the warp. Measure the amount of fabric which was made in 10 cm of warp and divide this by 10 which gives the percent yield in the fabric.

Fabric Consistency

The importance of fabric consistency in the areas of pick count and yield control is mentioned earlier. Papermaker fabrics, as shipped, are effectively an endless belt. Modern paper machines run at speeds around 2000 meters per minute (6562 feet per minute). If a fabric is 30 meters long, that means that a little defect comes around on the paper machine 66 times every minute! It can mean hundreds of blemishes in every roll of paper.

Defects have a tendency to project above the plane of the cloth, even after they are repaired. They, therefore, have less resistance to wear on the machine and can quickly turn into holes. The wood fibers can no longer be evenly distributed, the spacing may not be the same as the rest of the fabric and fiber support differs in this area. As a result, the repaired area will tend to leave a wire mark on the paper sheet.

Preliminary Heatsetting

After weaving, fabrics are subjected to heatsetting at high temperatures. The purposes of heatsetting are to relieve the internal stresses in the flat woven fabric, to increase fabric modulus and to improve dimensional stability of the fabric. For heatsetting flat woven fabrics, the fabric ends are temporarily seamed and the fabric is rotated around rolls on heatsetting machines (heatsetting "tables") as shown in Figure 2.45.

Typical loads during preliminary heatsetting are 0.4–0.8 gram/denier. During heatsetting, the forming fabric is stretched in the MD direction and also tentered along the CD direction as shown in Figure 2.46.

As the fabric is woven, the filling has very little crimp compared to the warp. During heatsetting, warp crimp decreases and filling

FIGURE 2.45. Heatsetting of forming fabrics.

crimp increases which is called crimp transfer or crimp interchange. Figure 2.47 shows crimp transfer during heatsetting of forming fabrics.

Hot air or infrared heating techniques are used for heatsetting of forming fabrics. Hot air heatsetting is relatively faster. It is also more suitable for multilayer fabrics.

The warp count during heatset can be calculated for single and two layer fabrics as follows:

New warp count

$$= \text{(initial fabric width before the heatset}$$

$$\times \text{raw warp count)}$$

$$\div \text{fabric width on heatset table} \qquad (2.4)$$

Initial fabric width can be taken as any distance, i.e., it does not have to be the whole fabric width. However, the same distance has to be measured on the heatset table as well.

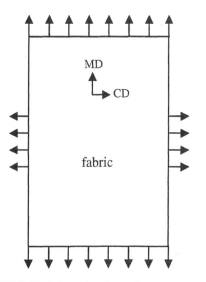

FIGURE 2.46. Schematic of stretching and tentering during preliminary heatset.

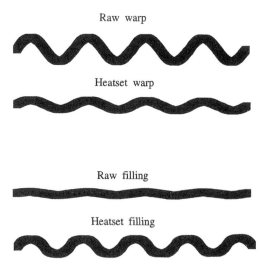

FIGURE 2.47. Crimp transfer during heatsetting.

FIGURE 2.48. Fringing.

Seaming

Most forming fabrics are woven flat on the looms. However, since the fabric is used like a continuous belt on the paper machine, the free ends of the flat woven fabrics must be seamed. Seaming is done after heatsetting, i.e., after the crimp is established in the warp and filling yarns.

In the seaming process, filling yarns at each end of the fabric are removed which is called "fringing" (Figure 2.48). Then, the fringed warp ends are brought together and re-woven using filling yarns.

Figure 2.49 shows typical seam styles for single layer forming fabrics and Figure 2.50 shows examples of seam styles for two layer forming fabrics.

In a special type of seaming technique called "pin seam," the warp ends are made into loops and a pin is used to connect the ends together as shown in Figure 2.51. Pin seamed fabrics are easy to install on the paper machine. However, drainage may be different in the seam area which is critical for mark sensitive papers.

Seaming is one of the most time consuming and tedious processes of forming fabric manufacturing. A considerable amount of the manufacturing cost of a forming fabric

comes from seaming. There has been extensive research and development work done by forming fabric manufacturers to automate the seaming process. Figure 2.52 shows manual and automatic seaming machines.

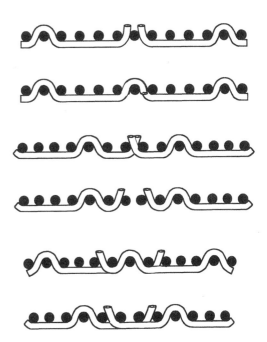

FIGURE 2.49. Typical seam styles for single layer forming fabrics.

4 harness

5 harness

7 harness

8 harness

FIGURE 2.50. Examples of seam styles for two layer forming fabrics.

single layer

two layer

FIGURE 2.51. Schematics of pin seams.

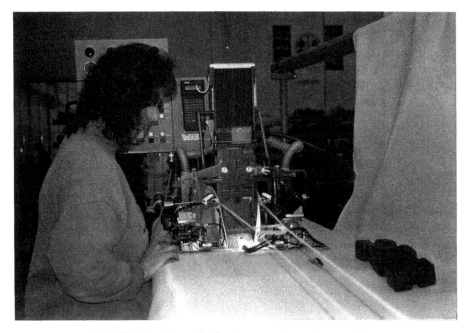

FIGURE 2.52. Manual (top) and automatic seaming machines.

Final Heatsetting and Finishing

After seaming, forming fabrics are subject to final heatsetting mainly to stabilize the seam area by relieving the stresses. Although the whole fabric is subjected to heat, final heatsetting is done at a much lower temperature than the preliminary heatsetting, therefore the properties of the fabric such as air permeability, crimp ratio, modulus, etc., which were set during the preliminary heatsetting, are not altered. During final heatsetting, the protruding ends of warp yarns in the seam area are shaved (Figure 2.53). The length and width dimensions of the fabric, which is already in a belt form, are established. The fabric edges are treated to prevent edge curl. A narrow band of resin is applied to the edges for strength and to prevent unraveling of warp yarns.

Fabric Treatments

Fabric treatments are intended primarily to reduce the tendency for fabric contamination, thus reducing drainage loss as a fabric runs for a period of time. In general, these treatments are tough film formers applied during the finishing process.

Most recent work has centered on specific treatment properties as they relate to surface charge and surface wettability. Surface charge is a significant factor as it relates to the zeta potential of the furnish. Interaction of electrochemical forces has an impact on the tendency for particular materials to accumulate on the fabric surface. Surface wettability becomes a factor because it tends to reduce the tendency for various materials to adhere to the base monofilament material. The layer of water carried on the surface of the monofilaments makes the removal of contaminant materials easier by means of normal showering.

On several applications, treated fabrics have reduced draws between the fourdrinier and first press as a side benefit.

Surfacing

If required by the application such as some fine papers, the surface of the fabric can be smoothed by surfacing which is also called sanding. The purpose is to eliminate the height difference between the warp and filling yarns and obtain a monoplane surface for good fiber support. Figure 2.54 shows a fabric surface after sanding.

After fabric is finished, it is boxed in proper containers to prevent any possible damage during shipment (Figure 2.55).

FIGURE 2.53. Fabric finishing.

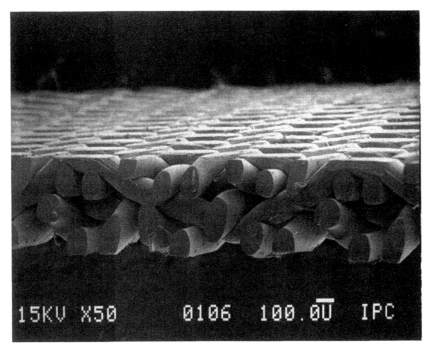

FIGURE 2.54. Surface of a two layer fabric after surfacing.

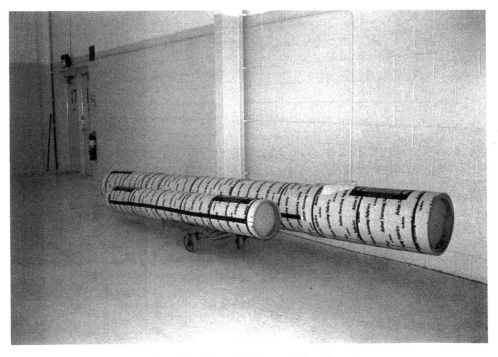

FIGURE 2.55. Boxed fabrics ready for shipment.

Potential Manufacturing Problems

Manufacturing of forming fabrics is a challenging process. Sometimes, it is not possible to get the desired fabric structure due to conflicting properties. High stiffness of monofilament yarns contributes to this. Heavy monofilaments are also the cause of extremely high weaving forces which require heavy duty, special weaving machines.

Figure 2.56 shows improper stacking in a two layer fabric, i.e., the top and bottom filling yarns are not stacked on top of each other properly. By changing the weave pattern, shed opening, weaving tensions, yarn materials or the position of the guide roll on the loom, this problem may be solved. It should be noted, however, that what may seem to be a problem in this case may be good for certain paper grades to obtain different properties in the sheet.

Figure 2.57 shows twinning problem in a two layer fabric. The top filling yarns are paired together which may result in nonuniform drainage and cause marking in the sheet. Again by changing yarn materials, weaving tensions or weave pattern, the problem may be eliminated.

The ridge in the fabric could be traced to a band of MD yarns that maintained a tension level different from that of the adjacent yarns. Differences in tension across the fabric can cause hills and valleys to appear parallel to the machine direction within the forming area of the paper machine wet end. Since the stock will tend to run off the hills and into the valleys, streaks of varying basis weight or mass will result in the sheet.

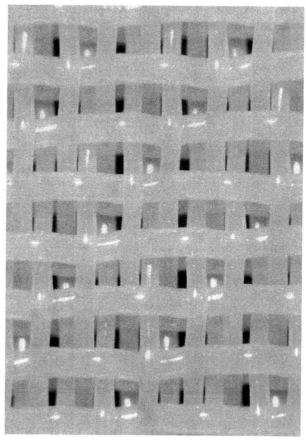

FIGURE 2.56. Improper stacking of top and bottom filling yarns in a two layer fabric.

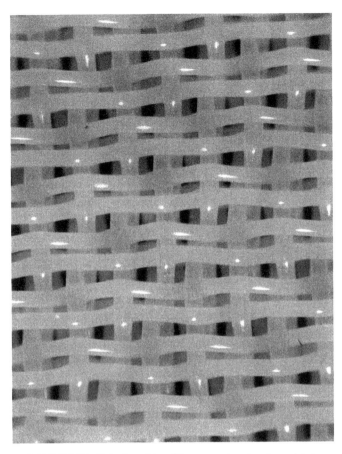

FIGURE 2.57. Twinning of top filling yarns in a two layer fabric.

2.2.4 *Properties and Testing of Forming Fabrics*

Many different factors are taken into account and are essentially built into the fabric when it is designed in order to satisfy the end use requirements. The following design and manufacturing parameters are calculated and recorded:

- material type
- mesh (warp and filling count)
- warp and filling sizes and types
- weave pattern
- plane difference
- percent projected open area
- air permeability
- void volume
- fiber support index
- drainage index
- number of holes per unit area
- hole size (length × width) and diagonal
- fabric thickness
- tensile strength and modulus
- yield as woven
- percent stretch in finishing
- maximum temperature and tension during heatset
- seam tensile strength

Mesh (Warp and Filling Count)

Warp count is the number of machine direction yarns per unit width of fabric and filling count is the number of cross machine direction yarns per unit length of the fabric. Generally, the finer the mesh, the finer the paper grade and the paper quality. Fineness has no direct correlation to fabric drainage. The absolute drainage of a fabric is related to the fabric's "openness" which not only is dependent on the weave pattern and mesh but also on the size (diameter) of the yarns used to weave the fabric. To maintain a level of "openness" for drainage, as fabrics are made finer and finer, the yarn diameters are reduced. Smaller diameter yarns result in reductions of wear resistance and stability.

Lunometers and counting glasses are used to count the number of warp and filling yarns in fabrics. As an approximate rule, a variation of ± 1 in warp count and a variation of ± 2 in filling count per inch are generally acceptable.

Plane Difference

In many instances, machine direction and cross machine direction monofilament yarns are not in the same plane on either side of the forming fabric. This topography is referred to as plane difference (Figure 2.58). In general, when the MD yarn is above the CD yarn, it is referred to as "warp runner" and when the CD yarn is higher than the MD yarn, it is referred to as "shute (filling) runner." When both are in the same plane, the fabric surface is "monoplane."

To obtain good fabric life, it is desirable to have the wear yarn over the other yarn to take the anticipated abrasion. For good fiber

"shute" runner

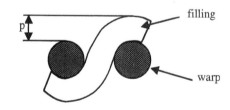

warp runner

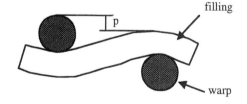

p : plane difference

FIGURE 2.58. Plane difference.

support on the sheet side, in many cases, it is desirable to be as close to monoplane as possible, thus giving support from both the CD and MD yarns to provide low wire marking. In tissue, plane difference is good for bulk generation.

With multilayer structures, some of which are inherently stable, material stiffness and heatsetting techniques are used to optimize plane difference levels. Optimization is possible for both the sheet and wear side for a particular application. Sometimes, forming fabrics that will be used for finer paper grades are polished with an abrasive roll. Polishing (sanding) removes the acute knuckle points on the fabric and provides a smooth, monoplane surface as explained earlier.

Percent Projected Open Area

It is a calculated number that describes, in a plane view, the percent projected open area. When comparing fabrics made with the same weave pattern and materials, it is a fair indicator of how each fabric will handle water. Open area is not a good indicator of drainage capability if changes are made to weave patterns, material or mesh. Open area does not take into account weave pattern, caliper, plane difference, material, level of crimp and wetted surface area. Open area is normally calculated based on mesh and yarn diameters.

% Open area

$$= [1 - (\text{warp count})(\text{warp size})]$$
$$\times [1 - (\text{filling count})(\text{filling size})] \times 100$$

$$(2.5)$$

For single layer fabrics, the percent open area is always larger than zero. For two layer fabrics, it ranges from 0 to a certain value. Figure 2.59 shows different levels of open area for different two layer forming fabrics. A 0% open area does not mean that there is no opening for water drainage. Figure 2.60 shows the top view and 30° angle view of a two layer fabric. Although the projected open area is zero, there is plenty of open area when seen from a 30° angle.

FIGURE 2.59. Different open areas in various two layer fabrics.

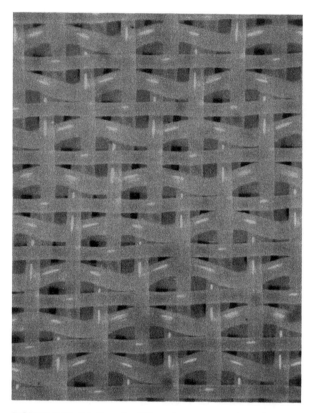

FIGURE 2.59 (continued). Different open areas in various two layer fabrics.

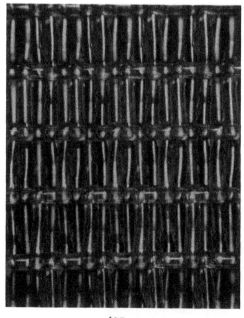

top

30° angle

FIGURE 2.60. Top view (left) and 30° angle view of a two layer fabric.

Air Permeability

It is a measure of air flow through a forming fabric at the standard pressure drop. The standard conditions are cfm/sq ft at 0.5 inch water column pressure.

Air permeability is of considerable value in predicting drainage characteristics of a fabric. It is, however, an air flow measurement, whereas the papermaker is interested in water flow. Within a given weave pattern and mesh range, air permeability, as a drainage indicator, may not correlate with actual machine data as sheet support characteristics also influence water handling capability. Nevertheless, air permeability is one of the most important properties of forming fabrics.

Air permeability is also a good indicator of fabric to fabric uniformity as long as weave and yarn diameters are constant.

Void Volume

Void volume, by definition, is the amount of space in a volume of fabric that is not occupied by solid material (Figure 2.61). Most single-layer and two layer fabrics have void volumes ranging from 60–70%. Both fabric thickness and internal structure affect absolute void volume.

Some studies have shown that having a high level of interconnected internal void volume could enhance dewatering over vacuum augmented boxes. On the other hand, a fabric with too high a void volume can lead to a wetter than normal sheet off the couch, due to the fabric having a high volume of air carried in the fabric.

Within a forming fabric, the void volume is not evenly distributed; some areas are more closed and it is the most closed area that best describes the water handling capabilities of a particular fabric. This area, called the "Minimum Average Hole Area (MAHA)" or "bottleneck," can be found by physically sectioning the fabric and measuring percent open area at the different planes through it.

The passage of water through a fabric is limited by the least permeable plane of a bottleneck within the fabric structure, but not by the projected open area. This bottleneck changes in shape and location when the material used to make the fabric is changed. It can also be altered by using different weave patterns. Void volume can be calculated as follows:

Total volume = length × width × caliper

Calculated void = total volume

  − (weight of sample/density of material)

% void volume

  = (calculated void/total volume) × 100

$$(2.6)$$

Fiber Support Index (FSI)

The concept of fiber support index was developed by Beran [13]. His work, based on a statistical model, predicts the fiber support characteristics of forming fabrics based on

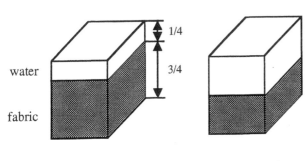
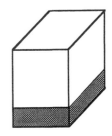

25% void volume   50% void volume   75% void volume

FIGURE 2.61. Void volume.

mesh, weave type and run configuration, i.e., long warp knuckle up or down (Section 2.3.2). Beran made four major assumptions:

- fabrics woven were planar grids with no thickness
- support from yarns is based on a line with no width
- yarns are straight and parallel
- yarns are uniformly spaced

Having made these assumptions, Beran then devised a rather complicated mathematical formula for his model. This was, however, simplified to his basic formula:

$$FSI = \frac{2}{3}(a \times N_m + 2 \times b \times N_c) \quad (2.7)$$

where

N_m = number of MD yarns per inch
N_c = number of CD yarns per inch
a = support factor for warp yarns
b = support factor for filling yarns on sheet side

Support factors a and b are calculated as follows:

$$a = \frac{\text{number of filling yarns under warp knuckle} + 1}{\text{number of harnesses}}$$

$$b = \frac{\text{number of warp yarns under filling knuckle} + 1}{\text{number of harnesses}}$$

Table 2.6 lists the single layer FSI factors, a and b. In general, the warp factor, which contributes to the fiber support number, decreases with the increase in number of harnesses.

FSI calculation examples for single layers:

- 80 × 75 mesh, 1/4 weave with long warp knuckle down

$$FSI = \frac{2}{3}[0.4(80) + 2(1)(75)] = 121.3$$

- 80 × 75 mesh, 1/4 weave with long warp knuckle up

$$FSI = \frac{2}{3}[(80) + 2(0.4)(75)] = 93.3$$

- 80 × 75 mesh, 2/3 weave with long warp knuckle down

$$FSI = \frac{2}{3}[0.6(80) + 2(0.8)(75)] = 112$$

FSI factors for two layer and two layer extra designs are determined by viewing fabric samples and establishing warp and filling profiles. For two layer designs, half of the total filling count is used since filling yarns are stacked. For two layer extra, 2/3 of the total filling count is used. For three layer factors, the factors for top layer (sheet side) only are used.

Examples of FSI calculations for multilayer fabrics:

- two layer (warp count = 162, seven harness, number of filling yarns under warp knuckle = 2, total filling count = 120, number of warp yarns under filling knuckle = 5)

$$FSI = \frac{2}{3}[162(3/7) + 120(6/7)] = 114.9$$

- two layer extra (warp count = 150, seven harness, total filling count = 135)

$$FSI = \frac{2}{3}[150(2/7) + 180(6.5/7)]$$
$$= 140.0$$

- three layer (top layer: plain weave, top warp count = 80, top filling count = 140)

$$FSI = \frac{2}{3}[(1)80 + (1)140] = 146.7$$

For fiber support index, a difference of between 5 to 10 is considered to be significant. There are several interpetations of Beran's work. In comparing index values, it is important that the same formulas are used. It should also be noted that Beran's work was done with single layer fabrics and the results are extrapolated to multilayer fabrics. There are

TABLE 2.6. Single Layer FSI Factors.

Weave	LWU		LWD	
	a	b	a	b
1,1	1	1	1	1
1,2	1	0.67	0.67	1
1,3	1	0.50	0.50	1
1,4	1	0.40	0.40	1
2,2	0.75	0.75	0.75	0.75
2,3	0.80	0.60	0.60	0.80
1,6	1	0.29	0.29	1
2,5	0.86	0.43	0.43	0.86
1,7	1	0.25	0.25	1
2,6	0.89	0.38	0.38	0.89

LWU: Long warp up.
LWD: Long warp down.

no known studies extending the theory to multilayers.

Wear Prediction

Several methods of wear prediction have been given in the literature. A general formula that has been used is:

Wear Prediction Number (W)

$$= 0.5\ N^{0.5}L^{0.5}D^2 \qquad (2.8)$$

where

N = number of knuckles/square inch
L = knuckle length
D = cross direction (CD) yarn diameter

The number of machine direction yarns divided by harness number gives the number of knuckles along the CD yarn per one inch. When this number is multiplied by the number of CD yarns/inch, the result is knuckles/square inch. The reciprocal of the machine direction yarns multiplied by the float length in the CD yarn results in the knuckle length.

Example:

For a four harness (1/3), 86×60 design with an 8 mil diameter CD yarn

$$N = (86 \div 4) \times 60 = 1290\ \text{knuckles/in}^2$$

$$L = (1 \div 86) \times 3 = 0.035\ \text{inch}$$

$$W = 0.5\sqrt{1290}\sqrt{0.035} \times 8^2 = 215$$

The Wear Prediction Number, W, is a relative number and only has merit in comparison with other fabric design wear numbers. Higher W indicates better wear resistance. For example, a design with $W = 215$ should show better wear characteristics than a design where the wear number is 200. The wear prediction number should be used as a "tool" in the overall evaluation of fabric designs for machine applications. It should be noted that material type is a key factor for wear resistance.

Drainage Index (DI)

The Drainage Index was developed by Johnson as a tool for rating fabrics based on relative drainage rates [14]. Drainage Index takes into account air permeability in cubic feet per minute, filling count, and filling factor as follows:

$$DI = b \times P \times N_c \times 10^{-3} \qquad (2.9)$$

where

b = support factor for filling yarns on sheet side
P = air permeability
N_c = filling count

For two layer fabrics, only half of the total filling count is used for calculation since only half of the filling yarns support the sheet. For double layer extra fabrics, 2/3 of the total filling count is used in the calculation. For three layer fabrics, only the top layer filling count is used.

DI calculation examples:

- two layer (air permeability = 320 cfm, total filling count = 120, seven harness)

$$DI = (6/7)320 \times 60 \times 0.001 = 16.5$$

- two layer extra (air permeability = 400 cfm, total filling count = 135, seven harness)

$$DI = (6.5/7)400 \times 90 \times 0.001 = 33.4$$

- three layer (air permeability = 400 cfm, top filling count = 70, plain weave)

$$DI = (1)400 \times 70 \times 0.001 = 28.0$$

In general, a difference of 1 or greater in drainage index is considered to be significant.

Hole Size (Length and Width) and Diagonal

Hole dimensions are significant for predicting the fiber bridging ability of the fabric.

MD dimension of the hole

$$= [1 - (\text{filling count})(\text{filling diameter})]$$

$$\div \text{filling count} \qquad (2.10)$$

CD dimension of the hole

$$= [1 - (\text{warp count})(\text{warp diameter})]$$

$$\div \text{warp count} \qquad (2.11)$$

Hole diagonal =

$$\sqrt{(\text{MD dimension})^2 + (\text{CD dimension})^2} \qquad (2.12)$$

Holes per unit area = warp count × beat count

$$(2.13)$$

Fabric Thickness

Fabric thickness is important since it affects the drainage characteristics. Too thick fabrics may cause water carryback. The standard test method for measuring thickness of textile materials is given in ASTM D 1777 [9].

Tensile Strength and Modulus

Fabric modulus is measured using ASTM Test Method D 885 [9]. Tensile properties of fabrics are measured using ASTM test method D 5034. Specifications of textile machines for tensile testing are described in ASTM D 76. The terminology of tensile properties of textiles are given in ASTM D 4848.

The tension measurement along the fabric width is done during manufacturing (on the heatset table) to determine if there is any length difference in the fabric along the machine direction.

Seam Tensile Strength

Seam strength is extremely important for proper functioning of fabric under high tension on the paper machine. Ideally, seam strength should be as good as fabric strength. However, seam strength is generally less than that of the fabric body. Seam efficiency is defined as

Seam Efficiency (%)

$$= (\text{Seam tensile strength}$$

$$\div \text{Fabric tensile strength})$$

$$\times 100 \qquad (2.14)$$

2.3 Applications of Forming Fabrics

The design of the forming fabric has an effect on sheet smoothness, retention characteristics and sheet integrity, which in turn affect the optical and physical properties of the sheet.

The forming fabric is just one parameter in the sheet forming process. The stock furnish, consistency, freeness, pH and fiber distribution are some of the other factors that will affect the final sheet properties. The retention of the fines and fillers must be optimized during formation. The sheet formation must be uniform. Fiber alignment in various directions (including the Z-direction) must be carefully controlled to ensure that there are not significant differences. The Z-direction changes will cause curl which will put constraints on other parts of the paper machine.

Sheet formation is affected by the bridging of the open areas in the forming fabric by the long fibers that are placed onto it. This bridging will subsequently affect the retention of the fines and fillers that are inherent in the sheet. Fabric hole size and shape affect the sheet characteristics. By changing the weave pattern, mesh and beat count, different hole sizes and shapes can be obtained as shown in Figure 2.62. In general, optimum bridging takes place over a square open area. This allows the fibers to interact and form a lattice that will in turn affect the fines and fillers retention.

In a fabric structure, the point of initial fiber bridging in the fabric is the point where a plug of fiber-filler-fines begins to form. As the fabric structure is changed by altering yarn diameters, the depth in the fabric of initial fiber bridging is changed. Using finer yarn diameters and higher mesh counts while maintaining or increasing the percentage of open area decreases the depth of the probable point of bridging. This in turn results in a plug of less total mass because it is not formed as deeply in the fabric. This shallower plug mass comes closer to matching the average mass of the sheet. In other words, the height of the hills on the fabric side of the sheet becomes less. A lower hill means more uniform calendering and a smoother sheet.

In Figure 2.63, $d_1 > d_2$. The fiber length a is the same in both cases which is equal to the distance between two adjacent yarns. As it can be seen from the figure, the depth of the fiber bridging is less in the case of small diameter yarns because $b > c$.

Wire mark is the result of optical and physical characteristics of the sheet caused by point to point density differences and surface differences caused by fabric topography. By minimizing the plug density effect and the height of the knuckle above the plane of the fabric, the wire mark will be minimized. Surfacing

of fabrics to obtain monoplanarity is done to reduce knuckle height.

Differences in tension across the fabric can cause hills and valleys to appear parallel to the machine direction in the forming area of the fourdrinier table. Since the stock will tend to run off the hills and into the valleys, streaks of varying basis weight or mass will result.

Forming fabrics have an average running life of three to six months on the paper machine. However, it should be noted that the life of a fabric depends on many things including the installation. Even a small damage to the fabric during installation may make the fabric useless, reducing the life to zero days. On the upper extreme, some forming fabrics, especially coarse fabrics, can run more than a year on the new, modern machines. A vast majority of fabrics are removed due to damage; very few are actually worn out.

2.3.1 *Fabric Design Considerations*

There are two major sets of considerations for forming fabric applications: paper type and paper machine design as shown in Table 2.7. Each of these factors in itself is extremely important. A survey of the machine is required prior to the manufacture of the first fabric. The survey should be updated periodically (Chapter 5).

Paper Considerations

The grade of paper is the first consideration. Table 2.8 lists recommended typical designs for common paper grades. Some machines run several different grades of paper and, generally, one design is required to run them all. It is then important to know what the predominant grade is as well as the most costly grade. The fabric may have to be designed to favor the highest quality grade, yet can only be designed to run the remaining grades adequately.

The type of furnish is important in determining how large a hole size can be tolerated. The makeup of hardwood, softwood, recycled and rag, as well as the combinations made from these greatly affect the decision

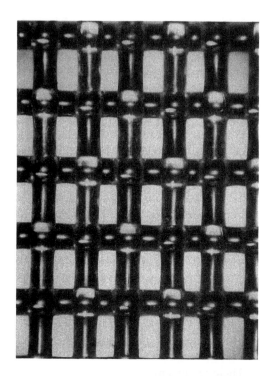

FIGURE 2.62. Various hole sizes and shapes (rectangular, square and trapezium).

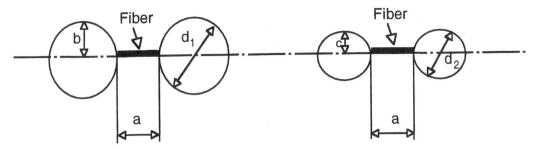

FIGURE 2.63. Fiber bridging.

on the number of cross machine yarns in the design as well as the fineness of the design. Due to the rising cost of pulp and the trend to make things recyclable, more and more recycled furnish is being used. This presents a finer fiber to the furnish and is more apt to bleed. Finer mesh designs are required for a furnish containing recycled material. The weight range as well as the predominant weight is important and requires the same considerations as the grade of paper to be run.

Paper Machine Considerations

Type and manufacturer of paper machine indicates what basic machine is being dealt with and along with width and length, indicate if a design has to be coarse or fine. Knock down type machines generally have more problems with fabric sizing than do cantilever machines. Twin-wire machines have two design considerations (Section 2.5.2). The speed range as well as the predominant speed is critical. Most generally the fabric will be designed to run best at the predominant speed and run adequately at the various speed changes.

In fourdrinier machines, the table configuration indicates how drainage takes place on the table as well as the sheet formation. A table consisting of all rolls could require a more open fabric design than a combination of rolls and foils or all foils (Section 2.5.1).

The operating tension and the available take-up are extremely critical. Long machines with short take-up, operating at high tension, will experience difficulty in fabric installation. This becomes a more serious problem if a fine mesh fabric design is required to run the product. The solution would be to install adequate take-up, but generally a design trade-off for a higher modulus coarser mesh fabric is made to compensate for the lack of take-up, resulting in quality somewhat lower than what would be expected from a finer fabric.

2.3.2 *Design Selection*

Proper choice of fabric design can provide benefits in minor variations and proper manufacture of the chosen fabric will relieve more serious problems. Fabric history is a major consideration in eliminating trial and error in fabric design. It establishes a good sense for

TABLE 2.7. Information Needed to Select the Proper Fabric Design for a Given Application.

Paper	Paper Machine
• Grade of paper to be run	• Type and manufacturer
• Type of furnish (recycled stock content)	• Size (width and length)
• Weight range of product line	• Speed range
• Wire mark considerations	• Type of pickup (open draw, suction or shoe)
• Type of forming (pressure or velocity)	• Table configuration, rolls versus foils, etc.
• Fabric history	• Operating tension and available take-up
• Similarity to other successful applications	

TABLE 2.8. Typical Recommended Designs for Common Paper Grades.

	Mesh (warp count × shute count); Air Permeability (cfm)			
	Single Layer Four Harness	Single Layer Five Harness	Double Layer	Three Layer
Corrugated Med.		46 × 37; 600	126 × 90; 550	78 × 72; 500
Linerboard				
lightweight		45 × 45; 580	126 × 96; 500	78 × 70; 400
medium/heavyweight		46 × 37; 605	126 × 90; 550	78 × 72; 550
Bleachboard		62 × 46; 650	144 × 135; 470	78 × 70; 400
Bag	58 × 48; 600	64 × 40; 550	144 × 135; 470	78 × 70; 400
Bond		92 × 70; 560	144 × 135; 400	72 × 80; 450
Duplicating		92 × 70; 560	144 × 135; 470	72 × 80; 450
Writing		92 × 70; 560	180 × 126; 450	72 × 90; 400
Directory		92 × 70; 560	144 × 135; 470	72 × 80; 450
Newsprint		92 × 70; 560	180 × 126; 450	72 × 70; 500
Rotogravure		92 × 70; 560	180 × 140; 400	72 × 80; 450
Coated book		92 × 70; 560	144 × 135; 470	72 × 80; 450
Publication		92 × 70; 560	144 × 135; 470	72 × 80; 450
Glassine		92 × 70; 560	144 × 135; 470	72 × 80; 450
Wrapping		92 × 75; 515	144 × 135; 400	72 × 90; 400
Kraft		84 × 75; 675	144 × 120; 525	78 × 70; 450
Fine		92 × 70; 560	144 × 135; 400	78 × 75; 400
Tissue		86 × 82; 680	200 × 165; 500	80 × 120; 550
Towel		84 × 75; 675	144 × 120; 600	80 × 120; 500
Napkin		84 × 75; 675	144 × 120; 600	80 × 120; 500
Wadding		84 × 75; 675	144 × 120; 600	80 × 110; 550
Carbonizing		110 × 90; 500	180 × 140; 400	78 × 75; 300
Cigarette		110 × 90; 400	180 × 165; 325	72 × 90; 400
Nonwovens		36 × 32; 700	84 × 66; 600	78 × 75; 400

fine tuning designs. The similarity to other successful applications is important. A successful design on a specific machine is a plus factor to be utilized with all of the previously mentioned points.

The design of the fabric for a paper machine must take into account grade, furnish, machine layout and operation. The furnish and grade influence the selection of design; machine layout and operation govern the actual drainage characteristics required within the design parameters of mesh and yarn diameters.

The design selection process encompasses two major concerns: the design of the fabric itself and the run attitude.

In a fabric design, there are some independent variables that control various dependent variables as shown in Table 2.9. The weave type specifies the number of harnesses (plain, four harness, five harness, etc.), warp and filling profiles (number of cross-over points and knuckle length) as well as the number of layers in the fabric (single layer, two layer, two layer extra, three layer). Any change in one of the independent variables will affect the controlled parameters. Each is weighed for its effect on the design.

The second part of the selection process

TABLE 2.9. Independent and Dependent Fabric Design Parameters.

Independent Parameters	Dependent Parameters
Weave	Open area
Mesh	Air permeability
Strand diameter	Hole size
Strand material	Hole shape
	Modulus
	Caliper
	Drainage index
	Fiber support index

deals with selection of the run attitude. Run attitude describes whether the fabric is run with the cross machine knuckles toward the sheet or toward the machine elements as shown in Figure 2.64. Wire mark considerations are important in determining the run attitude of the fabric.

There are advantages and disadvantages to running in a specific attitude as shown in Table 2.10. The advantages for a specific attitude are generally disadvantages for the opposite attitude. The type of formation is also important in determining the run attitude of the fabric. Pressure formation (Section 2.5.1) can lead to release problems. It also affects the cleanliness of the fabric as well as the drainage ability of a fabric over a long period of time.

The effect of running attitude on fiber support (fiber bridging) and drainage is shown in Figure 2.65. Long CD knuckle to the sheet side gives shorter distances for fibers to bridge resulting in higher fiber support. In the case of the long CD knuckle down, the protruding ("proud") knuckle is in machine direction, which is the direction of water flow. Therefore, the proud knuckles do not disturb the water flow as much. When the long CD knuckle is up, the proud yarns are perpendicular to the water flow direction. Water hits these yarns and loses its velocity and drains faster.

The fiber supporting characteristics of the fabric are primarily dependent on the following factors:

- number of support points
- size of support points
- distribution of support points
- orientation of support points
- surface topography
- hole size and shape

For grades that utilize intense dewatering forces (linerboard) and/or light in weight (tissue), fabrics are generally run long MD knuckle down. This allows better drainage and sheet release. The main disadvantage of forming on the long cross-direction knuckle is that the life potential is not as great as it would be if these CD yarns were the wear members. In fine paper fabrics, the hole is generally longer in MD direction.

Figure 2.66 shows a two layer fabric along with its top and bottom surfaces. The knuckle impressions of both surfaces are also shown which indicate the surface topography for initial fiber bridging.

In the event that a fabric is supplied in one run attitude and it becomes desirable to turn the fabric inside out, it is possible to do so before installation on the machine. Figure 2.67 shows a typical procedure for turning fabrics inside out with the least amount of damage.

In general, there is little choice of run attitude with multilayer fabrics. Single layer fabrics may make paper on either side.

Effect of Harness Numbers on Two Layer Fabric Properties

Assuming the same warp and filling counts and the same size yarns and that fabrics are woven in approximately the same weave styles, the harness number has several effects on fabric properties and performance on the paper machine.

Since there are fewer number of crimps in the warp of higher harness number design, it should have a higher modulus and, therefore, is less susceptible to stretching on the machine.

There is less interlacing of warp and filling in higher harness number design, which results in a higher internal void volume. The higher void volume causes higher air permeability. Consequently, as compared to a lower harness number design, the higher harness number design can have a higher filling count for better sheet support or it can have larger filling yarns for better wear for the same air permeability. However, larger filling yarns will make the fabric thicker in caliper, which might make it more difficult to remove the water.

Since there are fewer number of crimps and less interlacing of warp and filling in higher harness number design, it is not as rigid or has as much lateral stability as a lower harness number design. Therefore, it is more susceptible to skewing if the machine is misaligned. Fabric rigidity enters into guiding and fabrics

LONG CD KNUCKLE DOWN   LONG CD KNUCKLE UP

"X" diagrams:

Machine
Direction
(MD)

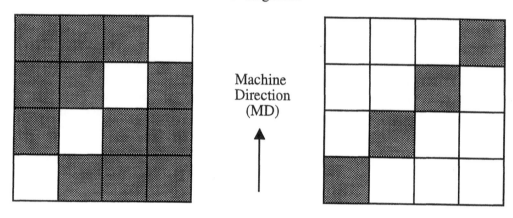

Schematic of top view (paper side):

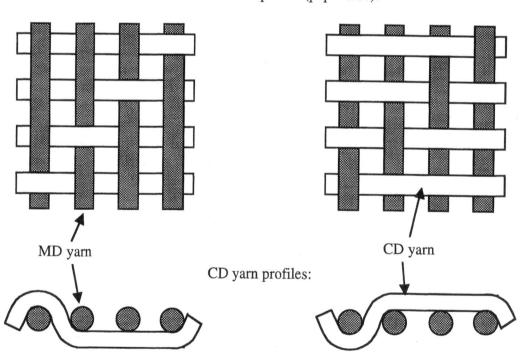

MD yarn   CD yarn

CD yarn profiles:

FIGURE 2.64. Fabric running attitudes for 1/3 single layer fabric.

98

TABLE 2.10. Effects of Fabric Running Attitude for Single Layer Fabrics.

Advantages	Disadvantages
1. Long Cross Machine Knuckle to the Sheet	
1. Earlier formation • good release • easier drainage • lower load 2. Sheet rides higher • less bleed tendency • easier cleaning 3. Better fiber bridging • better sheet formation due to good fiber bridging	1. Shorter life—less wear knuckles 2. Less stability—due to early wear on warp strands 3. More pronounced wire mark 4. More edge curl potential
2. Long Cross Machine Knuckle to the Machine Elements	
1. Better wear resistance, longer life • more strand surface against machine elements • not wearing the warp strands 2. Better stability • less edge curl potential • smoother surface 3. Less wire mark	1. Greater loads—more wear surface 2. Poorer sheet formation due to less fiber bridging 3. Poorer release—due to sheet following fabric 4. More bleed potential—fibers passing through elongated holes 5. More difficult to clean

which are too firm have more of a tendency for guiding problems. In general, the angle of the twill line is not affected by the harness number and, from this aspect, the guiding would be the same.

2.4 Service for Forming Fabrics

In today's industry, major paper machine clothing manufacturers generally provide service to paper companies. Services are designed to assist the papermaker in obtaining a maximum level of efficiency from the paper machine at the lowest cost per ton (see also Chapter 5, Paper Machine Auditing).

2.4.1 *Maintenance of Forming Fabrics*

Optimizing forming fabric life, thereby reducing machine clothing costs, can be achieved by preventive maintenance, good housekeeping and a good understanding of the care and handling of fabrics.

The proper maintenance of forming fabrics and of the machine as it relates to fabrics, can increase fabric life and productivity. Periodic checks of the machine and proper care and handling can maximize a fabric's useful life and machine performance.

It is necessary to prepare the machine area prior to fabric installation. The following steps or precautions should be taken:

• clean area for fabric unrolling
• remove all obstacles
• wash machine prior to installation
• pull fabric evenly during installation (prior to rotation)

Paper Machine Checkpoints

Inspecting the wet-end of the paper machine between forming fabric changes can drastically help to reduce damage during installation and improve fabric life. The following checkpoints have been compiled as a result of the most common points which are overlooked and are the major contributor to early

long CD knuckle down:

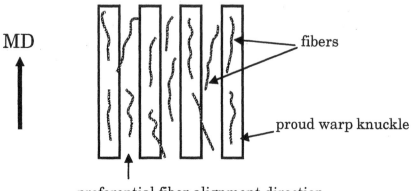

MD

fibers

proud warp knuckle

preferential fiber alignment direction

long CD knuckle up:

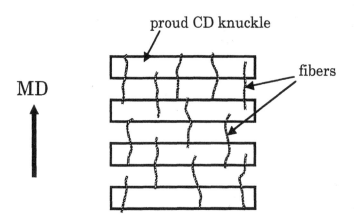

proud CD knuckle

MD

fibers

FIGURE 2.65. Effect of fabric running attitude on fiber support.

forming fabric removal due to damage or wear. It should be noted that some of these suggestions are machine specific and therefore should be followed if applicable.

(1) Check foils and forming board edges for rough, thin or chipped areas and for proper alignment. These can cause fibrillation to the underside of the fabrics. The end result is a marked sheet or severe damage to the fabric causing its premature removal. Poor alignment will cause the fabric to guide improperly.

(2) Check table rolls and return rolls for tight bearings, burrs and proper alignment. Tight bearings and burrs will cause considerable damage to the underside of fabrics as well as increase the energy required to drive the fabric. Tight bearings will cause the fabric to slip over the roll surface and fibrillate, closing the holes and restricting the drainage. Burrs on the roll surface will pick at the underside of

impression   top view   bottom view   impression

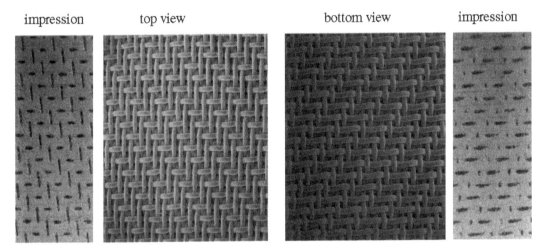

FIGURE 2.66. Two layer fabric knuckle impressions.

the fabric causing a fibrillation streak. This could result in sheet mark and possibly fabric ridging.

(3) Check shower nozzles for proper operation. All showers must be in proper operating condition with no plugged nozzles (Section 2.4.6). A plugged nozzle will cause a dirty streak in the fabric. The dirty streak will restrict the drainage and show up as poor sheet formation. All high pressure needle showers must oscillate when in operation and must only be operated when the fabric is moving. Fibrillation streaks will occur when high pressure showers fail to oscillate during the fabric run and will cut a hole in the fabric if left on when the fabric is stopped. Shower pumps should automatically shut down when the fabric is stopped.

(4) Check return roll doctor blades for grooves and nonuniform pressure. Tight doctor blades will slow a roll down, causing fibrillation to the underside of the fabric. Grooves in doctor blades will allow stock buildup, causing streaks in the fabric and possibly in the sheet. This condition can also cause a fabric to ridge.

(5) Check suction box covers for evenness and correct cut. Suction box covers should be maintained with a regular

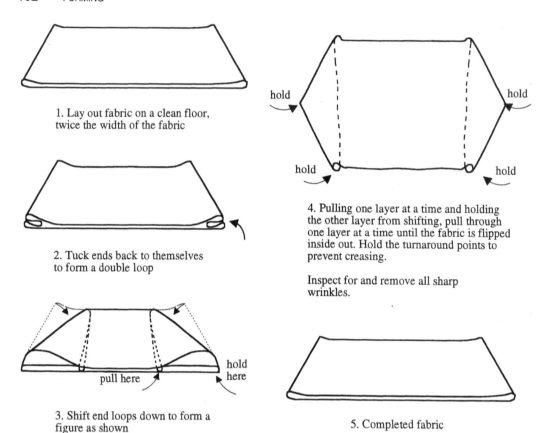

1. Lay out fabric on a clean floor, twice the width of the fabric

2. Tuck ends back to themselves to form a double loop

3. Shift end loops down to form a figure as shown

4. Pulling one layer at a time and holding the other layer from shifting, pull through one layer at a time until the fabric is flipped inside out. Hold the turnaround points to prevent creasing.

Inspect for and remove all sharp wrinkles.

5. Completed fabric

FIGURE 2.67. Turning fabrics inside out.

grind surface. If the surface is too smooth, the fabric will tend to seal off, causing higher loads, bleed and premature wear. If the chamferred edge is worn off, it could possibly cause fibrillation to the underside of the fabric. This condition becomes more serious as the fabric wears in its later stages, whereby the underside of the fabric is monoplane.

(6) Check couch roll, breast roll and wire turning roll for corrosion and pitting. This condition can cause fibrillation on the underside of the fabric leading to poor drainage, premature removal and poor sheet quality.

(7) Check fabric alignment before applying tension and rotation. An improperly aligned fabric can be ridged when tension is applied and the machine rotated. It is also important that the machine alignment front to back is plus or minus,

a maximum of 1/4″ (0.63 cm) differential.

(8) Maintain optimum slice clearance. Be sure slice is set properly and also clean. It is common for stock to build up on the underside of the slice. This can be knocked loose during cleanup when a fabric is being installed and can cause serious damage if it becomes lodged between the fabric and the slice during a startup. It is better to try to get minimum clearance and insure good cleanliness under the slice with good showering. Delivery will be better with clearance 0.100″–0.125″ for fine paper up to 0.200″ for brown papers.

(9) Smoking around fabrics during installation or shutdowns should be avoided. Extreme care should be taken when welding near a fabric. A hot spark from a pipe, cigar or cigarette can burn a hole

in a fabric faster than it can be reached to put out. Hot sparks from a weld unit can also burn holes in the fabric as well as embedded particles in the fabric which will cause holes.

(10) Cover all sharp edges during installation of fabrics. Be sure to cover all sharp protruding edges to prevent damage to fabrics during installation. If the fabric is to be socked on, it is recommended that an old fabric be used to cover the floor as well as to cover as much of the front edges of the machine as possible. Minor damage such as wrinkling can be removed by gently folding a wrinkle back in the opposite direction. Cross machine wrinkles will generally disappear when the tension is applied to the fabric. Some damage can require a refinishing of the fabric.

(11) Check guide palm for free movement. Make sure air is turned on. Be sure the guiding surface is not gouged or cut by the fabric edge.

Fabric Tension

Maintaining the proper fabric tension is important in optimizing fabric life and web quality. If a fabric is too loose, premature wear can result from slippage at the drive roll. Poor release is often seen, since the web tends to ride deeper into the mesh, on a slack fabric. In addition, sheet picking may occur if the seam ends are projecting above the plane of the fabric surface. Excessively high tension can cause stretching in the machine direction and narrowing in the cross machine direction. If the tension is high enough, the seam ends can be pulled from their locked position at the seam joint.

It is recommended that a tensiometer, an instrument for measuring tension in forming fabrics, be used to maintain proper tension. It is important that the fabric be run at the tension it was designed for and manufactured at.

The following method is recommended for taking fabric tension reading.

(1) Position
- with tensiometer on inside of fabric no closer than one foot to an outside roll
- no closer than two feet to an ingoing nip
- minimum of 15″ in from the fabric edge

(2) Direction
- Always align long direction of tensiometer parallel with fabric run.

(3) Tension
- Align tensiometer correctly with fabric.
- Gently press tensiometer down until the pointer stops.
- Read indicator. Using the conversion chart convert to PLI (pounds per lineal inch).

Rolls and Table Elements

Fabric life can be significantly reduced by worn or defective rolls and table elements. Thus, a preventive maintenance program, which involves periodic checks of all surfaces which come in contact with the fabric, is highly recommended. Periodic checks will identify potential problems in the early stages, which will not only help to eliminate these problems, but will increase productivity and fabric life.

It is imperative that accurate records be kept of periodic checks, in order to document defects and excessive wear. A defect documentation chart, with diagrams of the machine components, is very helpful to track data as they are collected. Table 2.11 shows the inspection frequency of various parameters.

TABLE 2.11. Recommended Inspection Frequency of Fabric and Machine Parameters.

Checkpoint	Frequency
Nick and burrs	Every fabric change
Bearings	Once per month
Wear	Once every two months
Alignment	Once every two months
Roll covers	Once every two months

2.4.2 *Potential Operational Problems of Forming Fabrics*

There are several areas of potential problems with fabrics on paper machines.

Abrasion

- definition: striations on a wear surface. Usually a narrow groove, channel, fine streak or line associated with a number of other parallel lines. Lines run parallel to machine direction.
- classification: dependent upon physical appearance of striations on wear surface. Abrasion can be classified as light, moderate and heavy.
  —light abrasion: wear surface nearly smooth. It would have striated marks that would be barely visible or shallow in appearance.
  —moderate: wear surface has deeper, more pronounced striations. Tendency of wear surface to appear more wavy and nonsmooth.
  —heavy: wear surface characterized by deep striations. Displays nonuniform, very jagged appearance.
- cause: rough stationary machine components such as suction box covers, fiberglass roll covers, sand, corrosion of breast and couch rolls.

Bow and Skew (Figure 2.68)

The condition in which the CD yarns lie in the fabric in the shape of an arc is called fabric bow. Bow is defined as the greatest distance, measured parallel to the selvages, between a CD yarn and a straight line drawn between the points at which this yarn meets the two selvages. It is expressed as a direct measurement of AB in Figure 2.68, or as a percentage of the fabric width:

$$\% \text{ Bow} = AB/CD \qquad (2.15)$$

Fabric bow can be symmetrical or non-symmetrical. The causes of bow are fabric profile, stock buildup, improperly installed bowed roll and roll deflection. There are several possible actions to correct fabric bow:

- Install a bowed roll.
- Clean stock buildup.
- Reduce fabric tension.
- Install larger diameter rolls.

The condition in which the CD yarns in a fabric do not lie perpendicular to the MD is called skew. Skew is defined as the distance, measured parallel to and along the edge, between the point at which a CD yarn meets one edge, and perpendicular from the point at which the same CD yarn meets the other edge. It is expressed as a direct distance (EF in the figure) or as a percentage of the fabric width:

$$\% \text{ Skew} = (EF/EG) \times 100 \qquad (2.16)$$

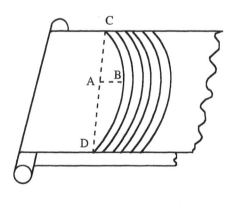

bow

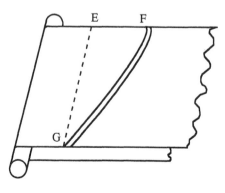

skew

FIGURE 2.68. Fabric bow and skew.

Causes for skew:

- roll misalignment
- stock buildup on roll
- nonuniform roll diameter
- suction box or foil misalignment
- fabric slackened on one edge

Burring

- definition: burring occurs in plastics as a slight extension or flattening of the wear surface edge occurring opposite the fabric run direction.
- classification: dependent upon physical appearance of edge on wear surface, burring can be classified as light, moderate and heavy.
  —light: very little extension of the edge. Edge is nearly in line with edge of wear surface.
  —moderate: more pronounced extension of wear surface. Tendency of extension to curl slightly.
  —heavy: severe extension of wear surface. Extension displays moderate to heavy curl.
- causes: roll slippage, off-speed table rolls, high load operating conditions, high speed applications.

Stretching

Stretching is the extension of the fabric on the paper machine. Low fabric modulus and/or excessive run tension are the cause of fabric stretching. There are two sets of variables which affect the stretching of fabrics on a paper machine: manufacturing variables and paper machine variables.

Manufacturing Variables

(1) Nature of materials: Most forming fabrics are made from polyester. Polyester fabrics will stretch a given amount based on the load applied. Using higher modulus yarn materials improves the situation.
(2) Type of weave: The type of weave has a great deal to do with how much modulus can be incorporated into a fabric. A five harness generally will have a better modulus than a four harness and a four harness better than a three harness. As yarn crimp in the MD is increased, the modulus decreases. Modulus is a measurement of the fabric elongation at a given load (Section 2.2.4).
(3) Yarn diameter: Larger yarns have a greater resistance to stretching than smaller yarns.
(4) Temperature and time at temperature: The temperature at which a fabric is manufactured as well as how long it is at that temperature has an effect on how much modulus is incorporated into the fabric. Proper heatsetting increases fabric modulus drastically. Using higher stretch tension during heatset improves the situation. This is because the crimp in the MD yarn will be decreased. The straighter the MD yarn, the higher the fabric modulus. Yarn has a higher modulus than the fabric. Crimp in the yarn causes modulus to drop under load.
(5) Yield: Yield is the final length of a machine direction yarn in the fabric, compared to its length before starting manufacture of the fabric. Generally a fabric having higher yield has less tendency to stretch.

Paper Machine Variables

(1) Operating tension: Fabrics are sized to be at a given length for a given tension. The fabric length will follow the tension if it is increased or decreased.
(2) Drainage components: The drainage components on the table affect the tension to which a fabric is subjected.

Fabrics are processed to modify the amount of stretch they will yield for a given load. Figure 2.69 shows typical stretch amount a fabric will yield when subjected to a specific load. Referring to the figure, to find the fabric stretch, select a desired load, trace to modulus

INCHES OF STRETCH/100' BASED ON MODULUS OF FABRIC

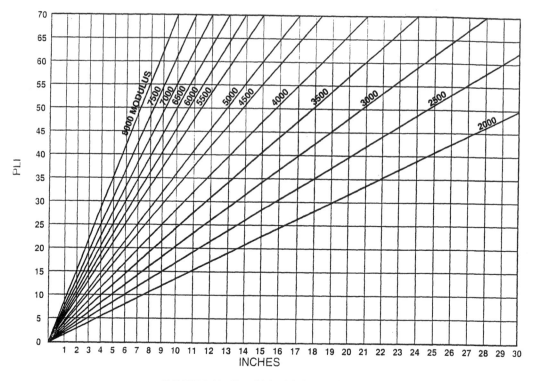

FIGURE 2.69. Fourdrinier fabric stretch chart.

and drop perpendicular to find fabric stretch in inches. For example, at a running load (tension) of 25 pli, a fabric with an 8000 pli modulus will stretch approximately 3.5″ per 100′ whereas a fabric with a 2500 pli modulus will stretch 13.5″ per 100′.

Then the total fabric stretch (TFS) is calculated by:

$$\text{TFS} = (\text{machine length in feet}/100')$$

$$\times \text{(inches)} \qquad (2.17)$$

Ridging

Ridging is caused by five major conditions.

(1) The flatter a fabric is, the less susceptible it will be to ridging.
(2) Damage: A fabric can be ridged by severe damage in the machine direction as well as through stock buildup on rolls.
(3) Loading: Loading is extremely critical. If

the power ratio between the couch and wire turning roll is incorrect, a fabric will tend to ridge.
(4) Machine misalignment: Machine misalignment is a prime suspect when a fabric creases at an angle as opposed to a straight machine direction line.
(5) Table shake: Excessive table shake can cause a fabric to ridge.

Using a crayon tracing technique, it was shown that ridges in forming fabrics could be the cause of ridges in finished rolls of paper. If a ridge develops in a fabric, it can usually be removed, or lessened, if done properly. When attempting to remove a ridge, the fabric should be held under tension.

Fabrics are generally flattest at their heatset tension. If the fabric is being run at higher tension, ridging can be reduced by reducing tension. If run tension is less than heatset tension, flatness will be improved by increasing tension.

One method to remove a ridge is with the use of a household iron with variable heat settings. Using a thermometer on the flat surface of the iron, set the temperature to 350°F. If the fabric is wet, blow it dry with an air hose before attempting to remove the ridge. While rotating the fabric, move the iron in a back and forth motion over the ridge. To prevent puckers, it is important to keep the iron moving. It is important to use caution in the seam area, to prevent the seam ends from backing out. Lower the setting of the iron to 250°F when removing the ridge in the seam area.

An alternate method of removing a ridge is with the use of a hot air gun, or steam hose. However, extreme caution must be used so that the temperature does not exceed 350°F. As with the iron, keep the heat gun, or steam hose, moving at all times.

Bleed-Through

Bleed-through is a condition where fines or fillers pass through a fabric and collect on machine elements inside the fabric loop. Associated with bleed-through are the conditions of stapling, where fines become embedded in the fabric mesh and cannot be knocked off by showering; and carryback, where the fines collect on the return rolls and doctor blades.

Causes of bleeding, stapling and carryback:

(1) Fabric too open
  - A fabric in which the hole size is too large will allow fibers to filter through and be lost upstream. While manifestations may not be readily evident in the sheet produced, the condition loads up the system with fines which, in many cases, eventually becomes a problem. This is especially true with furnishes consisting of a high percentage of hardwood or recycled fiber.
  - A fabric having excessive drainage will allow the sheet to set too rapidly or seal. If this occurs, residual water cannot be effectively removed at the suction boxes and fines will be pulled through and be deposited on the covers. Some of the fines may be trapped in the mesh after the sheet releases and these may then cause stapling and/or carryback.

(2) Fabric too closed: A fabric having insufficient drainage does not allow enough water to filter through before passing over the suction boxes. Removal of the excessive water at the boxes without a proper mat being formed allows fines to be sucked through and collected on the boxes. Here again, some of the fines may be trapped in the mesh after the sheet releases and these may cause stapling and/or carryback.

(3) Fabric attitude with the long warp knuckle either on the sheet side or machine side may influence whether or not carryback occurs, especially on open draw machines. Fabrics run with the long warp knuckle on the machine side allow the fiber to ride higher on the fabric and the sheet releases more readily. As a consequence, carryback occurs less frequently with fabrics run in this attitude.

Remedies to reduce bleeding, stapling and carryback:

- Drop some of the flat boxes and redistribute the vacuum.
- Cut down the vacuum on the boxes to move the wet line ahead.
- Use fresh water in the knockoff system. It is very possible that the bleed problem is not bleed, but contaminated shower water. This is easy to tell by the amount of buildup on a fabric when the fabric is stopped and the low pressure showers remain in operation.
- Lubricate the leading edge of the flat boxes through a low pressure shower.

Guiding

If not properly guided, fabric may move off the machine during operation. Figure 2.70 shows the schematic of guiding. When the

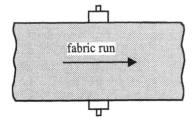

proper guiding

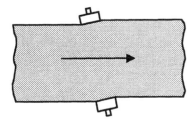

guide left

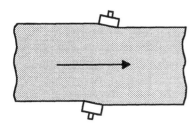

guide right

FIGURE 2.70. Schematic of guiding.

guide roll pivots to the left, the fabric will be shifted to the left. When the guide roll pivots to the right, the fabric will be shifted to the right. Guiding problems are generally caused by:

(1) Low tension: Guide palm does not pick up movement.
(2) Insufficient wrap on the guide roll: A minimum of 25° is recommended as sufficient wrap on the guide roll (Figure 2.71). Lead-in and lead-out distance are critical. Figure 2.72 shows the recommended lead-in and lead-out setup.

(3) Improper guide palm sensitivity
   - too sensitive: If the guide palm is too sensitive, it may pick up the normal oscillation of a fabric and transfer to the guide roll causing the fabric to shift unnecessarily.
   - Insufficient sensitivity: If the guide palm is too insensitive, it may not make the necessary corrections if the fabric walks on the machine.
(4) Nonuniform fabric tension. Nonuniform tension across the width may cause a fabric to tend to walk off the machine. Nonuniformity can be caused by a problem in the manufacture of the fabric, making one side longer than the other or a machine misalignment having the effect on the fabric as though one side was longer than the other.
(5) A strong twill pattern in the fabric may also cause guiding problems. The possible solutions to this are:
   - Break the twill pattern and even out the protruding knuckles.
   - Reverse the twill pattern to be opposite to the hole pattern on rolls.
(6) Tension difference in MD direction along the fabric width. Fabric moves toward the shorter side. Nothing can be done on the fabric to solve this problem. The fabric movement needs to be aligned using guide rolls.

Fibrillation

Fibrillation is a shredding of the monofilament which can occur uniformly across the entire width of the fabric or nonuniformly in streaks (Figure 2.73).

Uniform Fibrillation

Uniform fibrillation can be caused by:

- extremely high or low tension: High tension will tend to rub the underside of the fabric excessively against the machine elements. Low tension will

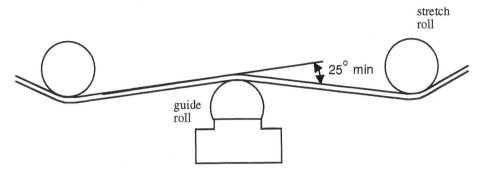

FIGURE 2.71. Guide roll angle of wrap.

allow the fabric to slip causing the turning rolls to fibrillate the underside of the fabric.

- insufficient wrap on the return rolls: Insufficient wrap will allow the fabric to slip across the rolls.
- defective table roll bearings: This creates heavy drag loads.
- high vacuums at the flat boxes: High vacuum will cause high drag loads.
- incompatible flat box material: Incompatible material will wear the fabric prematurely.

Nonuniform Fibrillation

Nonuniform fibrillation can be caused by:

- metal roll corrosion or pitting
- burrs on suction boxes, foils or forming boards
- chipped or porous areas in hard box covers

- hard box cover segments out of alignment
- foreign matter caught in slice: This problem is common in startups. Dry stock will break loose during a cleanup and get caught between the fabric and the slice, causing damage to the fabric.

Holes

Holes are very common cause for fabric removal. Causes of holes include:

- foreign matter from the machine that has built up due to long intervals between fabric changes
- hot sparks from cutting and welding being done near a machine
- hot ashes from cigars, cigarettes and pipes
- tools and implements dropping from pockets of personnel working nearby
- rust from inside the fabric run

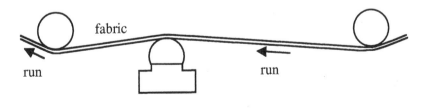

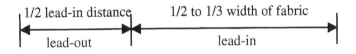

FIGURE 2.72. Fabric lead-in and lead-out ratio.

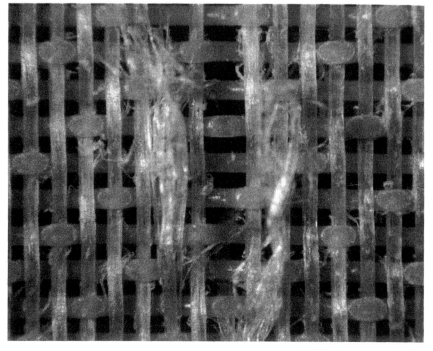

FIGURE 2.73. MD yarn fibrillation.

- foreign material knocked into fabric run by wash-up hoses (broken concrete from the floor, rust from the frames)

Wear

Forming fabrics may be worn due to friction between the fabric and various machine elements. Figure 2.74 shows levels of a two layer extra fabric wear. Improper graduation of flat-box vacuums is a large contributor to wear. The wear prediction of new fabrics is given in Section 2.2.4.

Determination of Single Layer Fabric Wear

As a guide in determining the amount of wear that has taken place in a single layer fabric, the following calculation may be used (Figure 2.75).

Percent Worn:

$$W = \frac{C - C_1}{0.58d} \times 100 \qquad (2.18)$$

Worn out:

$$C_2 = C - 0.58d \qquad (2.19)$$

where

W = percent worn
C = original caliper of the fabric
C_1 = caliper of the worn fabric
C_2 = caliper when approximately worn out
d = diameter of the wear yarn given by the fabric manufacturer.

The wear rate of a fabric decreases considerably as it continues to run, especially when the maximum point of potential wear life is being approached. Even at this maximum point, the fabric may retain sufficient strength to sustain the normal tensions encountered in operation. Therefore, in the final analysis, other factors should be considered in determining whether the fabric should be removed because of wear. The major effects of fabric wear are:

- a dramatic change in drainage that usually causes excessive bleed-through

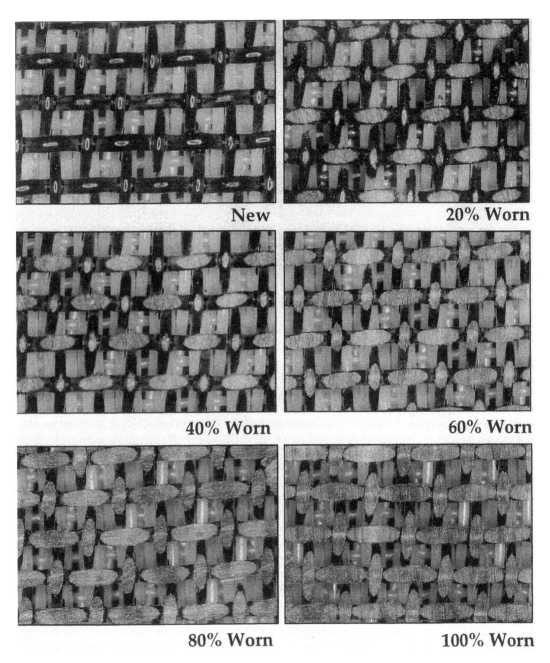

FIGURE 2.74. Various degrees of a two layer extra fabric wear.

FIGURE 2.75. Calculation of single layer fabric wear.

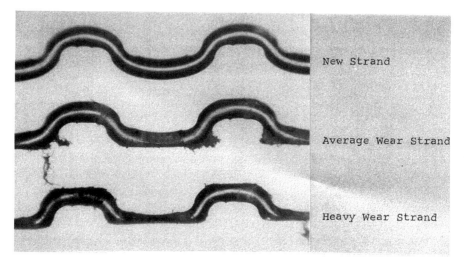

FIGURE 2.76. Comparison of a new yarn with worn yarns.

- development of pin holes in the fabric, which reflects in the quality of the sheet
- fabric stretching

Figure 2.76 shows a comparison of a new yarn to an average wear yarn and a yarn taken from a heavily worn edge area. The fibrillation and degree of top wear can also be seen.

Seam Picking

Occasionally, a seam end may be pulled from its locked position in the seam joint, and project above the plane of the fabric surface as shown in Figure 2.77. This may cause picking of the fibers in the web, which may result in pinholes, or poor web release from the fabric. This problem can generally be eliminated by gently sanding the seam ends until they are flush with the fabric surface. Starting with a 320 grit abrasive paper, sand in a machine direction, back and forth manner, to eliminate the formation of burrs on the fabric knuckles. Finish with 400 grit abrasive paper, in the same fashion. A simple pick test can be performed by rubbing over the seam ends with a cotton ball, or a handful of synthetic fiber, such as polyester or rayon. The seam ends should not snag any of the fibers when rubbed across them.

Edge Sealing

If the fabric edges begin to fray, they can be resealed while the fabric is on the machine. The machine, however, should not be running when sealing edges. A standard soldering gun with a flat tip can be used. For best results, the tip should be bent into a slight V. This will help guide the gun when sealing the fabric edge. Trim off any rough edges with a scissors prior to sealing. When sealing the edges, caution should be used not to melt into the fabric. Begin by sealing in a fast motion along the edge. If the edge does not seal, slow the motion down slightly. After the edge is heat sealed, an adhesive can be brushed on.

Sheet Release

Monoplane fabrics with long MD yarn up may have sheet release problems. Sheet does not release itself to be transferred to the press section (Section 2.7). The solutions are to use long top CD yarn or not to have monoplane surface.

2.4.4 Fabric Cleaning

Forming fabrics may be cleaned with most solvents and cleaning agents used in the paper industry. Inasmuch as these cleaners and their

FIGURE 2.77. Seam picking.

usage varies from mill to mill, it is not possible to specify standard cleaning methods. The following points should be considered when cleaning forming fabrics:

- Concentrated acid, concentrated alkali, phenolic compounds, and strong bleaching solutions should be avoided.
- Due to their toxicity and flammability, caution should be exercised when handling organic solvents.
- A metal wire brush or stiff bristle brush should not be used for scrubbing. Also, the solvent should not affect the bristles in the brush used.
- High pressure showers to rinse off cleaning solutions should be used only if the fabric is rotated.
- High pressure steam is not recommended. When used, the hose nozzle should not come in contact with the fabric and should be moved constantly.

The question of chemical resistance of forming fabrics is of high importance for many applications. As it cannot be answered generally, evaluations are necessary on an individual basis.

The chemical resistance decreases with higher temperatures. Besides this, the concentration of chemicals is important in many cases. Often there is in the end use only a short term influence of a certain product or temperature on the monofilament. This can be of high importance in estimating the lifetime. Last but not least, the mechanical influence also has to be considered. Projected open area in a fabric makes it easier to clean.

2.4.5 Fabric Contamination and Contaminant Resistance

Contaminants such as pitch and tar stick to forming fabrics causing contamination. When recycled paper is used, the situation gets worse due to latex. The polyester yarn is susceptible to contaminant buildup due to the chemical groupings in the molecules. Contaminant buildup fills the fabric causing drainage and release problems. One way to prevent contamination is to treat the surface of yarns or fabrics with anticontaminant finishes or coatings which reduce the fabric's attraction for contaminants. Coatings can be used on either the individual yarns or the finished fabric. Coating can be applied during manufacturing of the fabric or on the paper machine. Coating also may "weld" the MD and CD yarns at crossovers preventing the mechanical attachment of the contaminant to the fabric. It is claimed that coated fabrics increase production due to speed increases, fewer sheet breaks and less downtime due to changing or cleaning filled fabric. It is also reported that coating increases fabric life and improves fabric stability and stiffness. However, as the coating wears out or washes off due to high pressure showers, the performance characteristics of fabrics may change over time.

Figure 2.78 shows wetting angle comparison of two different monofilament yarns. The standard filling yarn has a contact angle of approximately 25°. The contaminant resistant yarn shows a contact angle of 47°.

Mechanisms of Contamination

(1) Chemical bonding between yarn and contaminant. To prevent bonding between the yarns and contaminants, yarns with contaminant resistant properties are used. Contaminant resistance in a yarn can be achieved in three ways:

- using contaminant resistant (CR) additives in the yarn polymer: This method is the most efficient because the contaminant resistance property is inherent in the yarn.
- coating the yarn with a contaminant resistant solution: The drawback of this system is that the coating may be lost partially or totally during operation of the fabric.
- combination of the two methods

(2) Mechanical locking of contaminants at yarn crossovers: Fabric coating is used as a remedy to this kind of contamination. However, fabric coating requires capital investment and environmental issues may also be a concern.

Recycled stock is the major source of contamination. In general, recycled fibers are shorter than virgin fibers. Due to short fibers in recycled stock, higher mesh fabrics are used to increase fiber support index and improve retention. The fabric should provide a surface that is easy to bridge with short recycled fiber.

regular yarn
contact angle = 25°

contaminant resistant yarn
contact angle = 47°

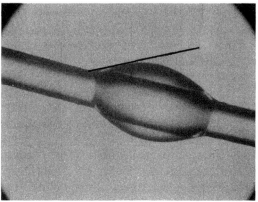
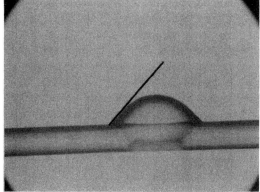

FIGURE 2.78. Wetting angle comparison of regular and contaminant resistant materials.

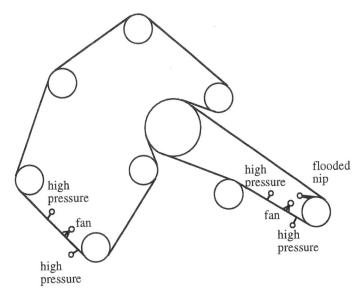

FIGURE 2.79. Shower positions on twin-wire formers.

Fabrics used in recycled furnish should allow easy cleaning with existing shower systems. Low caliper in general allows ease of cleaning.

Fabric showering is an effective method to prevent contamination buildup on the fabric during operation. High pressure showers are used for this purpose. Due to the high pressures involved, fabric stability and seam strengths may become an issue in fabric design. With more open area (higher permeability) and straight through drainage, the showerability of fabrics is improved.

2.4.6 *Fabric Showering*

Properly operating cleaning showers are necessary to remove fibers and contaminants from the fabric during operation to optimize fabric life. Showers are used on wet-laid machines for fabric cleaning, sheet knockoff, trimming and lubrication. In any showering system, it is critical to maintain uniform coverage and open flowing nozzles. Another consideration in shower design should be that nozzles cannot fall into the fabric run, creating a major fabric/roll accident.

The major types of showers are high pressure needle and fan showers, flooded nip showers and knockoff showers [15]. High

pressure fan showers are generally used in tissue manufacturing. Figure 2.79 shows the typical positions of the showers on a twin-wire former. Table 2.12 summarizes the forming section showers.

High Pressure Needle Showers

High pressure showers are the most effective device for contaminant removal on the wet-end of the machine. The high energy jet can cause fabric vibrations, resulting in premature failure if not located close to a support roll. The shower should be located as close to the outgoing nip of a support roll as possible. High pressure needle showers are needed to penetrate the voids in the fabric to optimize cleaning.

There are different opinions about shower distance from the fabric surface. One should consider the fabric structure and production needs before making this decision. Published data indicate that very close showering distances can clean contaminated fabrics well, minimizing damage. As the shower is moved away from the fabric, it begins to entrain air which results in a nonuniform pounding of the surface. While this can be utilized to remove contaminants, it should be reserved for inside

TABLE 2.12. Forming Section Showers.

Application	Nozzle Type/Size	Spacing	Distance	Pressure	Gallons/Inch
Flooded nip (machine side)	40° Fan stationary	3″	8″–16″	100–150 psi	RVV × 1.1
Inside high pressure (machine side)	0.040″ Needle oscillating	3″	6″–8″	350–500 psi	.18–.23
Outside high pressure (sheet side)	0.040″ Needle	3″	2″–4″	200–500 psi	.18–.23
Fiber/sheet knockoff (machine side)	40° Fan stationary	3″	4″–7″	200–400 psi	1.0–2.0
Suction roll	0.040″ Needle oscillating	3″	4″	350–500 psi	.21–.25
Wire return roll	40° Fan stationary	6″	8″	40 psi	.07–.09

single layer showering or multilayer structures.

Oscillating showers are typically 3 inch spaced needle orifices with a 6 inch stroke. Oscillating speed must be set to machine speed to insure total fabric high pressure cleaning. The speed of shower is calculated as follows:

$$OS = \frac{MS}{FL} \times NS \qquad (2.20)$$

where

OS = oscillating speed (m/min)
MS = machine speed (m/min)
FL = fabric loop length (m)
NS = nozzle size (m)

The high pressure showers should have an automatic kick-off system to shut off when the machine shuts down. There should also be a relay that will not allow the high pressure showers to be started until the machine is put into the run mode.

Pressures to be used in the high pressure showers are dependent upon the amount of cleaning and the speed at which the machine is running. At faster speeds, higher pressures are needed to penetrate the voids. Conversely, operating at high pressures and low speed can and will fibrillate the fabric. Optimizing pressures need to be determined.

Inside High Pressure Shower

Inside high pressure showers are generally used on forming fabrics utilizing a blend of secondary fiber and virgin fiber or 100% virgin fiber. This shower is very effective in cleaning contaminants from the void volume of the fabric. It is critical to have an oscillating needle jet-type shower with a recommended operating pressure of 350 to 500 psi. The nozzle size is generally 0.040 inch installed at a distance of 6 inches to 8 inches from the fabric. When very fine single layer fabrics are required, inside showering is recommended.

Sheet Side High Pressure Shower

This shower is more effective in removing fiber and other contaminants from the surface of the fabric. For two layer fabric design applications with no projected open area, it will be necessary to have the ability to adjust the showers to a 30° angle (CD) with respect to the fabric (Figure 2.80). The 30° nozzle allows penetration of the high pressure showers through the angular voids, which is a characteristic of two layer design fabrics (Figure 2.60). It should be noted however that this technology is relatively old.

Since seam ends on single layer fabrics are generally terminated to the sheet side, it is recommended that operating pressure be kept

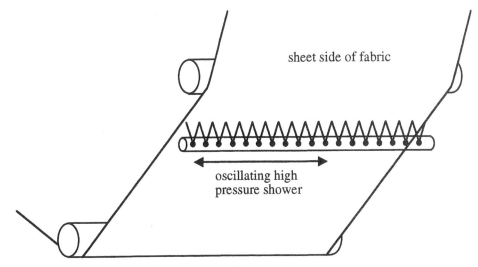

sheet side of fabric

oscillating high
pressure shower

FIGURE 2.80. Two layer showering.

at 250 psi or lower. Yarns smaller than 0.005 inch can break at high pressure showers above 250 psi. When running a multilayer fabric, because of design structure difference and depending on yarn sizes, pressure should be set from 200 to 500 psi and run perpendicular to the fabric surface. Oscillating needle shower utilizes nozzle sizes of 0.040 inch and is generally set at a distance of 4 inches or less to the fabric.

To effectively prevent the fabric from plugging, most cleaning applications require both inside and outside high pressure oscillating showers.

Suction Roll Showers

On suction roll machines, a high pressure needle shower is located perpendicular to the roll surface. The nozzles should be spaced at 3 inch centers, 4 inches from roll surface, at a pressure of 350–500 psi.

Flooded Nip Showers

On a standard fourdrinier with wire turning roll, the flooding nip shower is placed on the inside of the fabric run at the ingoing nip of the wire turning roll (Figure 2.81). They should be of a fan type and provide complete coverage into the nip. The nozzle

distance, from the nip, should be 3–4″ away. The jet from the fan nozzle should hit the roll slightly before the nip. The angle of the fan shower jet should be approximately 30° with respect to the fabric. It is commonly recommended that this shower nozzle be of a self-cleaning type. The pressure and flow rate are largely dependent on the void volume within the two-layer constructed fabric. When expelled through the opposite surface, the water will affect the separation of the fabric and the sheet. The main criteria used to determine rates are machine speed, number of nozzles, forming fabric caliper and void volume.

In high speed twin-wire formers, flooded nip showers have dual purposes of internal cleaning and sheet knockoff. Sheet knockoff can occur running a gallons per minute rate 75% of the fabric's total running void volume at speeds above 3000 fpm. This result is created by the centrifugal force as the water is thrown through the fabric due to the wrap on the roll. Flooded nip cleaning flow is 10% more than the calculated running void volume. The roll chosen for this shower should have a minimum wrap of 30%. A stationary 40 to 45° fan nozzle shower should be installed to the roll/fabric nip and is much more effective in a pressure range of 100 to 150 psi. The flooded nip shower should be located 12 to 16 inches from the roll/fabric nip to allow

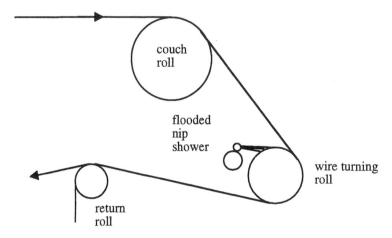

FIGURE 2.81. Flooded nip shower on a standard fourdrinier with wire turning roll.

water to directly enter the nip. In several instances, knockoff showers have been eliminated by application of the flooded nip with good results. Many shower manufacturers reduce water volume as speed is increased above 5000 feet/min.

Minimum flooded nip shower requirement to fill fabric void:

$$Q = 0.055 \times C \times W \times MS \times V \quad (2.21)$$

where

Q = water requirement to fill fabric void (GPM)
C = caliper of fabric (inch)
W = fabric width (inch)
MS = maximum machine speed (fpm)
V = void volume factor (usually 0.6)

Normally, Q is multiplied by 1.1 to obtain water requirement for cleaning.

Fiber/Sheet Knockoff (Fan) Showers

The sheet knockoff shower or fan shower is used to supply fairly large quantities of water at a relatively low pressure across the full width of the fabric. The application of higher pressure significantly improves the efficiency of this type of shower. The effectiveness of lower pressure showers is questionable when moving to multilayer structures. The bottom surface of multilayer shears the fan jet, creating machine direction flow before water passes through the plane of the fabric.

This shower is generally a stationary 40 to 45° fan on 3 inch nozzle spacing at a pressure of 200 to 400 psi. Since distribution of water by a fan nozzle is uneven, a possible second pipe with offset nozzles is recommended. The maximum impact force is obtained by placing the nozzle 4 inches to 7 inches from the surface of the fabric.

The actual impact of the shower depends on the distribution of water across the face of the fabric. This is controlled by the fan angle which is chosen. Table 2.13 shows the effect of changing fan angle and fabric distance on coverage.

Shower Maintenance

Periodic inspection of the shower system is recommended at every fabric change. Excessively high pressure can fibrillate the individual yarns in the fabric, leading to a shortened fabric life. Low pressure can prevent optimum cleaning, possibly decreasing its drainage capacity. On abrasive fiber grades, such as glass, a decrease in fabric life can result from fibers embedded into the mesh. Table 2.14 lists the recommended inspection frequency of showers.

It is critical to inspect shower nozzles visually on each machine down. Plugged nozzles

TABLE 2.13. Effect of Changing Fan Angle and Fabric Distance on Coverage.

Fan Angle	Distance to Fabric (inch)				Impact Factor
	4	5	6	7	
30°	2.0	2.5	3.0	3.8	.16
40°	3.2	4.2	5.0	6.0	.12
50°	4.0	5.2	6.5	7.4	.10
60°	4.8	6.2	7.5	8.8	.08
80°	6.8	8.8	10.4	12.0	0.5
	Surface coverage (inch)				

can cause MD bands of varying tension inducing drainage problems and premature loss of the fabric. Shower nozzles should routinely be replaced on a 3–12 month cycle because tips erode over time, changing the water distribution pattern. All shower pipes should be set up to allow flushing on the run. Self-cleaning nozzles are recommended where possible.

Forming Section Doctors

Doctors are an integral part of cleaning and conditioning of forming fabrics. Contaminants naturally transfer to the smoother roll surface from the rougher forming fabric surface and must be doctored away from the roll or they will build up and cause operational problems. Roll doctors are generally made out of poly or fiberglass. Generally, doctors do not oscillate in wet-end installations, but is an option provided by some vendors. It is recommended that all rolls have a wire return roll shower to aid in doctoring efficiency and provides lubrication between the roll and doctor.

Brush Cleaning

Brush cleaning has been tried on several machines and is utilized to a limited extent as

TABLE 2.14. Recommended Inspection Frequency of Showers.

Checkpoint	Frequency
Plugged nozzles	Every fabric change/on run
Pressure check	On run
Purge shower header	Every fabric change

a final cleaning device. Fiber must be flushed from the surface of the fabric before contacting the brush to avoid brush plugging problems. A properly operating brush will allow stickies to be removed from the machine white water system.

2.4.7 Returned Fabric Analysis

Returned fabric analysis, which is extensively done in the industry, gives valuable information about the performance of fabrics. This information can pinpoint any potential problem areas on the machine. It is also useful to improve fabric design and structure.

A typical returned fabric analysis includes the following:

- wear profile and degree of wear
- air permeability profile
- abrasion
- burring
- fibrillation
- contaminants
- edge condition
- seam condition

Air permeability test done on the returned fabric is to determine how fabric wear has affected drainage quality. To do this, it is necessary to remove as much contamination in the returned fabric as possible. Some contaminants such as dried binders, wax or tar-like specks commonly found in secondary furnish cannot be removed without destroying the polymer yarns. However, many contaminants found in returned fabrics such as pulp, calcium carbonate, paper dyes, ink or grease can be easily removed. For these contaminants, a mild alkaline liquid (containing potassium hydroxide, water, tetra-sodium pyrophosphate, sodium heptagluconate, tetra-potassium pyrophosphate) solution can be used with a water content of 90%.

First, the air permeability of the test sample is measured as it is received. The test sample is then completely submerged in the solution. The time the fabric is submerged depends on the amount of contamination. The average time can range anywhere from 20 minutes for lightly contaminated test samples to 2 hours for heavier contaminated fabrics.

After the test sample has been soaked, it is rinsed with water and scrubbed with a soft bristle hand brush. The test sample is then blown dry with a compressed air nozzle. The test sample is then retested for air permeability. At this point, any loss in the cleaned test sample can be attributed to:

- burring, which causes the fabric mesh to close up resulting in less air flow
- the development of a wear pad, which increases the area in which the air must travel around also reducing drainage quality

The air permeability measurements, taken before and after cleaning, are compared to the new value and reported in the form of percentage retained.

2.4.8 *Wet-End Surveys*

It is important to measure the performance of paper machine clothing on paper machines. Some paper machine clothing manufacturers provide extensive technical support that may include:

- showering surveys
- crew training
- startup assistance
- lab reports
- chemical analysis
- roll abrasion studies
- drainage studies

Machine surveys are commonly done in the paper industry to assess the performance and productivity of paper machines and papermaking process (Chapter 5). The collected data are analyzed and used to make recommendations to the papermaker to improve his process. Appendix B gives a guide for forming section troubleshooting.

2.5 Forming Section Configurations on Paper Machines

The most important part of choosing a paper machine is probably selecting the forming section. This is because sheet forming mechanism has a significant effect on sheet properties. Since the invention of the first paper machine in the 1790s, there have been major improvements and inventions in forming sections of the paper machines. As a result there are virtually endless varieties of paper machines in existence today. A comprehensive coverage of the paper machines is out of the scope of this book. A brief introduction to each major group will suffice for the convenience of the reader.

There are five main types of formers in industry today: fourdriniers, twin-wires, two-wire formers, multiwire formers and Crescent formers. Although multicylinder machines are not manufactured anymore, many are still in use.

2.5.1 *Fourdriniers*

Fourdrinier machines are considered the conventional paper machines. However, their use is declining. Figure 1.10 shows the schematic of a typical fourdrinier machine.

There are many elements on a fourdrinier machine to help with the formation and drainage. These elements include forming board, foils, table rolls and suction boxes. These elements are located under the portion of the fabric where sheet is formed and transported. There are other machine elements that are used to control the fabric such as stretch rolls, guide rolls and fabric turning rolls.

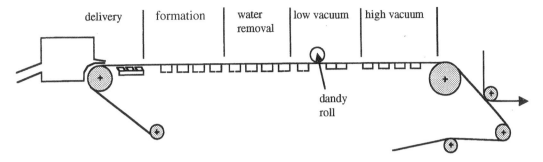

FIGURE 2.82. Sheet forming process on a fourdrinier machine.

Operation of the Fourdrinier

Fourdrinier machines are very useful to explain the formation of the sheet. A rather detailed explanation of the operation of a typical fourdrinier paper machine is given below.

The operation of a fourdrinier wet-end is very complex. There are many interrelated components and actions, each of which is important to good operation and quality paper. Attention to the details of the whole operation is essential to having a well run machine.

In modern fourdriniers, the sheet forming process on the machine can be divided into several regions as shown in Figure 2.82: stock delivery, sheet formation, water removal, low vacuum and high vacuum. To get the best possible sheet of paper to the couch, several different conditions are established on the wet-end of the machine to form the sheet and to remove water. Each of these conditions has an impact on the efficient operation of the machine and the quality of the sheet being made [1].

Stock Delivery onto the Fabric

The delivery part of the process is made up of the headbox and slice, the breast roll and the forming board. All are related to the performance of the forming fabric.

Headbox

The headbox is one of the key parts of the machine as it determines the uniformity and stability of the basis weight of the sheet being made. It also has a major impact on the formation and strength properties of the sheet. It affects the way the forming fabric performs with regard to sheet physical properties and sheet finish and it can help even out problems with uniformity of the sheet which may originate in the stock preparation area, the approach flow system, and the headbox manifold.

Headbox consistency is one of the primary concerns for good formation. The fibers must be dispersed as uniformly as possible so as to minimize floc formation in the headbox.

Slice

The slice controls the angle of impact of the jet onto the fabric. The angle is governed by the relation of the top lip to the bottom lip or apron (Figure 2.83). If the top lip is moved downstream of the apron, then the jet is said to have a tendency toward pressure forming. If the top lip is moved upstream into the headbox from the apron, then the jet is said to have a tendency toward velocity forming.

Within practical limits, sheet formation is always improved as headbox consistency is lowered and the slice operated at the largest opening that the table, fabric and furnish are capable of draining.

The ratio L/b describes the relative location of the top lip with respect to the edge of the apron or bottom lip. It relates directly to the angle of impact of the jet. While it is an important number to be aware of, it is more important to observe the actual landing spot and

pressure forming   velocity forming

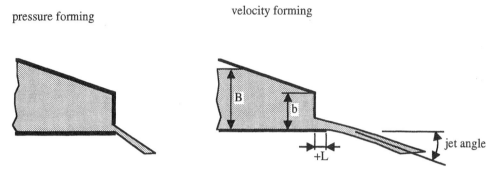

FIGURE 2.83. Control of jet angle with slice opening.

attempt to have it in the right position for the best formation and drainage. *L/b* has a minor effect on fiber orientation except as it relates to excessive turbulence and stock jump that may occur with too much pressure forming. In practice, pressure and velocity forming are defined based on the *L/b* ratio as follows:

$$\frac{L}{b} \cong 0.5 \text{ pressure forming} \qquad (2.26)$$

$$\frac{L}{b} \geq 1 \text{ velocity forming} \qquad (2.27)$$

In pressure forming, the jet angle to the fabric is steeper. In velocity forming the jet angle is low and the jet travels relatively parallel to the fabric a longer distance. In modern machines, velocity forming is preferred. The jet should land on the fabric just before the leading edge of the forming board.

Pressure Forming

Pressure forming tends to move the impact point upstream toward the headbox, breast roll and slice. If the contact is too close to the breast roll, the nip vacuum may act upon the initial fiber mat. The drainage force can pull fiber, filler, fines and water through the forming fabric. As speeds get higher, the drainage forces get greater, deflecting the fabric downward and creating severe flow disturbances on the table. Fibers can become tangled in the mesh of the fabric, creating high draws at the couch. The higher the early drainage forces, the worse the wire mark.

Pressure forming is very sensitive to any defects or irregularities in the slice opening or headbox flow patterns. These defects become set into the fiber mat because of the high drainage rates and are virtually impossible to remove or correct on the table. Pressure forming also tends to seal the sheet on the fabric, making efficient drainage very difficult to achieve.

Velocity Forming

Velocity forming is more desirable because there is more latitude to impact the jet correctly onto the fabric and there is a greater possibility of evening out flow irregularities from the slice or headbox if they exist.

Velocity forming, at the extreme of landing the jet completely on the forming board, will eliminate any possibility of drainage until the slurry is over an open area of the board where gravity alone will provide drainage force. Delivery may be smooth but the large amounts of water being carried down the table may give skating and other formation defects.

Ideal delivery lies somewhere between the two extremes of pressure forming and velocity forming. If the jet is directed so that it lands 1/4″ to 1/2″ before the forming board and the balance lands on the board thus draining 10–20% of the flow between the slice and the trailing edge of the forming board, some of the initial drainage will penetrate the forming medium, filling the interstices or holes of the fabric. Water will replace the residual air in the fabric, giving a uniform layer of water in the fabric rather than an unknown mixture of air, water and fiber.

The slice delivery is important in making a success of a fabric. With pressure forming, the fibers will be driven toward the bottom of the holes, making drainage difficult and creating draw problems at the couch. Using velocity forming, the fibers will not penetrate the fabric as easily at contact and will have a chance to become partially cross machine direction oriented over the forming board. As they turn, the fibers have a better chance of bridging across adjacent MD yarns, staying much further up toward the fabric surface and thus keeping the drainage paths more open.

The jet speed to wire speed (i.e., forming fabric speed) ratio is very important. This ratio can be changed to obtain a rush or drag effect during formation. Rush/drag changes are made to develop grade qualities. For example, if MD tensile is of top importance then the sheet may be dragged to a great extent to align the fibers in the machine direction. Other properties such as burst may need more rush to get some cross-direction orientation. When formation is of great importance, it is best to drag the sheet slightly to prevent the rolling of fiber which sometimes comes from a rush. The rolling fibers will make fiber clumps which result in poor formation. Rush/drag has essentially no effect on the angle of impact.

Breast Roll

In setting up the forming table, the center line of the breast roll is generally used as the point of reference from which the table layout is made. The roll must be very accurately positioned horizontally and vertically so that the slice lip and all other table elements are precisely aligned to one another. The geometry of the slice, breast roll and forming board arrangement determines the area where the jet will contact the fabric.

The breast roll is directly beneath the headbox. Since this is a roll with a high amount of wrap, it must be large enough and strong enough to prevent deflection from the tension forces generated by the fabric. It should have a properly adjusted doctor.

Any roll on the paper machine is an example of a very powerful drainage force. Foil blades are a second example of the same force.

A vacuum is generated in any nip where a roll surface separates from the fabric. This vacuum induces drainage through the fabric. In general, the amount of vacuum induced in the nip is dependent on the machine speed. The higher the machine speed, the higher the vacuum. The drainage rate, Q, is calculated as follows:

$$Q = \frac{DV^m}{R^2} \qquad (2.24)$$

where

D = diameter of the roll
R = drainage resistance factor of the furnish
V = speed of the machine
m = coefficient related to the furnish being used

The nip where the fabric leaves the breast roll is the first area on the table where drainage of the headbox slurry can occur. This drainage should be avoided because of its effect on the sheet.

Forming Board

As machines became faster and wider, the diameter of the breast roll became larger. The span of unsupported fabric from the center line of the breast roll to the next supporting roll became longer and the forming medium sag or deflection became greater under the force of the jet and the lack of support. The forming board reduced the sag by enabling the support point to be moved in closer to the roll, and improved the landing zone conditions for the jet by allowing some control of the drainage of the sheet (Figure 2.84).

The board is used to provide a well-defined landing place for the jet. It also controls the drainage at the entry zone of the table. Drainage at the forming board is largely controlled by gravity and is dependent on the open area of the board. The greater the open area, the greater the drainage. Doctoring by the blade nose also contributes to dewatering.

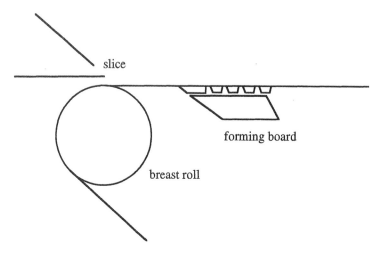

FIGURE 2.84. Forming board.

The forming board body must be extremely rigid. The water forces that result from the jet impact are very high. If the body is not rigid enough, there will be vibrations which will affect the sheet quality. The body also supports the blades and keeps them correctly aligned for uniform drainage. Usually there are three, five or seven blades on a forming board. Ceramic blades have replaced early forming boards made of polyethylene. Most forming board blades have no built-in drainage angle (they are flat). Drainage over the board comes from the velocity of the jet as it impacts the fabric before and over the early part of the lead blade. No drainage can take place over the blade itself and gravity provides the balance of the drainage. Since elapsed time over the open area is short and the height of the water column is not high, little drainage can occur. What does occur is related to the ability of the blade nose to doctor water from the underside of the fabric. There are some signs that the more blades (and noses) that there are, the more dewatering will occur.

The forming board nose can be located such that it is easy to adjust the top slice lip to get good jet impact. The forming board also may need to be located at the point on the table where the desired jet angle will put the jet in the right position with respect to the forming board.

Sheet Formation Region

In this region, the sheet is formed with the help of turbulence and shear.

Turbulence

In the stock approach system and the headbox, the consistency of the furnish is run at as low a level as possible to disperse the fibers as completely as possible. Turbulence generation in these systems ensures dispersion. On the forming table, dispersion must be continued and turbulence is the primary tool for maintaining dispersion. Shear is the primary means of providing turbulence for dispersing the fibers. Shear is simply causing controlled flow disturbances which disperse fiber flocs or bundles.

Shear

As the jet exits the headbox, shear may continue for a very short time due to surface friction between slurry and the slice surfaces. In addition, the way the jet impacts the fabric may be very turbulent depending on the entry zone setup. At speeds up to 2000 feet/min, dispersion may be accomplished by using the shake. On some machines, the shake may affect only the breast roll and on other machines, part of the table may also be shaken. The shake creates a cross machine direction shear in the fiber flow. As speeds increase from the very low to the 2000 fpm range, the time between the jet landing on the fabric and the sheet set point being reached becomes very short, decreasing the effectiveness of the shake.

As the fabric travels down the table, the foils create shear by deflecting the fabric downward. The change of direction of the fabric causes the water to attempt to travel vertically upwards from the fabric. This movement creates shear. If the angle is high and the consistency low, droplets may actually separate from the surface and travel through the air.

Collapse of ridges also provides a powerful source of shear and turbulence. A dirty slice or a damaged slice will exhibit streaks on the surface of the sheet at the defect. Slices can be purchased serrated which generate a CD series of streaks or ridges which collapse and reform. These streaks can be likened to valleys and peaks of hills. The hills collapse and are replaced by peaks. This cross-direction flow provides shear. At foil noses, peaks become valleys and valleys become peaks. This will

continue down the table for some distance depending mainly on consistency and spacing. Formation showers have the same effect as serrated slices.

Formation Shower

These showers are used to help form the sheet properly on the forming fabric. They are particularly helpful on heavier weights and may be harmful on lightweight sheets if the pressure and volume are not controlled to prevent complete penetration of the sheet on the fabric. The showers are stationary and use low pressure water to create ridges in the stock furnish on the forming fabric.

Quite often, these showers are simply a drilled hole shower which will work quite well with white water if the burrs are removed inside the pipe and around the circumference of the holes. The shower is positioned 5 to 10 inches above the fabric over the forming board or first foil unit. The shower is angled downstream 10–15° from perpendicular (Figure 2.85). Suspending the shower from the machine room crane for trials will allow easy adjustment for determining the optimum operation. Permanent brackets can be designed to locate the shower at the best location.

Water Removal Region

In this section of the table, formation is essentially completed and the consolidation

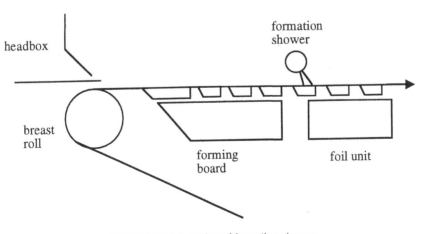

FIGURE 2.85. Location of formation shower.

of the sheet begins. The foils do the work of water removal and the downward flow of water causes the fibers to be brought into closer contact with each other.

As the consolidation of the sheet is taking place, it is important to diminish the turbulence level constantly. If the sheet is disrupted by high turbulence this far down the table, it will not reform into an acceptable sheet.

Foil Precautions

Plastic foils will wear under the action of the fabric passing over the surface. Since the amount of water removal is dependent on the amount of time that vacuum can act on the underside of the fabric, a worn foil will not be able to dewater a fabric as well as a new foil.

If the dewatering rate appears to be reduced, it is always a good idea to check the condition of the blades. Ceramic blades or partial ceramic blades will retain their shape for long periods of time. However, great care must be used when handling these blades as they will chip and break rather easily. Damage to fabrics may occur if blades are chipped.

If a pair of new foils are put onto a foil unit on each side of a worn unit, the worn foil will not contact the fabric, no nip will be formed and no doctoring or dewatering can take place. The same thing can happen if the total table is not accurately aligned. All foils should be at the plane of the fabric and level in all directions.

Low Vacuum Region

Dandy Rolls

Dandy rolls were originally used to impart special marks in the sheet for decorative and identifying reasons. Many dandies are used today to improve the formation of the sheet. This improvement comes from a shear action of the roll relative to the sheet and the fabric. The shear is intended to rearrange fiber in the sheet, breaking up large flocs and reducing their size. In order for this shear to be effective, it must take place when the sheet has enough water in it to permit the disruption of

floc and the rearrangement of fiber. If the sheet is too dry, the dandy will have no effect.

Special flooded boxes are sometimes used under the dandy to help with rearranging the fiber to improve formation and watermark clarity if a watermark is used. In the absence of these boxes, some dewatering can take place at the dandy but this is not a major function of the dandy.

It is generally accepted that 3% consistency is the highest consistency where good dandy operation will take place. This is somewhat grade dependent. It is best to run consistency trials to determine the optimum effectiveness of the dandy. In most cases, the dandy surface speed will exceed the wire speed by up to 1.5%.

Spray from the outgoing nip of the dandy can cause water spot defects in the sheet if the drop hits the sheet after it is too dry. Sometimes an internal fog shower with a water based defoamer in the spray will reduce or eliminate the spray.

Vacuum

Foils are capable of drying the sheet out to about 1.5% to 3.0% consistency, depending on the grade of paper being produced. Vacuum assistance of the foil drainage is necessary for further drying.

In the forming section, the sheet is being compacted and consolidated to create smaller and smaller flow paths to promote efficient dewatering. After the foil section, the sheet is a completely saturated fiber mat. As long as there is free water in the fiber mat, vacuum is useful in removing that water.

To remove the saturating water, low level vacuums are very effective. After the water film or saturating water has been removed, then high vacuums will remove much of the rest of the water.

Low Vacuum Equipment

The typical low vacuum unit consists of a stainless steel box which can be fitted with foil blades. There are several different designs of foils which can be used on the box. Each supplier of boxes has his own theories on what works best. All styles of units have deckle pieces which completely close the top of the box with a vacuum seal.

A vacuum inlet is provided to permit the application of vacuum to the underside of the fabric in the spaces between the foils. In many cases, this inlet is independent of the water drain of the box although a few units exist where the water comes out the same opening through which the vacuum is applied. In all cases, the water drain must be set up so that water can drain freely from the box without air leaking back in.

For best operation, the vacuum source should be different than the flat box vacuum source. One of the best arrangements is to use a low vacuum level blower with the vacuum line being connected to the inlet of the blower.

Low Vacuum Operation

Low level vacuum boxes are effective at removing water which is saturating a fiber mat. The water acts as a seal, allowing the sheet to drain into the box, evacuating the free water from the fiber mat. The water flow not only dries the sheet but the flow velocity creates drag on the fibers that gradually compacts the mat into a denser sheet. The effectiveness of the low vacuum level is gone when air breaks through the sheet. This happens at the dry line.

Early on the table, the sheet is liquid enough that the deflection caused by the foil action and the applied vacuum provides additional turbulence for better formation. Late on the table, the sheet is no longer mobile enough to have any formation improvement. At this point, the effect is solely to remove water.

Low vacuums dry the sheet to 7 to 9% consistency, again depending on the grade. Vacuum augmented foils typically operate at 1 to 20 inches of water vacuum.

Low Vacuum Troubleshooting

A vacuum foil body must drain water freely to work the way it should. There should be sufficient clearance under the seal leg. As a rule of thumb, the surface area of the imaginary cylinder beneath the seal leg should be at least equal to the cross-sectional area of the seal leg.

The vacuum lines connected to the unit should be large enough in diameter and as short as possible to ensure the total effect of the vacuum source is carried to the unit. Small diameter piping will cause large pressure drops in the piping. There should also be no low spots in the piping to create a water pocket which will cause vacuum surges and uneven water removal in the machine direction.

A vacuum gauge or a manometer tube should be installed on each unit. The maximum effective vacuum that can be applied to the unit is equal to the distance from the water in the seal pit to the bottom of the vacuum inlet. If this distance is exceeded, the vacuum level in the unit will vary and cause uneven drying.

High Vacuum

The water remaining in the sheet after the dry line is the water remaining on the fiber surface, in the fiber crossovers, and the water in the fibers themselves. The efficient way to remove the first two is to use high velocity air to strip the water off the surfaces and out of the crossovers. The last must be removed in the presses and dryer sections. Flat boxes with various types of open area (slots or holes) and high vacuums generate the high air velocities necessary to accomplish the stripping of water.

A flat box must be rigid enough to withstand the vacuum source applied and support the cover to prevent cover collapse. The number of slots and the width of the slots depend on the grades being produced, speed of operation and sequence of boxes on the table.

With polyethylene covers, the plastic of the cover will wear away and somewhat protect the fabric. The cover wear will however be uneven and eventually uneven water removal takes place over the cover. Profile problems and drying streaks will result.

Ceramic covers have a lower coefficient of drag than polyethylene covers. A good cover will last indefinitely and cause no fabric wear. However, the edges of the holes and slots must be very smooth to prevent fabric wear. Nicks, chips or rough margins will eat fabrics up.

High Vacuum Operation

Flat boxes are the greatest source of wear to a forming fabric. Covers for flat boxes have either drilled holes or narrow slots over which the fabric travels. The high vacuums cause the yarns to be distorted and abraded on the hole edges or by abrasive particles trapped between the fabric and the cover.

Typical fabric wear patterns show high wear rates at the end of the slots on both the drive side and the aisle side of the machine. This wear area is generally the place where the fabric will wear and split in the machine direction. Several options such as lubrication showers, graduated vacuums and staggered slot ends will reduce the problem. It can hardly be eliminated.

Fabric wear can also be reduced by the use of wear resistant yarns containing nylon or other special materials, by applying special resin treatments to the fabric and also by applying fabric wear beads to the high wear area of the fabric outside the trim squirts.

High power loads on the drive motors on the wet-end can be caused by poor flat box operation. In any vacuum application under the forming fabric, higher dryness and lower loads will result from graduating both the low and high vacuums from low to high, reserving the highest vacuum for the last box. This method of operation will also help reduce fabric wear.

The flat boxes should dry the sheet within the range of 16 to 18% if they are operated efficiently. Flat boxes operate generally at 1 to 6 inches of mercury vacuum.

Steam boxes are used over flatboxes on some grades to heat the sheet water, lowering the water viscosity, making it easier to remove.

Couch

Some styles of fabrics will cause a loss in couch efficiency. Thick, open bottom fabrics

will carry air or permit radial air leakage into the couch box, bypassing some of the high velocity air under rather than through the sheet. However, the most common cause of low drying efficiency is the couch itself, either from worn seal strips, poor internal lubrication, or a vacuum source not operating at design standards.

The couch is the final opportunity to dry the sheet to the maximum prior to transfer to the presses. On most machines, a 4 to 6% increase in consistency can be expected over the couch.

Couch rewet results from excess water at the couch, particularly the water from internal and external couch showers. It is helpful to install a flow meter on the internal shower and seal water lines. By reducing the flow rate until a slight drag increase is noted and then increasing slightly will ensure the minimum of excess water and reduce rewet possibilities. In addition, locating the internal suction box so that there is a very small airflow into the couch at the point of sheet release will remove the entrained water in the fabric, keeping that water from rewetting the sheet.

Lump breakers are used to compact the sheet over the couch. By compacting or densifying the sheet in this area, pore size is reduced which increases air flow velocity for more efficient dewatering.

2.5.2 *Twin-Wires*

Modern twin-wire machines were developed to overcome the following disadvantages of fourdrinier machines [16]:

- two sidedness of the sheet
- lack of fine scale formation
- nonuniform profiles
- length and speed limitations

Twin-wire forming is the type of forming in which sheet is formed between two rotating forming fabrics. Although the idea of forming paper between two screens dates back to 1875, the credit for the first modern twin-wire machine is given to Daniel Webster of Consolidated Paper who had a working model in 1953.

Twin-wire formers can be classified as gap formers and top wire formers.

Gap Formers

In a gap former, the slurry is injected from the headbox nozzle between two converging forming fabrics as shown in Figure 2.86.

Depending on the dewatering mechanism, there are three major types of gap formers:

1. Gap roll formers: roll wrap is the major dewatering mechanism (Figure 2.87). Since there are no stationary parts, fabric life is usually longer in these machines compared to fourdriniers. Other advantages of gap roll formers are low drive power, reliability, speed capability, reduced two sidedness, less linting and improved printability. The disadvantages are higher headbox consistencies, potential for pinholes, lower internal bond and grainy formation.

Neglecting the centrifugal and vacuum forces, the following simplified formula has been accepted to determine the water removal force in gap roll formers:

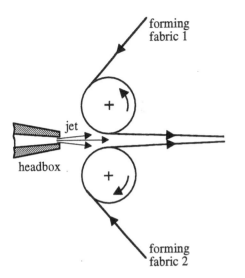

FIGURE 2.86. Schematic of gap former.

Since then, many variations of twin-wire formers have been developed. Delivery is through a nozzle which may have layering capabilities. A detailed discussion of these machine types is out of the scope of this book. Therefore, only a brief description of the machine types will be given below.

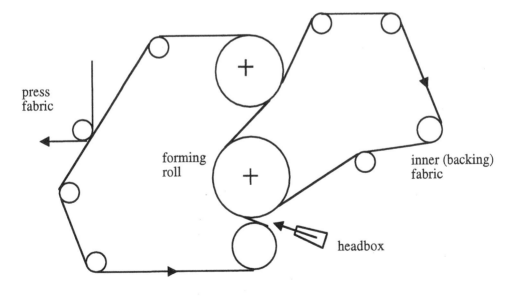

FIGURE 2.87. Schematic of a typical gap roll former ("C" wrap).

$$P = \frac{T}{R} \qquad (2.25)$$

P = dewatering pressure
T = outer wire tension
R = radius of curvature of the forming roll

Gap roll formers were initially used for newsprint grades. Today, their uses include brown paper, newsprint and high speed, lightweight sheet manufacturing such as tissue. The two common tissue formers are called "C formers" or "S formers" with respect to the shape of the drainage zone. The "C former" has a "C" shape drainage zone as shown in Figure 2.87. The forming roll is either solid for one-sided dewatering or perforated with suction for two-sided dewatering.

Figure 2.88 shows the schematic of an "S former" with an "S" shape drainage zone. Forming roll can be solid or with suction. In either case, two-sided dewatering takes place in this machine due to its configuration. The outer, conveying fabric also provides primary drainage.

2. Gap blade formers: blade action is the major dewatering mechanism (Figure 2.89). The drainage zone can be straight or curved. If the drainage zone is curved, then Equation (2.25) can be applied for curves that are approximately 500 cm in radius.

Gap blade formers have better formation, higher drive requirements and lower retention compared to gap roll formers [16].

3. Gap roll/blade formers: both wrap and blade actions are used for dewatering (Figure 2.90).

Top-Wire Formers

In top-wire formers, a forming section with an additional fabric is placed on top of the fourdrinier table, which carries the bottom forming fabric (or conveying wire). There are a variety of top-wire formers which are called retrofits, hybrids, preformers or on-top formers. The common characteristic of top-wire formers is that they all have a preforming zone of "open-wire" papermaking like a fourdrinier machine. Following open-wire preforming, dewatering can be done with roll, blade, roll/ blade or adjustable blade. Figure 2.91 shows

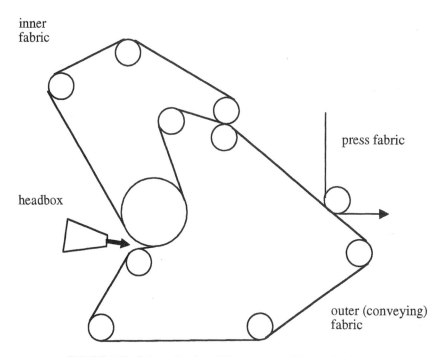

FIGURE 2.88. Schematic of an "S" wrap gap roll former for tissue.

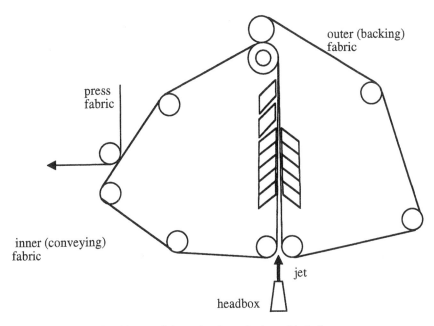

FIGURE 2.89. Schematic of a typical gap blade former.

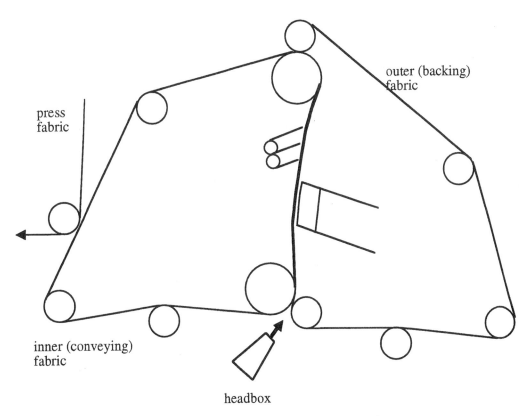

FIGURE 2.90. Schematic of a typical gap roll blade former.

131

hybrid roll preformer

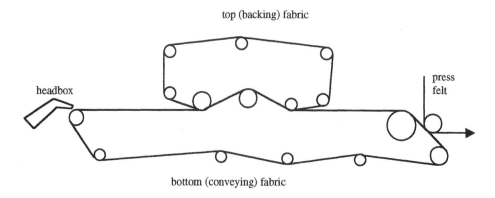

hybrid blade preformer

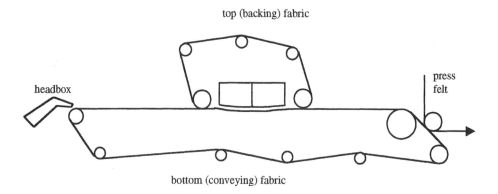

FIGURE 2.91. Examples of top-wire formers.

examples of top-wire formers. These formers are generally used for lighter weight sheets. It is reported that they improve two sidedness, reduce linting and improve formation compared to fourdrinier machines. Compared to gap formers, preformers have less streaking, good formation with improved first pass retention, better cleaning and drainage capability.

2.5.3 *Two-Wire Formers*

In two-wire formers, two sheets are formed separately and combined later on the forming section. Therefore, there are two headboxes on the paper machine as shown in Figure 2.92. These machines are usually used for heavier weight grades.

2.5.4 *Multiwire Formers*

In multiwire machines three or more fourdrinier tables are used to form the sheet. As a result, there are at least three headboxes and three forming zones as shown in Figure 2.93. Proper bonding between the sheet layers is critical. Bonding is controlled by the consistency of the various layers of fibers being combined.

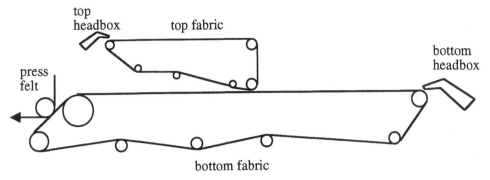

FIGURE 2.92. Schematic of a typical two-wire former.

2.5.5 *Crescent Formers*

A Crescent former is basically an inverted "C" gap roll former that is used for tissue manufacturing (Figure 2.94). Felt and forming fabric are used in the Crescent former which are unique to this machine.

2.5.6 *Cylinder Machines*

Cylinder machines were especially suitable for heavyweight multiple board grades. Although cylinder machines are still running in some applications such as tube stock, core stock, jigsaw puzzle and other grades that are too heavy to be practical for fourdrinier production, their design has become obsolete and therefore they are not manufactured anymore. Figure 2.95 shows the working principle of these machines. The sheet is formed on the wire mesh of a rotating cylinder mold, partially submerged in a vat of fiber stock. Aided by a couch roll, the fiber mat transfers from the cylinder to a making felt, which is a woven fabric combined with nonwoven batt. The flow of slurry into the vat can be in the same direction of or opposite to the rotation direction of the vat. Several cylinder vats are placed in a sequential manner which allows formation of a multilayer sheet structure. These machines were slow and had some formation problems such as uneven basis weight.

2.6 Manufacture of Specific Paper Grades

The sheet formation principles of different paper grades such as tissue, newsprint, fine paper, brown paper, etc., are basically the same. Nevertheless, there are some differences in manufacturing of some grades. This section explains the major manufacturing

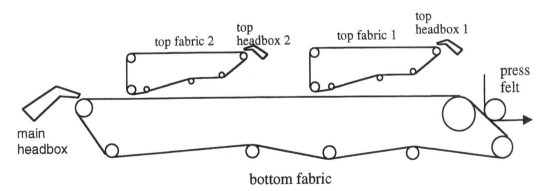

FIGURE 2.93. Schematic of a typical multiple forming machine with three forming fabrics.

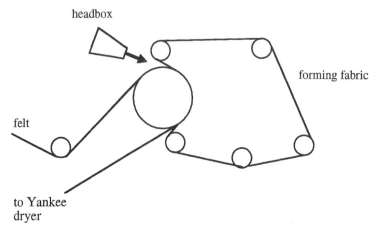

FIGURE 2.94. Schematic of a Crescent former.

characteristics that are specific to each paper grade. The manufacture of nonwovens is quite different than the other paper grades. Special methods and equipment have been developed to manufacture nonwovens. A more detailed description of these methods is given.

2.6.1 Tissue Manufacturing

The tissue market covers a variety of low-weight sheets used primarily in sanitary (facial and bathroom tissue, towel, napkin, etc.) applications. A much smaller segment includes specialized grades such as condenser, carbonizing and wrapping tissue [17]. Most sanitary tissue fabrics run CD knuckle proud to orient the fibers in the CD direction. The hole is longer in the CD direction.

Due to its low basis weight (as low as 5 g/m^2) or loose structure, special machines have been developed to manufacture tissue. One machine that has been used consists of a fourdrinier forming section and Yankee dryer. The Yankee dryer is a large steam cylinder with a polished surface to dry the tissue. Tissue is pressed against the Yankee surface and transferred to it. Through-air dryers are used before or after Yankee dryer to preserve bulk and increase drying efficiency as shown in Figure 2.96 [17–19]. Modern tissue formers include "C" wrap, "S" wrap and Crescent former (Figure 2.97). The wet tissue sheet must be supported in forming, pressing and drying sections of the paper machine.

In many cases, creping is done to increase the softness of tissue paper. As a result of

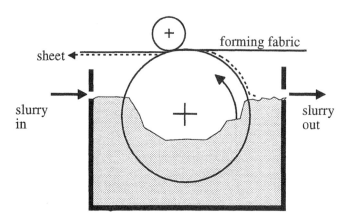

FIGURE 2.95. Schematic of a cylinder vat.

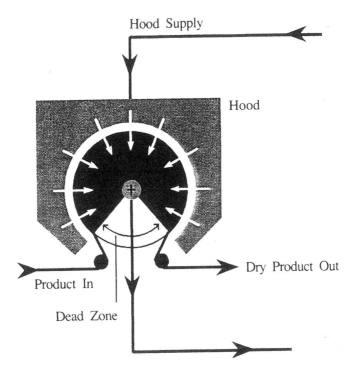

FIGURE 2.96. Through air dryer (courtesy of Valmet Paper Machinery).

creping, the MD and CD strengths of the paper are reduced. Extensibility is increased which increases the rupture resistance of the sheet. Creping also increases the bulk of paper.

Creping is done on the Yankee dryer cylinder using a doctor blade (Figure 2.98). The web follows the dryer surface to hit the doctor blade. The degree of crepe has an effect on the bulk. The degree of crepe is defined as the difference in peripheral speed between the dryer and the pope divided by the speed of the dryer. Around 50% crepe, the bulk reaches a maximum. Crepe formation is affected by doctor blade geometry, adhesion between the

paper and dryer surface and mechanical properties of the paper [20].

2.6.2 *Newsprint and Fine Paper Manufacturing*

During formation of newsprint and fine paper, it is preferred that fabric speed is higher than the jet speed.

During printing, newspapers ran in the direction of reading which is the machine direction on the paper machine. There is considerable amount of tension on the sheet

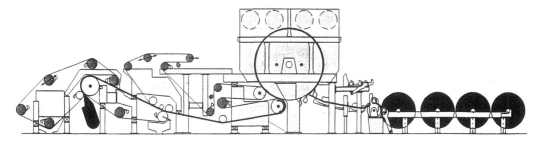

FIGURE 2.97. Schematic of a tissue machine with Yankee cylinder (courtesy of Voith Sulzer).

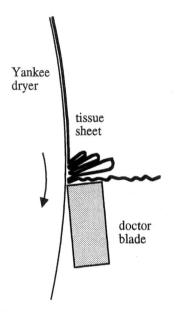

FIGURE 2.98. Schematic of creping process.

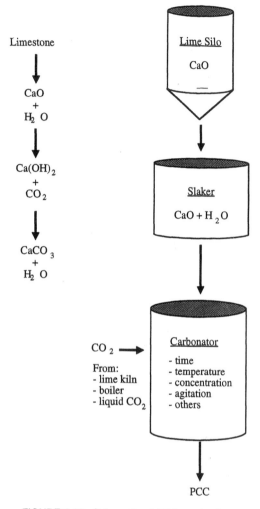

FIGURE 2.99. Schematic of PCC production.

during printing; therefore, the paper should be strong in this direction. This is achieved by orienting the fibers in the machine direction. In fine paper fabrics, the hole is longer in MD direction.

The major additives in the fine paper grades are fillers which give improved opacity and other optical properties to the sheet. The two major types of fillers are clays and carbonates, both of which are very small particles compared to fibers. To be effective, they must be retained in the sheet.

Acid versus Alkaline Papermaking

Fine paper manufacturing can be done either in acid or alkaline environment. Alkaline papermaking results in longer paper life. Therefore, the use of alkaline papermaking is increasing. Ground Limestone (GL) is used in alkaline papermaking.

Contents of Precipitated Calcium Carbonate (PCC):

(1) calcite
   - rhombohedron (barrel shaped)
   - scalenohedron (rosette clusters)
(2) aragonite
   - needle shaped (3:1 aspect ratio)

Figure 2.99 shows the schematic of PCC production. The acid and alkaline papermaking systems are compared in Table 2.15. Table 2.16 lists the effects of alkaline papermaking. For better abrasion resistance, nylon yarns are used along with polyester in forming fabrics that are used in alkaline conditions. In general, PCC produces less fabric wear than GCC (ground calcium carbonate).

2.6.3   Nonwoven Manufacturing

A nonwoven product is made directly from a web of fiber. Although paper also fits this

TABLE 2.15. Acid and Alkaline Papermaking Systems.

	Acid	Alkaline
pH	4.5–5	7.5–8
pH control	Alum	Calcium carbonate Soda ash
Fillers	TiO_2 Clay Extenders	Calcium carbonate TiO_2 Clay, extenders
Size	Rosin	Alkyl ketene dimer (AKD) Alkenyl succinic anhydrides (ASA)
Cationic starch	Tertiary amine	Quaternary amine
Points of addition		
Fillers	Machine chest/blend chest	Machine chest/blend chest
Size	Machine chest	Stuff box
Cationic starch	Machine chest/blend chest	Machine chest/blend chest

definition, paper is not considered to be a nonwoven product [21]. Due to manufacturing and product similarities between nonwovens and paper, a nonwoven product is usually considered as an intermediate form between paper and conventional textile fabrics.

Comparison of major textile fabric and paper properties are shown in Table 2.17. Nonwoven fabric properties generally fall in between those of paper and conventional textiles. Therefore, nonwovens frequently compete with textiles or paper in already established markets. Table 2.18 compares the production speeds of various processes.

Paper is a relatively low priced commodity compared to nonwovens which may be a motivational factor for paper companies to move toward nonwovens.

Five major types of nonwoven processes that use forming fabrics are wet laying, air laying, spunbonding, meltblowing and spunlacing. All of these processes are very

TABLE 2.16. Effects of Alkaline Papermaking.

Terms	Effects
Deposits	Generally less than with acid conditions, some press problems with size pick out in press section.
Drainage	Since $CaCO_3$ does not contain hold water, drainage is more rapid under alkaline.
Draws	May require slight changes.
Drying	Alkaline sheets dry easier than acid sheets due to less water bonding.
Formation	Table set up is critical. Refining is usually less and furnish has more long fiber.
Pressing	Equal to acid conditions.
Retention	Extremely important for filler and sizing efficiency. Early retention very important. Retention aids a must.
Runnability	Providing good wet-end stability, runnability is equal to or better than acid. Deposits and holes are most common problem.
Sheet dryness at couch	Usually better under alkaline conditions. $CaCO_3$ and lower fiber mass drain easier.
Table set up	Generally set up to carry water down the table. Grading of flatboxes is critical. Forming zone also critical.
Fabric wear	Proper fabric design and table setup minimizes wire wear. Equal to acid configurations under proper setup.

TABLE 2.17. General Characteristics of
Paper and Textiles [21].

Paper	Textiles
Less expensive	More expensive
High production rate	Low production rate
Short fiber	Long fiber or filament
Simple structure	Complex structure
Smooth surface	Textured surface
Dense	Bulky
Inextensible	Some extensibility
Able to hold sharp folds	Noncreasing
Stiff, little drape	Flexible, drapes relatively easier
Relatively nonporous	Porous
Low wet strength	Wet strength is essential
Low tear strength	High tear strength

TABLE 2.18. Productivity of Various
Manufacturing Methods [21].

Process	Linear Speed (m/hour)
Weaving	5–40
Knitting	100
Mechanical web formation	3,000
Spunbonding	4,500
Papermaking	50,000

different from one another and may require different designs of forming fabrics.

Wet Laying

In the wet-laying process, a mixture of fibers and water are laid down on the moving forming fabric (Figure 2.100). The water drains through the fabric, leaving a web or sheet of entangled fibers. The fibers can be wood, cotton, glass, polyester, rayon or any mixture of these. The function of the fabric in this process is to provide a carrier that is adequate to support the fibers, keeping them from penetrating into and through the mesh. Good web release and exceptional wear resistance are required in the fabric. An example of a wet-laid product is fiberglass mat used in the manufacture of roofing shingles which is shown in Figure 2.101 along with the forming fabric used to make it. This mat is impregnated with asphalt and provides the strength in the shingle. Table 2.19 lists the requirements of forming fabrics for wet-laying processes. Adhesives are required to replace the hydrogen bonding formed in papermaking.

Air-Laying

In the air laying process, wood pulp is ground up, mixed in an air stream and fed onto the moving forming fabric (Figure 2.102). A variety of fiber types can be used, but wood pulp is very common. An example of an air-laid product is a wipe which is characterized by being very soft, bulky and absorbent.

Forming fabrics used in air-laying must have good fiber support for retention of fibers and relatively high air permeability to ensure fiber hold-down. Due to the very high static buildup in this process, antistatic cross-direction yarns are used.

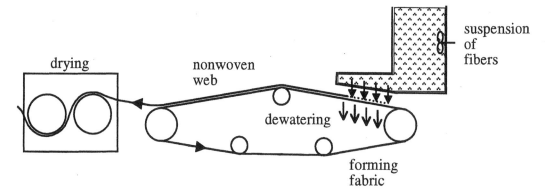

FIGURE 2.100. Schematic of wet-laying process.

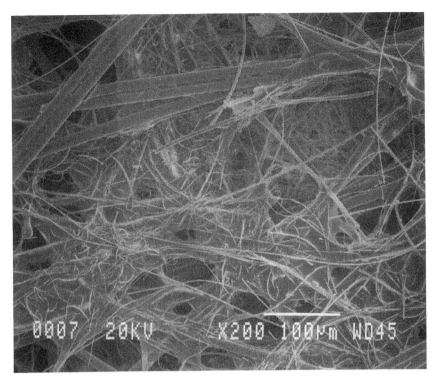

FIGURE 2.101. Structure of wet-laid fiberglass mat for roofing.

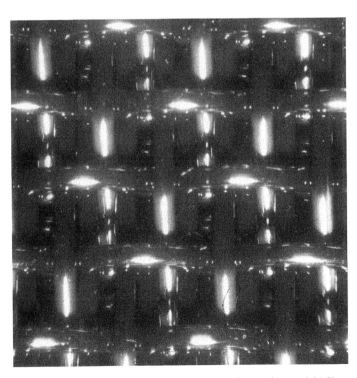

FIGURE 2.101 (continued). Forming fabric used to make wet-laid fiber-glass mat for roofing.

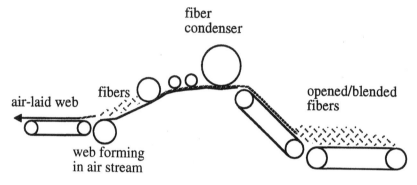

FIGURE 2.102. Schematic of air-laying process.

Forming fabric characteristics for the air-laying process:

- good fiber support
- fiber retention
- high air permeability
- medium mesh range
- two layer designs
- antistatic properties

Spunbonding

In the spunbonding process, pellets of polymer are melted down and forced through a spinnerette, a device with tiny holes like a shower nozzle (Figure 2.103). Continuous filaments are extruded through the spinnerette, are blown about and spread on the moving forming fabric.

The forming fabric used in spunbonding must have good fiber support characteristics to stop fiber penetration, provide good web

TABLE 2.19. Forming Fabric Requirements for Wet-Laying Process.

Binder Position	Forming Position
Stability	Stability
Durability	Durability
Good drainage	Good Drainage
Fiber retention	Fiber retention
Single layer designs	No wire mark
Good seam joints	Good seam joints
	Two layer designs
	Medium-fine mesh fabrics

release and not mark the sheet in any way. Figure 2.104 shows a polypropylene spunbonded medical product along with the forming fabric used to make it.

Forming fabric characteristics for spunbonding:

- stability
- durability

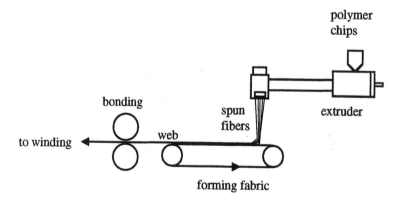

FIGURE 2.103. Schematic of spunbonding process.

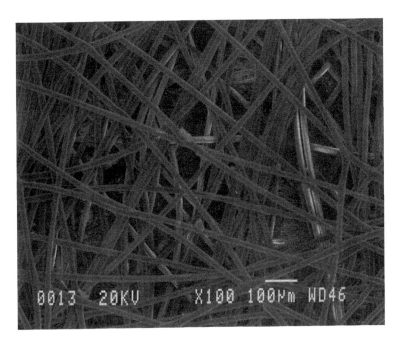

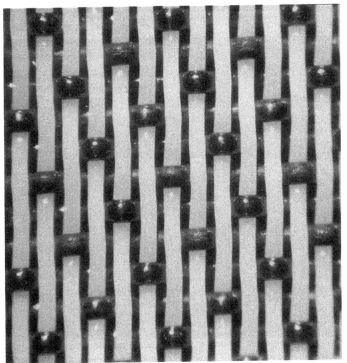

FIGURE 2.104. Polypropylene spunbonded medical product (top) and forming fabric used to make it.

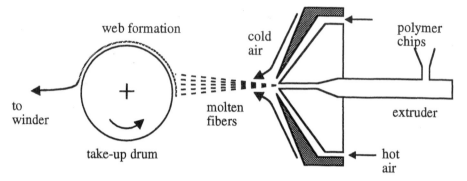

FIGURE 2.105. Schematic of meltblowing process.

- good fiber support
- high air permeability
- two layer designs
- medium-fine mesh range
- good seam joints
- anti-static properties on some applications

Spunbonded web characteristics are low weight, high stiffness and high tensile and tear strength.

Meltblowing

In meltblowing process, granules of polymer are melted down and blown with hot, high velocity air through extruder die tips. The filaments are then stretched until they break and are deposited onto the moving forming fabric (Figure 2.105). The fibers break into short lengths, as compared to the continuous fibers in the spunbonding process and are also much finer in diameter. Thus, the web produced is often used for wipes and filters.

The forming fabric for the meltblowing process must provide adequate fiber support, release and good stability. Figure 2.106 shows the fabric view from machine side with elastomeric polymer showing through. Figure 2.107 shows meltblown elastomeric polymer formed on a forming fabric.

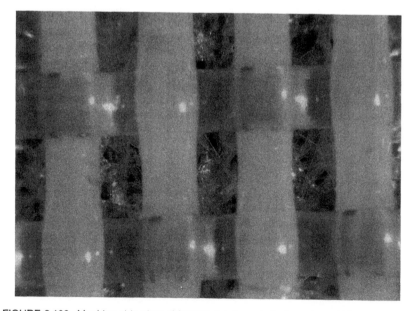

FIGURE 2.106. Machine side view of forming fabric and elastomeric meltblown polymer.

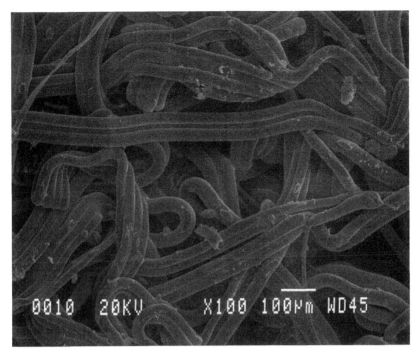

FIGURE 2.107. Photomicrograph of meltblown elastomeric polymer fabric formed on forming fabric.

Forming fabric characteristics for melt blowing process:

- large fabric design range
- single and two layer designs
- coarse-medium mesh range
- metal and synthetic fabrics
- good air permeability
- wire mark is less critical

Spunlacing (Hydroentanglement)

In this process, the finished product is produced by entangling fibers in a preformed web, generally carded, using high pressure, columnar water jets (Figure 2.108). As the jets penetrate the web and deflect from the forming fabric, some of the fluids splash back into the web with considerable force. Fiber segments are carried by the turbulent fluid and become entangled on a semimicron scale. In addition to bonding the web, spunlaced or hydraulic entanglement can also be used to impart a pattern to the web. Figure 2.109 shows a spunlaced rayon wipe and the fabric used to make it.

Forming fabric characteristics for spunlacing:

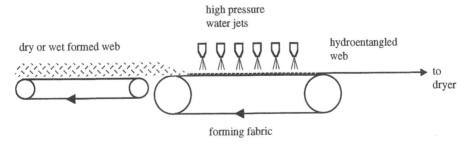

FIGURE 2.108. Schematic of spunlacing (hydroentanglement) process.

0008 20KU   X150 100μm WD45

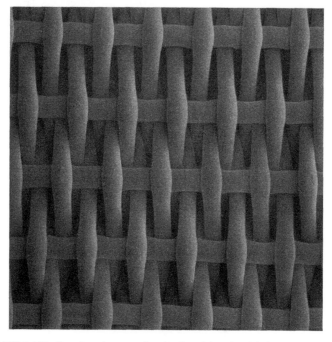

FIGURE 2.109. Spunlaced rayon wipe (top) and forming fabric used to make it.

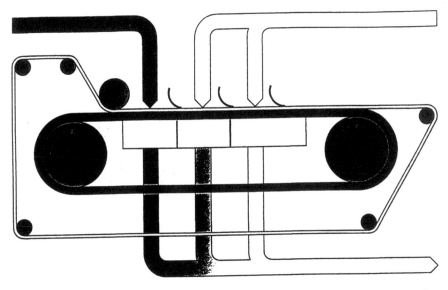

FIGURE 2.110. Schematic of filter belt application for cake washing (courtesy of Joy Manufacturing Company.)

- variety of designs
- stability
- good release
- fabric uniformity
- seam is critical

2.6.4 *Special Applications*

Fabrics with structures similar to paper machine clothing are also used in end-use applications other than papermaking. These fabrics are usually called "specialty fabrics." Examples of application areas for these fabrics are food processing, drying, dewatering and textile printing. Figure 2.110 shows the schematic application of a filter belt in cake washing.

Specialty fabrics are similar to forming fabrics or dryer fabrics (Chapter 4) for their structure and functions. The common characteristic of specialty fabrics is their coarse structure. They are woven under higher tensions than forming or dryer fabrics. Heatsetting is also done at higher tensions. The air permeabilities of common specialty fabrics range from 300 cfm to 1300 cfm.

Figure 2.111 shows the schematic application of a specialty fabric in textile drying

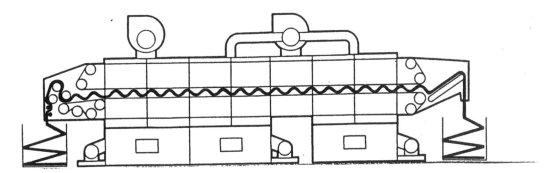

FIGURE 2.111. Schematic of textile drying application using a specialty fabric (courtesy of Santex AG Maschinenbau).

FIGURE 2.112. Leno design dryer fabric.

application. Figures 2.112 and 2.113 show two fabric designs that may be used for this type of application. Figure 2.113 is a leno design made of 100% Nomex® warp and Nomex® wrapped fiberglass filling. In the fabric shown in Figure 2.113, the warp yarns are grouped together to balance the weave and keep the fabric open. The fabric is made with 100% PPS monofilament.

Figure 2.114 shows a herringbone design that is used in recycled paper stock filtration. The design repeats on 12 warp yarns. Figure

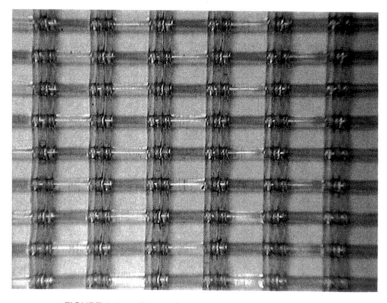

FIGURE 2.113. Fabric design with grouped warp yarns.

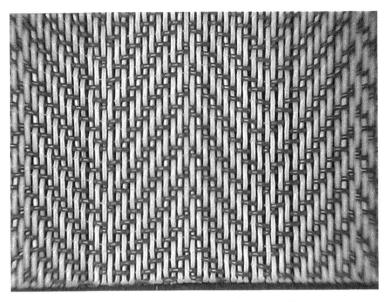

FIGURE 2.114. Herringbone point draw design.

2.115 shows a specialty fabric that is used for drying in milk powder production.

Some of these fabrics are woven endless while others are woven flat and seamed later. Seam styles include single wire clipper seam, coil seam, etc.

Coil Fabric

Figure 2.116 shows a unique polyester monofilament fabric designed for wet and dry formed construction board products. The design allows maximum water removal. The

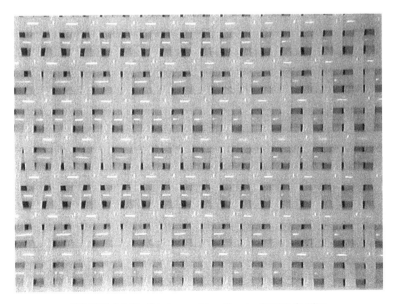

FIGURE 2.115. Fabric used in milk powder production.

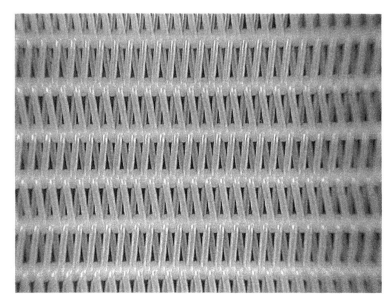

FIGURE 2.116. Coil polyester fabric.

durable coil construction offers good abrasion resistance and a smooth surface enhances doctoring properties for easy cake removal. The fabric can be supplied open ended without any type of conventional seam and is converted endless by merely inserting a synthetic filling yarn after meshing the ends.

Watermarks

Watermark is a pattern in paper made with a raised or indented design that comes in contact with the paper when it is approximately 90% water and 10% fibers. Watermarking has been known for almost 700 years. Watermarks first appeared in Italy around A.D. 1280 which depicted animals, fruits and flowers. Later two-color and light-and-shade watermarking were developed. One of the first commercial watermarks used in the United States was produced for Pennsylvania in 1777. Although it is not exactly known as to why watermarks were first used, they were probably used as a symbol of mysticism and religion. Today they are mostly used for aesthetic adornment and identification. Banks and insurance companies use them to prevent forgery.

Dandy rolls are used to make watermarks. With a wire mark dandy roll, while paper is still being formed, the stock passes under the dandy roll, and the design displaces some of the fibers to form the watermark pattern. As a result, the paper is more translucent and the watermark appears lighter than its surrounding area. Shaded mark dandy rolls utilize the intaglio process in which the design is depressed to form the watermark. As the dandy roll presses on the paper being formed, compression occurs everywhere but in the cutout design. Fibers accumulate more heavily in this depressed area, which results in a watermark slightly darker than its surrounding area.

2.7   Transfer of Sheet from Forming to Press Section

After the sheet is formed on the forming fabric, it is transferred to the press section for water removal and sheet consolidation. Figure 2.117 shows various configurations for sheet transfer from forming to press section.

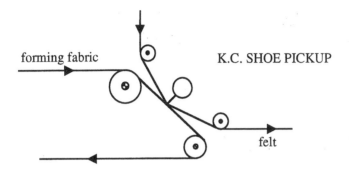

forming fabric   K.C. SHOE PICKUP

felt

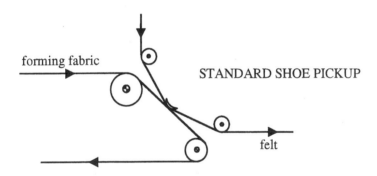

forming fabric   STANDARD SHOE PICKUP

felt

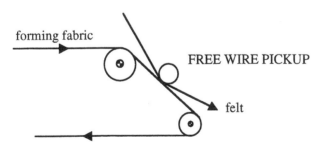

forming fabric   FREE WIRE PICKUP

felt

FIGURE 2.117. Sheet transfer from forming to press section.

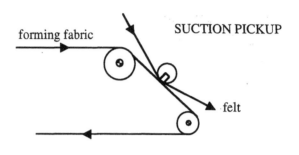

SUCTION PICKUP

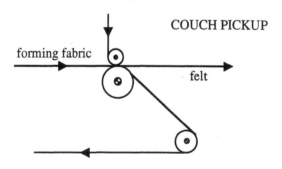

COUCH PICKUP

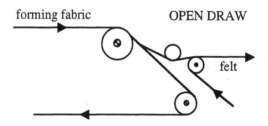

OPEN DRAW

FIGURE 2.117 (continued). Sheet transfer from forming to press section.

2.8 References

1 Beach, E., *Fourdrinier Wet End*, Asten, Inc., 1995.

2 Graham, L. L., and Ring, G. J. F., Introduction to Pulp and Paper Properties and Technology, Course Notes, The Paper Science Department, University of Wisconsin/Stevens Point, 1991.

3 Parker, J. D., *The Sheet-Forming Process*, TAPPI, 1992.

4 *Forming Fabrics: Design and Applications*, Asten, Inc., 1995.

5 Grundstrom, K. J. et al., "High-Consistency Forming of Paper," *TAPPI*, 56(7), 1973.

6 Fellers, C., "The Significance of Structure for the Compression Behavior of Paperboard" in *Paper Structure and Properties*, Ed. J. A. Bristow and P. Kolseth, Marcel Dekker, Inc., New York, 1986.

7 Beach, E. M., "Forming Fabric Design, Effect on Coated Sheet Properties," *TAPPI*, May 1979, Vol. 62, No. 5.

8 Adanur, S., *Wellington Sears Handbook of Industrial Textiles*, Technomic Publishing Co., 1995.

9 American Society for Testing and Materials, *Annual Book of ASTM Standards* (Volumes 7.01 and 7.02), published yearly by the ASTM, Philadelphia, PA.

10 Resistat® Conductive Fiber, BASF Corporation, Williamsburg, Virginia, 1995.

11 Certificate of Quality System Assessment, Shakespeare Monofilament Division, 1994.

12 American Association of Textile Chemists and Colorists, *Technical Manual of the American Association of Textile Chemists and Colorists*, published yearly by AATCC, Research Triangle Park, NC.

13 Beran, R. L., "The Evaluation and Selection of Forming Fabrics," *TAPPI Journal*, Vol. 62, No. 4, 1979.

14 Johnson, D. B., "Retention and Drainage of Multi-Layer Fabrics," *Pulp & Paper Canada*, Vol. 87, 1986.

15 McCumsey, K., *Showering Recommendations*, Asten, Inc., 1994.

16 Thorp, B. A., "Principles of Twin-Wire and Multiple-Wire Formation and Drainage" in TAPPI 1991 Wet End Operations Short Course, Orlando, FL.

17 Smook, G. A., *Handbook for Pulp & Paper Technologists*, Joint Textbook Committee of the Paper Industry, TAPPI and Canadian Pulp and Paper Association, 1989.

18 B. A. Thorp, and M. J. Kocurek, Ed., *Pulp and Paper Manufacture, Vol. 7*, Paper Machine Operations, Joint Textbook Committee of the Paper Industry, TAPPI and Canadian Pulp and Paper Association, 1991.

19 TAPPI Wet End Operations Short Course, Orlando, FL, TAPPI Press, 1991.

20 Hollmark, H., "Mechanical Properties of Tissue" in *Handbook of Physical and Mechanical Testing of Paper and Paperboard*, Ed. R. E. Mark, Vol. 1, Marcel Dekker, Inc., 1983.

21 Broughton, R. M., and Brady, P. H., "Nonwoven Fabrics" in *Wellington Sears Handbook of Industrial Textiles*, Technomic Publishing Co., 1995.

2.9 Review Questions

1 Why is the formation of sheet so critical for paper properties? Explain.

2 Explain how sheet is formed. What are the major forces acting on fibers during formation?

3 Explain the similarities and dissimilarities between textile structures and paper structures.

4 How can you control the orientation of fibers on the forming section of the paper machine? Can you orient the fibers in any direction in three dimensions (XYZ)? Why?

5 What kind of fillers are used in papermaking? What is the purpose of using a filler?

6 Papermaking is an inefficient process in the sense that approximately for every 1 kg of paper, 99 kg of water is required. Could you think of any other more efficient way to form the sheet with or without water?

7 Explain the functions of forming fabrics in sheet manufacturing.

8 What are the advantages and disadvantages of flat versus endless weaving of forming fabrics? Explain.

9 What are the major forming fabric designs? Explain the major differences among them for the following properties:
• drainage capacity
• fiber support
• fabric life
• strength

10 How can you change the percent open area of the forming fabric without changing the hole size?

11 Explain the major steps in forming fabric manufacturing. Why are very heavy

weaving machines required to make forming fabrics? (hint: think of forces involved during weaving of forming fabric).

12 What are the advantages and disadvantages of polyester and nylon yarns in forming fabric structures? Would a 100% nylon forming fabric work? Why?

13 Explain the extrusion process.

14 Calculate the denier and the weight per kg of the following yarns:
   • polyester diameter: 0.0055 inch
   • nylon 6,6 diameter: 0.030 inch

15 What is the relation between fabric tensile strength and yarn tensile strength? What other factors affect the fabric tensile strength?

16 Can air-jet weaving be used to make forming fabrics? Why?

17 How is pick count changed in weaving of forming fabrics?

18 What are the reasons for preliminary heatsetting of forming fabrics? Explain the effects of heatsetting on fabric modulus with reasons.

19 How can you improve the seam strength of forming fabrics?

20 Can monoplane surface be achieved with proper weaving conditions? How?

21 Calculate the holes/cm^2, hole dimensions and % projected open area of the following single layer forming fabrics:

Mesh	Warp Diameter	Filling Diameter
12 × 14	0.03″	0.03″
24 × 19	0.024″	0.035″
63 × 49	0.0095″	0.0125″
86 × 55	0.0067″	0.009″
95 × 96	0.0055″	0.006″
103 × 107	0.004″	0.005″

22 How is fabric permeability measured? How effective is the fabric permeability in predicting the drainage of forming fabrics?

23 When selecting a forming fabric for a special application, what kind of paper and fabric properties are need to be considered?

24 Define contaminant resistance? Explain the ways to make forming fabrics more contaminant resistant?

25 What causes fabrics to bow and skew? What can be done to correct the problem?

26 Explain the following terms:
   • pressure forming
   • velocity forming
   • rush/drag ratio

27 Why is turbulence is necessary in sheet formation?

28 In a gap roll former, if the tension on the outer fabric is 6 kg/cm, the forming roll radius is 75 cm, calculate the dewatering pressure (refer to Figure 2.87).

29 Explain the spunbonding process.

3

Pressing

The press section is the next part of the paper machine after the forming section. The function of pressing is to continue the water removal process which started in the forming section, consolidate the sheet, give texture to the sheet surface, support and transfer the sheet. The fabrics in this section are called press fabrics or press felts. The term "fabric" has a broader meaning; "felt" is a type of fabric which is made of individual fibers only, i.e., no yarn in the fabric structure. Nevertheless, the terms "press fabric" and "press felt" are used as identical terms in the papermaking industry. Although it depends on sheet grade and paper machine, typical sheet consistency at the beginning of the press section is 20% fiber and 80% water and at the end of the press section is 40% fiber and 60% water. At the end of the press section, the sheet is transferred to the dryer section.

During pressing, the sheet is compressed between one or two fabrics and two rolls in the press nip to squeeze water from inside the web and out of the felt fibers. Figure 3.1 shows this process in a plain press nip. Increased compression increases water removal [1–3].

The main functions of a press fabric are water removal from the sheet, sheet support and transportation, providing uniform pressure distribution, and imparting proper surface finish to the sheet. The fabric should provide proper protection for the sheet to resist crushing, shadow marking and groove marking. Other functions include transferring the sheet

from one position to another in case of closed draws, driving the undriven felt rolls and cylinder formers. The amount of water that the felt can absorb and the water flow resistance are affected by the void volume (volume that is not occupied by fibers or yarns) and air and water permeability of the fabric. Low flow resistance and ability to maintain void volume under load are important during operation. Important press fabric properties include strength, adequate void volume, required permeability, low compressibility, batt/base ratio, compaction resistance, abrasion resistance, contaminant resistance, heat and chemical resistance.

3.1 Water Removal Theory

The basis weight of the sheet and its drainage characteristics are critical factors in determining the mechanism for water removal. There are basically two types of nips that define the water removal process.

3.1.1 *Flow-Controlled Nips*

In these nips, water removal is primarily influenced by water flow resistance in the sheet. In flow-controlled nips, the resistance to fluid flow within the mat of fibers controls the rate at which water can be pressed. These

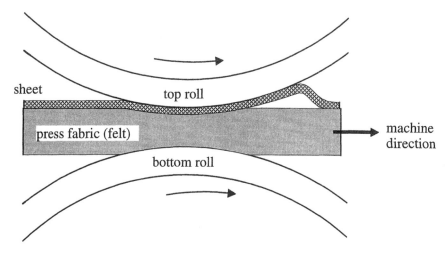

FIGURE 3.1. Schematic of simple press nip.

nips are characterized by high water loads, heavyweight sheets, slow draining and low freeness stocks.

Major symptoms of problems in flow-controlled nips are crushing and hydraulic flow mark. Water removal is aided by:

- soft roll covers
- large diameter press rolls
- double felting
- high sheet temperatures

Design Considerations

(1) Low flow resistance: In flow-controlled nips, it is very important that flow resistance be minimized. Typically, coarser batt deniers and higher permeability fabrics are used to facilitate water removal without crush or hydraulic mark.

(2) High void volume: Because of the higher water loads, high void volume structures are usually required. Generally multilayer or laminated fabrics are used to handle the water flow in the nip.

(3) Flow-controlled nips, where the sheet structure restricts water flow, may occur with lightweight, low freeness sheets, such as carbonizing, condenser and glassine: In this case, coarser batt denier would not be applicable, and void volume requirements would be less.

3.1.2 *Pressure-Controlled Nips*

In pressure-controlled nips, the mechanical resistance to compression within the sheet controls the rate of water removal. These nips are characterized by low water loads, lightweight sheets, fast draining and high freeness stocks.

Major symptoms of problems in pressure-controlled nips are vibration, poor consistency and physical impression mark. Water removal is aided by:

- hard roll covers
- small diameter press rolls
- pressing uniformity
- high sheet temperatures

Design Considerations

(1) Base pressure uniformity: Higher sheet contact during pressing is very important. High contact base designs are preferred.

(2) Mass uniformity: Because of the higher nip intensity and peak nip pressures, press bounce and vibration are potential problems.

(3) Batt stratification: Finer deniers are used to improve the pressure uniformity of the fabric. These are usually needled over coarser fibers for permeability control and resistance to filling.

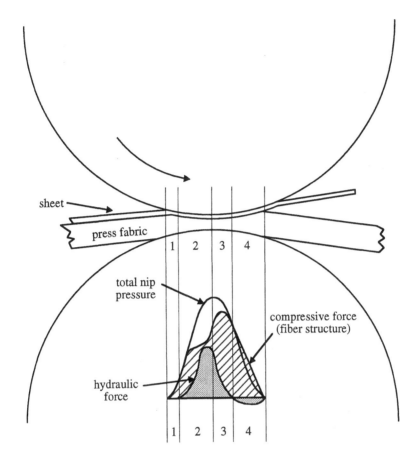

FIGURE 3.2. Transversal flow nip.

An understanding of the fundamentals of transversal flow pressing is necessary to optimize press and press fabric design. Figure 3.2 shows a typical transversal flow press nip in which press operation has been broken down into different phases based on mechanisms involved in water transfer. The nip is defined by two solid rolls with paper and fabric passing through the nip. Both the press fabric and paper are unsaturated entering the nip.

Phase 1 starts at the entrance of the nip where pressure develops in the sheet structure and continues until the sheet is saturated. There is no hydraulic pressure generated in this phase.

Phase 2 begins when the sheet becomes saturated and hydraulic pressure develops. Water is pressed from the sheet into the press fabric. If the press fabric becomes saturated, hydraulic pressure causes flow of water into the voids of the roll (grooves, holes, etc.). This phase extends to the mid-nip or the point of maximum pressure.

Phase 3 extends from the point of maximum total pressure to the point of maximum dryness. Maximum dryness occurs at maximum structure pressure and where the hydraulic pressure in the paper reaches zero.

Phase 4 begins where the paper and press fabric start to expand resulting in negative hydraulic pressure. Both the fabric and paper became unsaturated and rewetting occurs due to capillary forces and pressure differences between the press fabric and paper. During the expansion phase, the sheet and fabric compete for the boundary water.

The press fabric is a necessary component of this water removal mechanism. The press fabric provides a porous structure to which the water can flow from the paper in the ingoing nip and it should retain this water in the expansion phase of the nip. The ideal fabric should provide perfectly uniform pressure distribution, lowest possible flow resistance, and smallest rewet in the outgoing nip.

An analysis of nip conditions using Walstrom's model [4] can help identify the terms of our compromise. The following equation represents a model for outgoing moisture ratio for a nip.

$$MR_{out} = k - f_1 - f_2 - f_3$$
$$- P - RW - R_1 - R_2 \qquad (3.1)$$

where

MR_{out} = outgoing moisture ratio

k = moisture ratio of maximum paper dryness at zero flow resistance and uniform pressure distribution determined by compression characteristics of the paper

f_1 = increase in moisture ratio due to flow resistance in the fiber wall

f_2 = increase in moisture ratio due to flow resistance in the paper structure

f_3 = increase in moisture ratio due to flow resistance in the press fabric

P = increase in moisture ratio due to nonuniform pressure application

RW = increase in moisture ratio due to rewetting, redistribution of water between paper and fabric

R_1 = increase in moisture ratio due to rewetting of the press fabric from the pressure structure

R_2 = increase in moisture ratio due to rewetting after the nip

A closer look at this model suggests how press fabric properties influence water removal.

f_3, the effect of fabric flow resistance, is dependent on speed, capillary structure, incoming moisture, moisture change, temperature, and basis weight. It is worth noting that open press fabric structures, to reduce flow resistance, are in direct conflict with the property of uniform pressure distribution.

P, the effect of uniform pressure distribution, is likely determined primarily by the press fabric. The fabric should bridge the grooves, suction holes, etc., to exert a uniform pressure to the paper.

RW, R_1 and R_2 concern rewet and are dependent on press nip conditions, paper and fabric structure and their dryness.

The relative importance of each factor and the direction of fabric design compromise are determined by the following:

- water load
- nip pressure and width
- speed or nip dwell time
- sheet properties; freeness, furnish and weight
- temperature

At low basis weights, free sheets, and low speeds, flow resistance in the paper or fabric (f_2, f_3) is relatively negligible; P and rewetting dominate. Smooth, dense fabrics are preferred. At the other extreme—high basis weights, low freeness, and high speeds—flow resistance is highly important. Double felting and/or high-void volume fabrics can be of benefit. Pressure distribution and rewetting continue to be important.

In between the two extremes, all press variables need to be considered. Also it should be noted that the relative importance of each of these variables changes with water load. First presses with high water loads tend to behave more as flow-controlled nips, even on lightweights. Last presses have lower water loads and move toward pressure-controlled conditions. Table 3.1 shows the relative importance of fabric characteristics for pressure- and flow-controlled nips.

Figure 3.3 shows the distribution of hydraulic and compressive forces at the first, second and third nip in a press section. Hydraulic force is greater in early presses; compressive forces are higher in later presses. Table 3.2 lists typical sheet consistencies for various grades.

TABLE 3.1. *Relative Importance of Fabric Characteristics for Pressure- and Flow-Controlled Nips.*

Fabric Characteristics	Pressure-Controlled Nip	Flow-Controlled Nip
Caliper	↓	↑
Permeability	↔	↑
Void volume	↔	↑
Pressure uniformity	↑	↗

3.2 Press Fabrics (Felts)

3.2.1 *History*

Until the 1930s papermakers' felts had improved very little over the woven or pressed woolen fabrics observed in China by Marco Polo. In this pre-World War II era chemical treatments were developed to alter the woolen molecule. This improved the ability of wool to resist chemical, mechanical and bacterial degradation. In 1946 the introduction of staple man-made fibers contributed greatly to the strength and durability of wet felts. Early attempts to full or felt 100% synthetics resulted in fabrics that were too open or porous and unstable.

Since about 1944, the asbestos cement industry had been using fabrics made by a process known as needling. This was the process of laying a carded web of textile fibers on a woven base and needle-punching the web fibers into the base with barbed needles. It was considered that higher synthetic, unfulled fabrics might be made by this method to not only achieve desirable properties of woven, fulled felts, but also to make improvements in terms of life, finish and drainage. Needled fabrics were then produced in the synthetic ranges from 35% to 49%. In the late sixties, 75% to

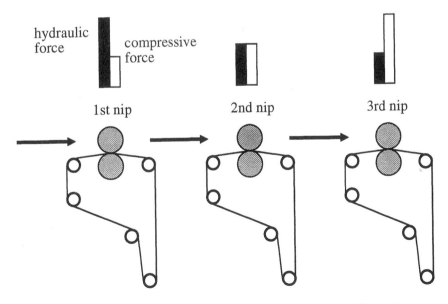

FIGURE 3.3. Distribution of hydraulic and compressive forces at different nips.

TABLE 3.2. Typical Press Dryness Number [5].

	Load		Exit Sheet Consistency, %
	kN/m	pli	
Newsprint			
First press	60	300	33
Second press	80	450	38
Third press	95	550	43
Bond			
First press	60	350	37
Second press	80	450	41
Third press	95	550	43
Corrugating medium			
First press	70	400	37
Second press	105	600	41
Third press	315	1800	46
Linerboard single felted			
First press	70	400	35
Second press	105	600	39
Third press	175	1000	42
Linerboard double felted			
First press	105	60	38
Second press	210	1200	43
Third press	350	2000	46

100% synthetic needled fabrics appeared and provided what some thought was the optimum in felt construction.

The introduction of the fabric press (inner fabric), whereby an incompressible fabric of monofilament or highly twisted multifilament synthetic is run between the felt and the press nip, resumed the evolution of the wet felt. The principle of the fabric press was built into the needled felt using a base fabric in mesh construction which provided significantly greater void volume within the total fabric structure. This improved the water handling capacity of the felt, allowing increased press loading. The fabric was easier to clean, and its tremendous strength and comparative rigidity provide remarkable improvements in felt life. The batt-on-mesh felt was, therefore, developed and felt-making took another step forward to the benefit of the papermaking industry.

3.2.2 Press Fabric Functions

Once the sheet has left the headbox, the general requirement of the paper machine is to increase the fiber consistency of the sheet from 0.2–1.5% to 92–96%. After the forming fabric, water removal costs far less in the press section than the dryer section. The value of efficient press fabric performance, therefore, can not be overemphasized.

Water removal, however, is not the only function of the press fabric. Its basic performance criteria includes other functions as well. In general, the press fabric must:

(1) Accept the water that is expressed from the sheet in the press nip.
(2) Provide the proper protection for the sheet to:
   • resist crush
   • resist shadow mark

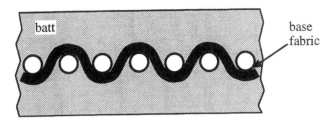

FIGURE 3.4. Schematic of a typical press fabric structure.

- resist groove mark
- resist base fabric mark

(3) Present the proper surface to the sheet so that the necessary degree of smoothness or finish requirement is imparted to the grade of paper to be manufactured.

(4) In case of closed draws, transfer the sheet from one position to another

(5) Provide desirable durability in terms of strength, resistance to mechanical abrasion, chemical degradation and fill-up.

3.2.3 *Construction of Press Fabrics*

Press fabrics, in general, consist of two components as shown in Figure 3.4: base fabric and batt.

The Base Fabric

Today, most press fabrics are made of 100% synthetics, primarily polyamide (nylon) polymers. Base fabrics are usually woven or constructed with cabled monofilament, plied multifilaments, spun yarns and/or single monofilaments (Figure 3.5).

Each yarn has properties that influence the operational characteristics of the press fabric. They are designed into the weave of the base fabric to affect sheet quality, water removal performance, runnability and ease of installation. Primary characteristics of the more commonly used yarns are given in Table 3.3.

The base fabric may be a single layer or multilayer mesh. The fabrics can be woven endless or woven flat and joined with a seam (Figure 2.3, Chapter 2). The weave of the base fabric is engineered to manipulate pressure uniformity, flow resistance, void volume, and compression properties. The basic classifications of press fabrics are conventional (endless) designs, stratified (laminated) designs and seam fabrics. Figure 3.6 shows the main types of base fabrics used in press fabrics.

Seam fabrics have shown a substantial increase in usage since their introduction in the late 1980s. The primary factors for the increase in use of seam products are safety, reduced installation time and, in some cases, improved press fabric performance. There are basically two types of seam products as shown in Figure 3.7.

Pin seam fabrics are woven flat with loops woven into the fabric in a separate process. These fabrics are characterized by crimp in the machine direction yarns. Woven loop seam fabrics are woven endless with loops formed during the weaving process. Like conventional endless fabrics, the woven loop design

FIGURE 3.5. Yarn types used in manufacture of base fabrics.

TABLE 3.3. Characteristics of Multifilament and Monofilament Yarns Used in Press Fabrics.

Multifilament	Monofilament
Very durable	Durable
Supple	Stiffer than multifilament
Compressible	Resists compression
Higher elongation	Less stretch than multifilament
Low resistance to chemical attack	Better resistance to chemical attack

has crimp in the cross machine direction. Both products can also be made with laminated fabrics on the surface.

The Batt

The process in which the batt is locked to the base fabric is called needling. The batt is carded into a uniform web first and then is applied in a series of layers onto the base fabric. The web and base fabric are fed through a zone where several thousand reversed barbed needles are needle-punched into the composite to lock the web to the base fabric. The batt is typically spliced at the start and stop of web application. Some processes can apply the web in a spiral method that eliminates the cross machine direction oriented splice. These processes are generally

applied on fabrics for critical press bounce and vibration prone positions.

The needling process can be engineered to affect the density, surface properties and permeability of the press fabric. The batt fibers that are used in the manufacture of press fabrics are purchased in standard sizes. The most commonly used unit to indicate the density of the fiber or yarn is the denier. Denier is actually a weight measure but it has become synonymous as an indicator of fiber fineness (Section 2.2.3). As long as the specific gravity of the polymer is approximately the same, denier can be used to compare fiber diameter. The relative sizes of the most common fibers used in press fabrics are shown in Figure 3.8. As can be seen from the figure, a 30 denier fiber is not twice as large as a 15 denier, and a 6 denier fiber is not twice as large as a 3

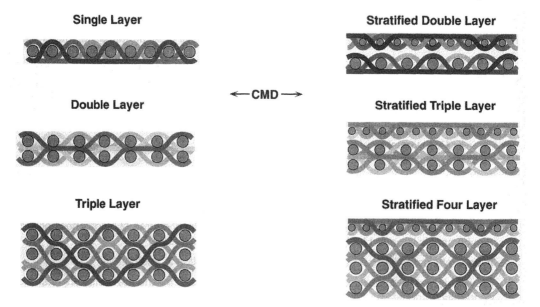

FIGURE 3.6. Major types of base structures used in press fabrics.

Pin Seam Fabric   **Woven Loop**

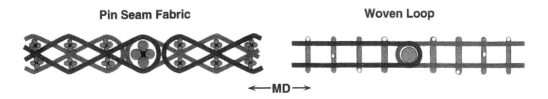

←—MD—→

FIGURE 3.7. Seam fabrics.

denier fiber. The relation between denier and diameter is explained in Section 2.2.3 for some common paper machine clothing materials.

Many factors influence the selection process of the fiber denier. These include:

- water handling requirements of the fabric
- break-in time
- available uhle box vacuum
- sheet control issues such as drop offs, blowing and sheet stealing
- fiber shedding propensity
- filling propensity
- bleed through propensity

Fibers with different deniers can be blended or applied in stratified layers for desired performance attributes. Figure 3.9 shows the most common means of batt application.

3.2.4 *Press Fabric Designs*

Conventional woven felts, which were woven with spun yarns became obsolete. Other major press fabric designs that have been used are explained below. Of these, the first two designs also have become obsolete.

Batt-on-Base Felt

A woven fabric, usually of spun yarns, is used as the base fabric on which fiber batt is needle-punched. Batt-on-base felts were made of either a combination of wool and synthetic fibers or 100% synthetic fibers (mostly nylon), which became the standard. The base fabric, which can have synthetic content between 50–100%, can be woven as endless belt or can be woven flat and joined later. The batt web has synthetic content between 20 and 100%. After washing and chemical treatment, the felt is stretched on the dryer to the operating dimensions. Singeing is usually done to prepare the surface and remove the loose fibers. Finally, the felt is dried at an overstretch length for easy installation. Figure 3.10 shows the random nature and structure of the top surface of a needled felt.

Batt-on-base felt designs have become obsolete and are not used anymore.

Baseless Felt

In this type of structure, no base fabric is used; felt is made by needle-punching only. The absence of base yarns reduces marking

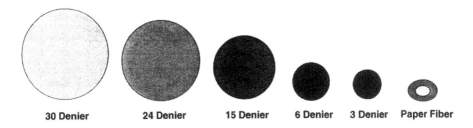

| 30 Denier | 24 Denier | 15 Denier | 6 Denier | 3 Denier | Paper Fiber |

FIGURE 3.8. Relative sizes of the most common fibers used in press fabrics.

Stratified Needling
Fine Denier Over Coarse

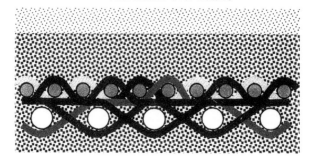

Blended Coarse And Fine Deniers

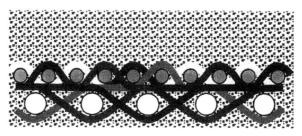

Single Denier Blend

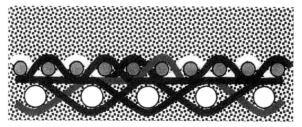

FIGURE 3.9. Most common ways of batt application.

and provides a uniform pressure distribution in the press nip. Nonwoven felts are suitable for suction, grooved and shrink sleeve presses, to provide a smooth sheet finish for fine paper and board machines. This type of design is outdated and not used anymore.

Batt-on-Mesh Felt

Needle-punched batt-on-mesh felt structure is similar to batt-on-base felt structure. Batt-on-mesh fabrics are generally made of 100% synthetic materials. The base structure may be a single or multilayer fabric woven with 100% monofilament, a combination of mono- and multifilament, all high twisted multifilament, all high twisted multifilament treated with resin for rigidity, or any combination of these yarns. Since these yarns are stiffer than spun yarns, batt-on-mesh fabrics have higher resistance to compression and compaction than batt-on-base felts. The base mesh is woven endless and the batt is applied on needle-punching machines. Figure 3.11

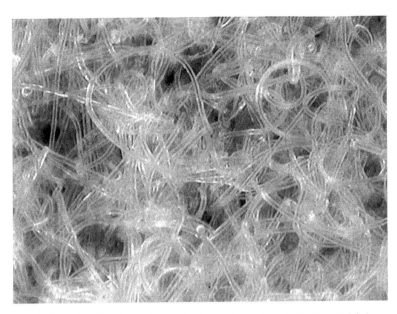

FIGURE 3.10. Random nature of the top surface of a needled press fabric.

shows a batt-on-mesh felt with single layer monofilament base fabric. Installation of batt-on-mesh fabrics may be more difficult than batt-on-base fabrics due to stiffer base structure.

Multifilament batt-on-mesh fabrics are suitable for suction, grooved and shrink sleeve presses. Resin treated multifilament fabrics are applicable on first and second suction presses for coarse paper grades such as kraft. Monofilament or mono- and multifilament combination base structural felts have excellent stability, resistance to wrinkles, and provide good finish for fine paper applications on suction pick-up positions, straight suction presses, and grooved or shrink sleeve presses.

Felts with No-Crimp Base-Fabric (Knuckle-Free)

In this type of design, the base fabric is not a woven fabric but is formed by two or more, completely separate uncrimped yarn layers. Figure 3.12 shows a two layer no-crimp felt structure. It is reported that the no-crimp base fabric structure provides improved sheet dewatering, good resistance to compaction and dimensional changes, resistance to filling and improved microuniformity of pressing. Eliminating knuckles in the base structure reduces the possibility of paper fines and dirt being trapped at the yarn crossover points. Moreover, it reduces sheet marking. Other advantages of knuckle-free felts include improved abrasion resistance, easy cleaning and smoother surface. However, due to lack of interlacing of yarns, void volume is drastically reduced and fabric stability may suffer.

Laminated (Stratified) Press Felts

In this relatively new press fabric structure, two or more base fabric structures are included in the fabric. The top base fabric is usually a single layer. The bottom base fabric can be a one, two or three layer integrally woven fabric (Figure 3.13). Laminated fabrics allow a wider range of base fabric pressure uniformity and low batt-to-base ratio which is critical for open and clean operation. As the machine speeds increase, the nip resistance time decreases and better surface contact between the sheet and press fabric becomes a requirement.

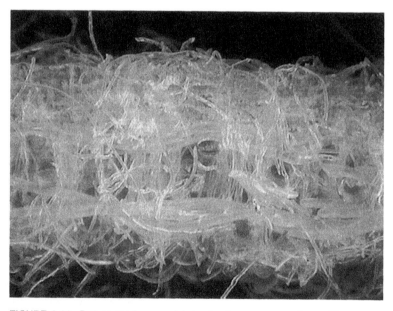

FIGURE 3.11. Batt-on-mesh press fabric structure: paper surface with exposed base fabric and cross-sectional view along the MD.

Continuous research is being done to develop new and improved press fabric designs. The latest developments include utilization of round, hollow yarns in cross machine direction. Hollow yarns flatten under pressure and immediately rebound after exiting the nip as shown in Figure 3.14. Flexing of yarns provides better sheet contact which in turn improves water removal and smoothness. Machine runnability and printability on the sheet are also improved. Another development is 100% monofilament (no batt) press fabrics.

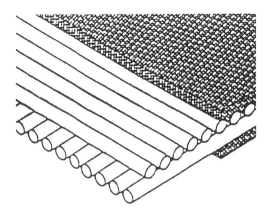

FIGURE 3.12. Two-layer no-crimp base press fabric [6].

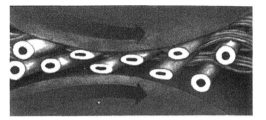

Process through the nip

Table 3.4 shows the base fabric weight, total fabric weight and air permeability ranges of typical single and multilayer fabrics.

3.2.5 *Manufacturing of Press Fabrics*

Due to batting, manufacture of press fabrics is different than forming or drying fabrics. Figure 3.15 shows the major manufacturing steps for press fabrics.

Batt fibers are blended, carded and usually preneedled first. Base fabrics are woven (except no-crimp base fabrics) similar to forming fabrics. They can be woven as a flat or endless belt as shown in Figure 2.3 (Chapter 2). The warp (MD) direction of an endless woven belt becomes the CD direction on the paper machine.

Flat woven fabrics have to be joined or pin-seamed to make an endless construction.

base fabric

FIGURE 3.14. Fabric with hollow yarns in CD.

Today, approximately 25–30% of the press fabrics are seamed. Usually pin seaming is used which makes installation of the fabric on paper machine a lot easier. Figure 3.16 shows a press fabric seam. Monofilament warp yarn ends are looped and woven back into the fabric at the two ends of the fabric. A monofilament pin passes through the loops connecting the two ends together after the installation of the fabric on the paper machine.

Endless fabrics are woven in a spiral with the filling yarn forming the length of the fabric. Fabric length will be twice the reed width. Woven loop seam fabrics are woven similar to an endless fabric; however, loops are formed at the ends of the fabric (Figure 3.17). Some looms for endless weaving are as wide as 31 meters (Figure 3.18).

After base fabric manufacturing, carded fiber web is needle-punched into the base struc-

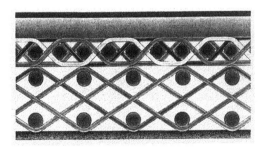

FIGURE 3.13. Cross section of a laminated press fabric.

TABLE 3.4. Typical Property Ranges of Single and Multilayer Press Fabrics.

Structure	Base Fabric Weight (oz/sq ft)	Total Fabric Weight (oz/sq ft)	Air Permeability (cfm)
Single layer	0.9–1.8	2.3–4.9	8–160
Two layer	1.9–2.7	3.7–5.8	12–140
Three layer	2.2–2.7	4.0–6.0	15–120
Laminated two layer	2.5–2.7	4.6–5.2	12–90
Laminated three layer	2.2–2.7	4.9–7.0	15–130
Laminated four layer	3.4–3.8	5.2–7.0+	15–130+

ture to form the batt. Batt laying can be applied from preneedled web (Figure 3.19) or as part of a direct lay batting system (Figure 3.20). Figure 3.21 shows the schematic of needling process and a barbed needle used for this purpose. The needle plate moves up and down into the batt and base fabric several hundred times a minute. In the process, the barbs on the needle grab and penetrate the batt fibers into the base structure. In modern press fabrics, different length and diameter fibers can be used to obtain stratified batt layers, the coarsest fibers typically being next to the base fabric and the finest fibers being on the surface. Special purpose fibers such as fusible fibers can be used to improve bonding.

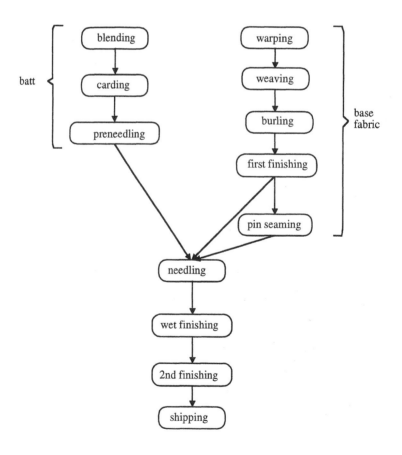

FIGURE 3.15. Manufacturing steps of press fabrics.

FIGURE 3.16. Pin seam press fabric.

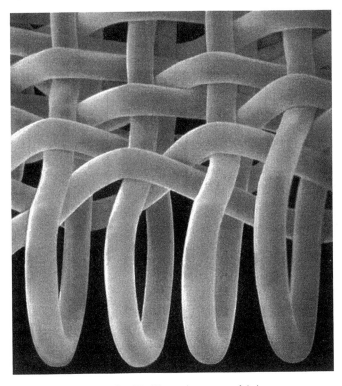

FIGURE 3.17. Woven loop seam fabric.

167

FIGURE 3.18. Weaving machine for endless base fabric manufacturing (machine width = 31 meters).

FIGURE 3.19. Preneedling.

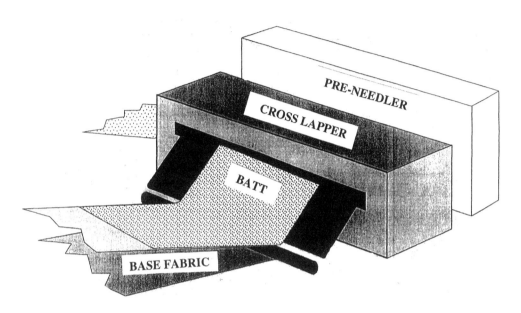

FIGURE 3.20. Direct-lay batting.

barbed needle

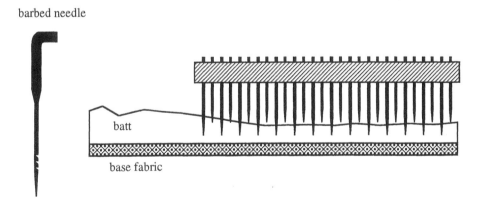

FIGURE 3.21. Schematic of needling.

FIGURE 3.22. Press fabric heatsetting.

The remaining major manufacturing steps are heatsetting for stabilization (Figure 3.22), washing (Figure 3.23), treatment and singeing. Treatments can refer to a chemical additive, a special process, or a specialty fiber that is designed to improve the properties of the press fabric. Most treatments are either colorless or reddish-brown in appearance. The most common uses for treatments are:

- improved flexibility
- improved batt adhesion
- shed resistance
- resistance to chemical degradation
- resistance to mechanical wear
- improved compaction resistance
- inhibit contamination
- inhibit wad burns

Singeing is done to remove loose fibers and properly prepare the surface. Press fabrics are sealed in packages prior to boxing in order to protect the fabric against moisture and dirt. The fabric remains in this package until ready for installation. Certain types of press fabrics are rolled and put into a paper tube to prevent possible creasing of the fabric.

3.2.6 *Properties of Press Fabrics*

Based on their structures (number and dimensions of base yarns, size of batt fibers, the smoothness of the surface finish), press fabrics are classified as superfine, fine, medium and coarse. The important properties of press fabrics are as follows.

Fabric Mass and Thickness

Fabric "mass" is defined as the fabric weight per unit area (oz/sq ft or g/m^2). Press fabric should have a uniform mass distribution across the width. Uneven mass or thickness (caliper) may cause press bounce or vibration. Thickness is important for felt wear and compaction characteristics which influence void volume and drainage of the fabric. The rate of thickness change during the operation depends on the fabric design. The thickness decreases quite rapidly during the early days of the operation. After a certain thickness loss, the fabric has to be removed from the machine. Thicker fabrics are also used to increase nip width and residence time.

FIGURE 3.23. Press fabric washing.

Batt-to-Base Ratio

Batt-to-base ratio is defined as the mass of the batt fibers divided by mass of the base fabric. As this ratio increases, compactness of the felt also increases, which is not very desirable for dewatering capability. However, a small batt to base ratio may cause marking on the sheet due to base yarns.

Air Permeability

Similar to forming fabrics, it is measured as the air volume (cubic feet) passing through of per unit area (square feet) of fabric per minute (cfm) at 0.5 inch water column.

Void Volume

Void volume is the volume that is not occupied by the yarns and fibers in the fabric. It is an indication of the amount of water that the felt can absorb. Void volumes are measured at various loads to predict the fabric's performance on different types of presses.

Flow Resistance

Properly designed fabrics should provide adequate drainage and filtration of water and therefore should have minimum resistance to flow. Resistance to water flow in the fabric is measured in three directions using water permeability testers: machine direction, cross machine direction and vertical direction. Low flow resistance indicates higher water acceptance capacity of the fabric.

Compressibility and Resiliency

Compressibility is a measure of compactness under load and resiliency is a measure of rebounding capability of the fabric after compaction. Compaction resistance is a desired property in press fabrics. A compaction resistant fabric can be either incompressible or compressible but resilient.

Uniformity of Pressure Distribution

Pressure from the rolls should be transferred to the sheet uniformly by the felt in order to prevent uneven sheet moisture profiles. Press fabrics with fine base structure and fine batt fibers provide more uniform pressure distribution. Too coarse base yarns, improper weave design or insufficient batt may also cause sheet marking.

Fabric dimensional stability, abrasion resistance, contaminant and chemical degradation resistance are important requirements of modern press fabrics. The fabric should be dimensionally stable throughout its operation. Some fabrics may get narrower after running some time due to heat from furnish and steam boxes. Smooth sheet transfer also requires dimensional stability. Abrasive inorganics in alkaline paper making and paper making chemical additives for mechanical conditioning have increased the importance of abrasion resistance in press fabrics. Increased use of secondary fibers has made contaminant resistance a necessity in today's press fabrics. Paper making chemical additives such as bleaching, oxidizing and cleaning agents require chemical degradation resistance.

3.3 Applications of Press Fabrics

3.3.1 *Installation and Operating Procedures*

Before Installation

- Clean press and other machine components thoroughly.
- Check all machine components such as uhle boxes and roll faces for rough surfaces.
- Ensure that stretch roll and hand guide are perpendicular to the machine.
- Check automatic guide for free movement and guide palm for cuts or grooves.

Installation

- Install fabric to run with the arrow and fabric number (if applicable) on sheet side.

- Keep fabric as dry as possible during installation.
- Spread fabric out evenly across width of the machine and make certain that the guideline is straight and parallel to the CD.
- Ensure that the fabric is flat and free of wrinkles and machine debris.
- Adjust stretch roll until fabric is snug.

Start-Up

- Load press to dead weight and jog slowly while checking that the fabric is flat and free of creases and wrinkles.
- Increase fabric tension to normal running level (14–22 pli) making certain that tension does not exceed 30 pli (5.4 kilogram per cm).
- Maintain tension during start-up and check frequently.
- Wet up by applying water evenly across the fabric using a full width shower, making certain that no puddling occurs.
- When the fabric is uniformly wet, load press to normal weight and turn on vacuum equipment.

Trimming

Trimming should be delayed, if possible, in the event that the width comes in during break-in. If trimming is necessary, the following procedure is recommended:

- The fabric should be trimmed on the non-guide paddle side.
- A guide mark should be made with a pen or marker on the edge to be trimmed.
- A razor or sharp knife should be held firmly on the guide-mark where the fabric is supported
- Jog the fabric a full revolution; then cut the trimmed loop and pull it free.
- If raveling occurs after trimming, a heat-sealing iron is effective at stopping the raveling.

If a seam fabric is trimmed, special caution should be taken to ensure that the pintle is not cut so that it can be stitched back into the fabric. This will keep the seam from opening on the edges.

3.3.2 *Fabric Shutdown Procedures*

When a machine is scheduled for a shutdown exceeding two hours and the clothing is to remain on the machine, the following procedures will help to extend fabric life as well as prevent start-up problems.

Wash Fabric and Follow with a Thorough Rinsing

It is important to remove contaminants from the fabric. If allowed to remain in the fabric, they would harden and be difficult to remove and may affect start-up.

- Reduce the machine speed to a crawl to increase the dwell time at the suction boxes and rolls.
- Shower on the felt cleaner and allow it to act on the fabric for 15–20 minutes. Ensure that the whole fabric has been treated.
- Reduce the press load to the lowest practical level and leave the uhle box vacuum on.
- Rinse the fabric using flooding showers, lubricating showers, or high pressure showers (use low pressures so the felt is not damaged)
- Set showers at normal operating pressures for 10–15 minutes to thoroughly rinse the fabric
- Shut off the showers.
- Use vacuum boxes to dewater the felts uniformly and to the lowest practical moisture levels.
- A fabric softener or wetting agent can be applied to improve wet-up of the fabric at start-up.

Relax Fabric Tension

Once the fabric has been cleaned, rinsed, and conditioned, the showers and vacuum should be turned off and the machine stopped. Release the fabric tension by backing off the stretch roll. This will prevent excessive tension from developing and roll deflection from occurring as the fabric dries.

Inspect Rolls, Showers and Uhle Boxes

- Inspect all roll surfaces for wear or damage which could cause abnormal wear to the fabric.
- Inspect and clean shower nozzles and uhle box covers.

3.3.3 *Repairing Damaged Areas of Fabric*

Occasionally during the normal operation of a paper machine, fabrics may become damaged by some object producing a hole or tear in the fabric. Due to the rising cost of downtime, it may be feasible to try and repair the damaged area temporarily until an extended downtime is scheduled. Sewing kits utilizing nylon thread or similar type yarn with curved needles may be supplied by the various paper machine clothing suppliers usually free of charge.

While there is no one method for repairing these areas, some suggestions are given below:

- The mended area should not be appreciably thicker than the fabric itself.
- In stitching, caution should be used not to pull the thread too tight causing the area to pucker, allowing for normal widening and stretching during machine operation.
- Two steps may be necessary for complete repair: closing the damaged area with stitching and then covering the closure with additional stitching.
- After stitching is completed, the ends should be left untied. A resin used for sealing edges of seam fabrics may be applied thinly to the roll side to prevent the yarns from coming loose.

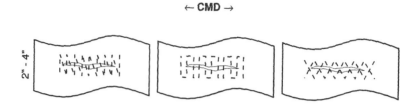

FIGURE 3.24. Damaged areas running in the cross machine direction and corresponding stitch types.

- Inspection of the damaged area as often as possible is recommended for safety purposes in case the fabric starts coming apart causing injury to personnel.

For damaged areas extending in the cross machine direction (Figure 3.24), the repair must be strong enough to prevent the fabric from ripping off the machine when run under tension. The stitch should extend vertically at least 2 inch on either side of the tear. The length will depend on the fineness of the fabric and the paper grade. Various stitching methods are shown in Figure 3.24.

If the tear is in the machine direction, the stitch will be predominantly back and forth horizontally. Since there will not be tension on the MD yarns seen in the cross machine direction shear, the stitch need to be only minimum of 1 inch on either side (Figure 3.25).

Depending on the size and direction of the tear, both vertical and horizontal stitching as a backup may be required on both types of tears. In addition, a "needle punch board" and batt can be provided for by punching the batt back and forth in the damaged area for grades such as fine paper.

3.3.4 *Guiding*

In general, a press fabric will line up at right angles to the surface it touches first. As shown in Figure 3.26, a guide roll angled or pivoted with the run of the press fabric will cause it to guide in the direction of the roll side that it contacts first. If the guide roll is angled in the opposite direction or pivoted away from the run of the fabric, it will cause it to track to the other side.

The above principle is valid for a normal guide roll wrap of 25° to 35° (Figure 3.27). Angles greater than recommended could cause a distance effect and skew the fabric. Angles less than recommended could result in poor fabric response to the guideline roll effect. The distance from the lead in roll to the guide roll should be approximately twice that from the guide roll to the lead-out roll. This ensures that the fabric response to the guide roll is effective.

Other Factors Influencing Guiding

Misalignment of a roll can cause guiding problems. The symptoms are typically characterized by a fabric that consistently tracks to

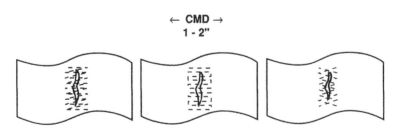

FIGURE 3.25. Machine direction tear and stitch types.

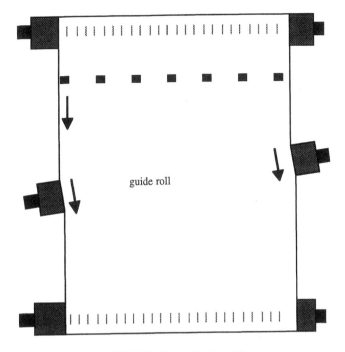

FIGURE 3.26. Press fabric guiding.

one side. It can occur with any roll but usually it is the stretch roll that is the culprit. With independent movement on each side of the stretch roll, it is inherently more likely to be skewed. The influence on the fabric is the same as by the guide roll. However, because of the large degree of wrap by the press fabric, the guideline will skew due to distance effect

that is created. The stretch roll should be checked periodically for misalignment.

Bowed Rolls

A bowed roll positioned in the same plane as the fabric run operates on the guiding principle, guiding the edges away from the center.

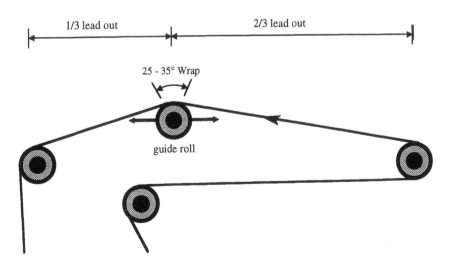

FIGURE 3.27. Lead-in and lead-out distances for guide roll.

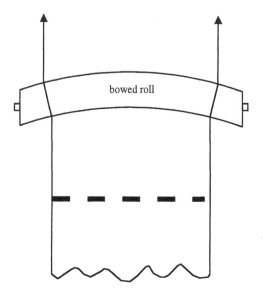

FIGURE 3.28. Widening effect of the bowed roll.

In this position, it provides the maximum widening or flattening effect on the fabric (Figure 3.28).

If the bow is turned into the plane of the fabric run, the widening and flattening effect will be lessened, proportional to the angle. A distance effect will be created, increasing proportional to the angle, causing the tradeline to lag in the center.

3.3.5 *Fabric or Guideline Distortion*

Press fabrics are designed to run with a straight guideline. A line that is bowed or skewed will distort the weave changing the permeability and other properties of the fabric. This could negatively impact water handling capability and result in a poorer response to conditioning and cleaning. The distortion could also cause the fabric to become unstable and possibly wrinkle.

Press fabric distortion is caused by two effects: distance-producing devices and speed-producing devices.

Distance Producing Devices

A distance-producing device is anything that causes a point on the fabric to travel a greater distance than any other point on the fabric during a revolution of the fabric. A cocked or skewed stretch roll, for instance, is such a device. However, any roll that is misaligned can cause the distance effect and fabric skew (Figure 3.29).

Side "B" would travel a shorter distance than side "A," and all points between "A" and "B" will travel progressively shorter distances per revolution than side "A." This will result in the "B" side skewing ahead. Adjustment to

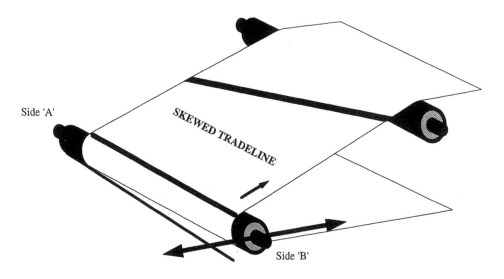

FIGURE 3.29. Schematic of a distance-producing device.

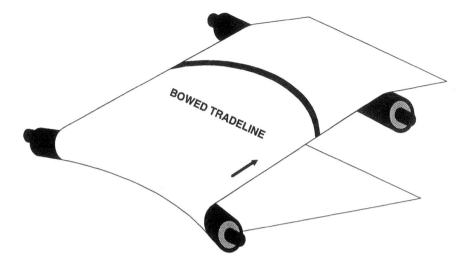

FIGURE 3.30. Roll deflection.

equal the horizontal positioning of both "A" and "B" side of the stretch roll will usually correct the skew.

Limber press fabric rolls or too high fabric tension causing roll deflection can cause the edges of the fabric to travel a greater distance than the center (Figure 3.30). The center travels a shorter distance resulting in the fabric and guideline running ahead in the center.

A bowed roll, such as Mt. Hope roll, that is turned into the press fabric run will also cause a distance effect in the center of the fabric. As the effect is greater length in the center, causing the tradeline to bow behind, the bowed roll is frequently used to correct a guide line that chronically runs ahead in the center.

Speed-Producing Devices

A speed-producing device is anything that causes a point on the fabric to travel at a different speed per revolution than any other point on the fabric. A crowned roll is such a device (Figure 3.31). If point "B" is the center of the crowned roll and point "A" is the edge of the roll, it can be seen that point "B" must travel at a greater speed to make one revolution in the same time that it takes point "A" to make one revolution. As a result, the center and all points of greater in diameter than point "A" are pushed ahead. The result on the press fabric is a guideline that is bowed ahead. The degree of bow is proportional to the difference in circumference.

Other Speed-Producing Devices (Figure 3.32)

Concave rolls have the opposite effect as a crowned roll. The edges are greater in circumference resulting in higher surface roll speed at the edges. The effect on the fabric would be to correct a chronic problem with guideline bowing ahead.

Leadered roll is a roll that has had the surface altered to create a greater diameter. This usually involves wrapping the roll with fabric

FIGURE 3.31. A crowned roll as a speed-producing device.

concave roll

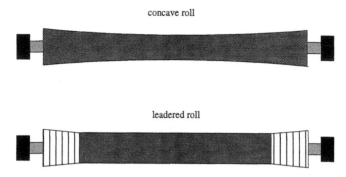

leadered roll

FIGURE 3.32. Concave (negative crown) and leadered rolls.

and/or tape, resembling a screw thread. Thread direction changes at the roll center line and must be set to widen, not narrow, the sheet or fabric running across it.

There may be other speed or distance producing devices that cause guideline distortion, but the point to remember is that distortion is caused by one of these devices. It follows then that one of these devices can also be used to correct distortion. Occasionally, it will be seen that fabrics from two different suppliers will show a different guideline distortion on the same position. Usually several devices on a position are affecting the fabric guideline, whose position is the net result of these several devices. The internal resistance to distortion of fabrics from different manufacturers may result in a different net effect.

3.3.6 Press Fabric Problems— Evaluation Guidelines

Table 3.5 gives a summary of common press fabric problems.

Dimensional Stability

Too Narrow

The question is whether the fabric started too narrow or became too narrow after running.

- Measure length and width (preferably at start-up and after running at least one day). If accurate measurement is not possible, determine sizes relative to fabric rolls and to amount of jack. Also note the tension (tight/slack).
- Determine if fabric was trimmed and how much.
- Measure tradeline distortion. Determine when it occurred. Is it bowed or skewed?
- Measure maximum running width; usually equal to fabric roll width.
- Measure minimum running width. This will usually be at least two inches outside maximum sheet width on either side or "wire" width for a pick-up fabric.
- Determine whether previous fabrics and succeeding fabrics exhibited the same problem.
- Determine operating temperature of press fabric or wet end.
- Determine whether steam boxes were used on press.

Too Wide

The same procedures apply for the items above, additionally:

- Check to see whether variable bowed rolls are used.
- Check to see if press fabric is worn excessively.

TABLE 3.5. Common Press Fabric Problems.

Blowing	Common on last press on machines that make lightweight fine paper and news at high speeds. The sheet rides on the fabric entering the nip. Usually during initial start-up the sheet will bubble or lift off the felt prior to entering the nip. This can cause a wrinkle in the sheet as it passes through the nip.
Drop-offs	Occurs on PU positions usually during initial start-up as the sheet rides on the bottom of the fabric from the pick-up roll to the first nip.
Bleed-through	Occurs primarily on machines that make highly refined (slow draining stock) and highly filled grades. Paper stock and/or filler material will collect on inside felt rolls causing guiding problems and vented rolls to plug.
Bounce/vibration	Typically occurs on highly loaded, hard roll presses (last press positions). The intensity of the nip makes the press sensitive to mass variations in the fabric which usually causes once or twice per revolution bounce. Most high frequency vibrations are caused by machine problems.
Shedding	Most critical on coated grades which can cause coater streaks or printing problems with the sheet of paper. It usually occurs later in life as the fabric is starting to wear and the batt fibers weakened by mechanical or chemical action.
Poor drying	A common problem on all grades in which the felt causes poor water removal in the nip, either from rewet or misapplication of design. It is usually indicated by high sheet draws and/or increased steam usage in the dryer section.
Difficult installation	Common problem on older machines in which the stiffness of the fabric makes it very difficult to bend around rolls and pack into tight spaces. Single monofilament CD structures and heavier designs make fabrics relatively stiffer.
Seam wear (mark)	Occurs with all seam products in which the batt over the seam wears prematurely, exposing the seam and base. This causes an impression mark in the sheet.
Shadow mark	Usually occurs in the first nip suction presses, where water loads are high. If flow resistance in a nip is too high, fines and filler material in the sheet will migrate to areas of low pressure causing a density variation in the sheet. It can also be caused by the suction couch roll.

Press Bounce or Vibration

Press bounce may be caused by either the press rolls or the press fabric. Improperly balanced rolls, worn rolls, corrugated rolls, or flat areas caused by leaving rolls nipped when the machine is down may cause bounce.

Answers to the following questions may reveal the cause of bounce:

- What is the machine speed? What is the frequency of bounce? (Once or twice per revolution of the press fabric? Where relative to tradeline? Once per roll revolution? Which roll?

Is it more of a high frequency chatter?)
- Was bounce present at start-up? When did it begin? Is it getting worse or better?
- Was bounce present prior to last press fabric change?
- Were any press rolls changed immediately prior to start of bouncing?
- Did the next press fabric bounce? If so, answer the first two questions again.
- Is the fabric length a multiple of a press roll circumference? (What are the roll hardnesses? What is press loading? What was the effect of cocking the tradelines?)

Water-Handling Problems

Running Wet or Speeds Down

Press water removal may be less than standard resulting in slow speeds, increased dryer steam, increased draws, crushing, or increased paper breaks. Profiles may also be poor. Causes of running wet or slow speeds can be stock-related, machine-related or clothing related. Analysis of the problem can be very complicated:

(1) When did poor water-handling occur?
  • Were start-up speeds down? How long?
  • Were speeds slow late in life? When?
(2) How much were speeds down?
  • What is normal speed?
  • What is normal break-in time?
(3) Were press loads normal? If not, why?
(4) Was water removal equipment operating normally? How much uhle box capacity was available? Measure air flows if possible. How did uhle box vacuum (inches Hg) compare to normal?
(5) Were furnish and grades normal? Was the sheet wet on the wire?
(6) Did problem occur after any change such as press fabric or forming fabric change?
(7) How did the steam use compare to normal?
(8) Did any crushing occur? In what press? Was it localized?
(9) How was press fabric tradeline? Distorted tradelines can close up fabrics.
(10) Is the problem localized? If so, where?
(11) What is the effect of cutting back lube or cleaning showers?
(12) Is nip condition flooded or dry? Is there a save-all pan? What is the condition of roll doctors or purge showers?
(13) How did the previous and succeeding fabrics perform?

Marking

Sheet marking is typically caused by either the press rolls, press fabrics or the forming fabric. The marks can be either a physical impression or a flow pattern.

Physical impression marks are almost always a result of nonuniform pressing by the press fabric. It can be caused by yarns in the base fabric that are too large, too widely spaced, or too rigid. If insufficient batt is used or if the fabric is worn, base fabric yarns may become exposed and mark the sheet. Physical impression marks will typically occur in the latter presses where nip intensities are usually high.

Flow pattern marks are usually a result of flow resistance in the nip that causes lateral water movement in the sheet. This causes a concentration of fines and/or filler material in the sheet that shows the vented pattern of the press roll. The area over the suction holes or the roll grooves represents zones of low pressure under load that can cause a density variation in the sheet that may be visually apparent. Flow marks usually occur in nips where water loads are high. Worn press rolls are usually the cause; however, filled fabrics or fabrics that run too wet can contribute to the problem.

Determination of the Causes of Sheet Marking

(1) Examine sheet samples in low angle light (best at about 20°).
(2) What is the pattern of the marks?
  • Can it be associated with the pattern of the press roll in the machine? Trace pattern from spare roll if possible or, if a grooved roll, count the number of grooves per unit length.
  • Are there large number of marks per unit length? This will distinguish a press fabric mark from a groove mark.
  • Is the direction of the mark diagonal? This will confirm or eliminate a forming fabric related mark.

- Is the mark on the top side or wire side? Which press fabrics and rolls contact which side? It needs to be determined which press fabric or roll is causing the problem.

(3) When did the marking begin?
  - Does the date coincide with a roll change?
  - Does the date coincide with a press fabric change?
  - Have any press fabric or roll changes occurred since the beginning of the sheet mark problem?
  - Did a roll change, press fabric change, or a forming fabric change eliminate the problem?

(4) What are the grinding schedules of the rolls in the machine? What is the remaining roll cover thickness compared to new?

(5) What are the loads? Are they normal?

(6) What are the roll hardnesses? Are they normal?

(7) How does the existing caliper of the press fabric compare with the initial or edge caliper?

(8) Is the press fabric surface soft; is the batt worn away?

(9) Is the press fabric carrying more water than normal?

(10) In all instances obtain a sample of both the objectionable mark and the sheet that is considered satisfactory.

Fabric Wear

Excessive press fabric wear has many causes which can be divided into three categories: mechanical, chemical, and hydraulic.

(1) Mechanical causes
  - improper roll crown/load
  - improper roll dubbing
  - bad press roll bearing
  - scissored rolls (misaligned press rolls)

- worn or rough suction box covers
- improper high pressure shower oscillation
- excessive shower pressure
- abrasives in the system
- worn or rough press rolls
- drag on the save-all

(2) Hydraulic causes: In some cases the press fabric is unable to handle water properly under load in which case hydraulic damage may occur to yarns and batt material.

(3) Chemical causes
  - reductive agents in the system
  - oxidative damage

Wrinkling

Uniform fabric tension across the entire width of the fabric is the best way to avoid press fabric wrinkling. The most common cause of uneven tension is uneven moisture distribution during press section start-ups. Proper wet up procedures will avoid slack areas or pockets that may develop wrinkles. Tradeline alignment can be an indicator of potential wrinkling. On exceptionally long draws, a supporting press fabric roll or bowed roll may be necessary.

In the event that wrinkling occurs, the press should be relieved and the fabric spread evenly across the width of the machine. Hot water or a steam hose may help flatten the wrinkle. The press should then be loaded lightly and started up slowly.

Drop-offs

Proper pick-up usually depends upon the presence of a uniform water film on the surface of the fabric to adhere the sheet to the fabric. Drop-offs can occur during the initial start-up or later in life. Many factors can influence the propensity for drop-offs:

- freeness of the stock
- press geometry
- sheet weight
- pick-up roll vacuum

Most drop-off problems occur during initial start-up of a new press fabric. In this situation the press fabric is usually designed too open or the batt component applied too coarse. Sometimes adding filler to the fabric or cutting back on uhle box vacuum will alleviate the problem. While the problem may be related to the design of the fabric, other factors could contribute to the problem:

- poor deckle trim
- improperly set pick-up roll vacuum deckle
- plugged suction pick-up roll
- improperly positioned pick-up roll
- worn forming fabric (poor sheet release)
- improperly set draws

Blowing

Blowing, like drop-offs, is influenced by the same factors and can either occur during the start-up of a new fabric or later in life. The majority of problems are usually initial and occur in the last press. Blowing is usually seen as a bubble or layer of air between the fabric and the sheet. The propensity for blowing is contributed by the press geometry. The sheet and fabric are in contact with each other entering into the nip. Air that is either entrained in the fabric or expressed backwards in the nip causes the sheet to lift off the fabric ahead of the nip. This can cause sheet wrinkles or breaks.

If the sheet blows when the fabric is new but stops after break-in, the fabric design may need to be modified to decrease the potential air volume in the fabric by increasing its density. Modifying the surface characteristics of the fabric may provide better contact between press fabric and the sheet. If the sheet does not blow when the fabric is new but blows as the fabric gets older, conditioning and cleaning efficiency may need improving or the fabric may need to be manufactured more open.

Sometimes, increasing sheet draws between the presses or running the fabric tighter can reduce or eliminate the blowing.

Crushing

Crushing may be caused by press loading too high for the flow resistance of the sheet. It also occurs when the press fabric can not accept the water at the rate it is squeezed into the fabric at the press. The result is a deformation of the sheet. On a new press fabric the problem may be too low a permeability or insufficient void volume within the press fabric. On an older fabric anything that reduces the void volume such as wear, filling or compaction contributes to crushing. Possible temporary solutions to minimize crush are to increase sheet temperature and to reduce press loads.

Troubleshooting guidelines for press section are given in Appendix C.

3.3.7 *Press Fabric Conditioning*

Press fabrics must be kept clean of filler materials that accumulate in their structure. Very seldom are press fabrics removed because they are worn out. They are generally removed because they are filled or compacted to the point where they no longer handle water uniformly or lack the drainage capabilities to maintain pressing efficiency. All fabrics should be continuously and mechanically conditioned from start-up to removal.

Good cleaning systems use hydraulic forces from high and low pressure showers to loosen and flush contaminants from the press fabric structure through a uhle box. A typical system includes a high pressure needle shower, a lubricating shower, and a uhle box and vacuum system (Figure 3.33). Chemical showers are sometimes used to facilitate the cleaning process.

Showers

High pressure showers provide hydraulic forces to loosen contaminants and minimize felt compaction. Continuous operation on the sheet side generally is most effective. Continuous application is recommended as it is easier to keep a fabric clean than to clean a fabric

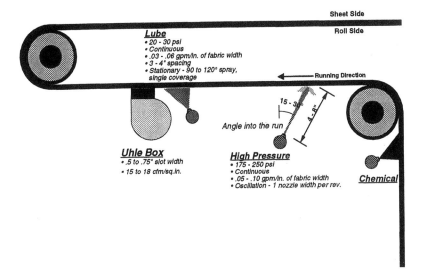

FIGURE 3.33. Press fabric conditioning system.

once it has become contaminated. Sheet side showers are more effective as the contaminants are usually filtered or trapped by the surface batt fibers and sheet side showers are more able to remove or loosen the contaminants. Roll side high pressure showers are usually not an effective means to clean the fabric. The pressures required to penetrate the base fabric and affect surface contaminants usually damage the press fabric.

Oscillator speed should be coordinated with machine speed so that the shower jet moves one nozzle width per fabric revolution. This provides complete coverage and cleaning of the fabric.

High pressure shower pumps must be interlocked with press drives so the shower will not operate when the fabric is not running, or if the oscillator fails.

Lube showers should be located immediately ahead of each uhle box on the same side of the fabric as the uhle box. These showers provide lubrication to reduce wear between the fabric and the uhle box cover.

Chemical showers provide a solvent or detergent that facilitates the cleaning process. The showers may be located either on the roll or sheet side and chemicals are applied at the point where the felt and felt roll converge.

This provides a flushing action that allows the chemical to penetrate the fabric thoroughly. Chemicals that are used continuously should be applied to the fabric as far ahead of the uhle box as practical to allow sufficient time for the cleaning solution to work.

pH of the shower water and cleaning solutions is critical in that there should not be a large differential between the stock and the showers. The pH of the sheet and showers should not vary more than ± 1. This is to prevent pH shock that could cause precipitation of filler material. This guideline does not apply to batch cleaning as stronger solutions can be used to clean the fabric as long as it is thoroughly rinsed afterwards to neutralize pH.

Temperature of the shower water should be within $\pm 10°F$ of the stock temperature. As with pH, this is to prevent upset of the sheet in the press section.

Dewatering

The typical dewatering system consists of one or more uhle boxes connected to a vacuum pump. The variables that can usually be controlled in vacuum dewatering are dwell time (slot width), pressure difference (inches

slotted

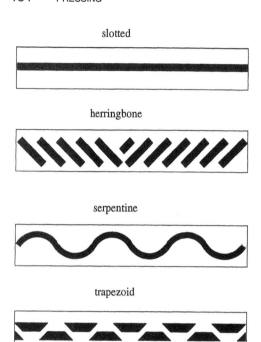

herringbone

serpentine

trapezoid

FIGURE 3.34. Common cover patterns.

of Hg) and pump capacity (cfm/sq in. of open area). Pump capacity is an important measure of dewatering energy and it should not be sacrificed to maximize one or more of the other variables. Uhle boxes do more than dewater fabrics. Showers loosen surface contaminants. The uhle box, not the showers, is the primary mechanism of fabric cleaning.

The ability to clean and dewater is related most closely to permeability. To optimize press fabric conditioning, a pump capacity of 15–18 cfm/sq in. is required. Uhle boxes on a single press fabric should be connected to a dedicated vacuum source. In nonisolated vacuum systems, air flow is proportionate to the fabric permeability.

Most uhle box covers are made from high density polyethylene. Other cover materials used are ceramic or ceramic composites. While ceramic covers are significantly more durable than polyethylene, they require greater care to prevent chipping and damage, and felt wear is usually increased. If ceramic is used, the radius of the trailing edge of the cover should be no less than 1/16 inch.

The most common cover pattern is single

or double slotted. Other patterns used are the herringbone, serpentine, and trapezoid (Figure 3.34).

The illustrated specialty patterns are used primarily on positions where seam fabrics are used. They are designed to minimize wear at the seam by providing support over the slots and reducing the degree in which the fabric is pulled into the slots. It is recommended that slot widths be no wider than 1 inch to minimize wear and shedding.

In order for a press fabric conditioning system to operate efficiently and effectively, special attention should be given to the design of the vacuum system. There are four common sources of problems with the vacuum system (Figure 3.35).

(1) Uhle box and piping diameter: Uhle box tubes should be sized to keep the linear air velocities through the pipe below 5000 fpm. Air velocities significantly in excess of this speed induce a pressure gradient from the blanked end of the uhle box tube to the outlet side. This creates a skewed moisture profile in the fabric.

(2) Convolution: Excessive turns and angles in the piping results in a loss of vacuum efficiency. This is important with systems that are marginal for vacuum.

(3) Infiltration: Atmospheric air that enters the system through improperly sealed joints or holes in the piping robs the fabrics of vacuum.

(4) Drop leg: An inch of Hg is roughly equivalent to a foot of water. If the drop leg to the seal pit lacks sufficient height, the inches of Hg at the uhle box can exceed the footage height of the drop leg. This can result in water being sucked from the seal pit back into the uhle box.

3.4   Service for Press Fabrics

3.4.1   *Service Laboratories*

Paper machine clothing manufacturers established customer service laboratories to enable the papermaker to learn more about the

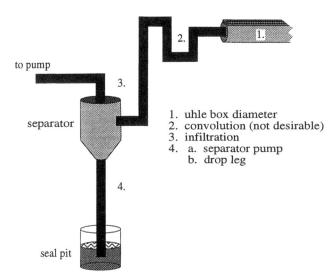

FIGURE 3.35. Typical vacuum system problem areas.

1. uhle box diameter
2. convolution (not desirable)
3. infiltration
4. a. separator pump
   b. drop leg

solution to press fabric problems. Working with returned, used fabrics, the laboratory often identifies the conditions that help determine the cause of the less-than-satisfactory fabric performance.

A comparison of the initial and residual measurements are made to determine the condition of the fabric. As a general rule, a laboratory analysis on a used fabric is a good tool for trend analysis and data comparison as well as to identify press section and machine-related problems.

Laboratory analysis, in general, does not provide enough information to understand application related problems. Application problems, such as initial blowing and drop-offs, that result from the design specification of the press fabric are usually diagnosed with data gathered from on machine performance compared to normal operation (Chapter 5). In some cases, the physical analysis of the fabric helps resolve application problems.

Chemical Analysis

Standard fill-up analysis determines the general type of plugging materials. These are divided into four categories:

- alkaline extractables—pitch/oil, starch, dyes, salts, dirt and detergents
- alcohol extractables—grease, oil, tar, wax, ink, pitch, and stickies (resins and latex)
- ash—alum, titanium, clay, iron, sand, talc and all inorganics
- cellulosic materials—paper stock and fines

Qualitative analysis is done for positive identification of any material in the four categories above. Infrared spectrophotography is used for specialized identification of foreign materials in the used fabric that are difficult to identify. Chemical degradation tests are done for acid, alkaline, reductive and oxidative damage.

Physical Analysis

Strength tests and profiles are done to determine the residual strength of used fabrics compared to original strength. This often reveals if the fabric could have performed longer. Profiles indicate possible problem areas such as undercrown, overcrown, misaligned rolls, worn suction boxes or any source of unusual machine wear across the width of the fabric. Weight profiles can confirm these problems as well. Caliper profiles reveal problems of uneven compaction and/or uneven physical wear. Air permeability profiles can reveal any filled or compacted streaks across fabric width.

3.4.2 *Technical Service*

In today's practice in the papermaking industry, technical service is the primary contact a clothing supplier has with its customer and product. Outside of direct communication with the paper machine superintendent usually performed by the sales engineer, the technical service engineer is the main source of running performance information.

In general, technical service is made up of the following:

- general visual observations of the press clothing and sheet
- physical measurements of the press clothing and sheet using specialized instrumentation
- a thorough analysis of the data to translate how well the clothing is performing
- a usable report documenting findings and making pertinent recommendations

The technical service engineer reviews the press section performance and records baseline information for future comparisons. Engineers use a variety of instruments to collect baseline information. These instruments are useful to the papermaker as troubleshooting tools and as capable aids in setting and maintaining machine variables. The most widely used instruments and measurements are listed below:

- tensiometer: This instrument measures fabric tension in pli.
- running caliper gauge: The running caliper gauge is used to measure the thickness of running fabrics, which is useful when predicting fabric removal and determining loading conditions.
- gamma gauge: The gamma gauge measures the mass of the fabric. The reading obtained with the gamma gauge can be used to help determine the water balance in the press.

- water content in the fabric: Microwave measuring instruments are used to measure the amount of water in the fabric. These measurements are used to evaluate the profiles of the press fabrics.
- draw measurement gauge: Tachometer devices are used to measure clothing and sheet draw in real time to determine draw schedule throughout the machine.
- vibration analysis: Synchronous time averaging with Fast Fourier Transform (FFT) analyzer can isolate vibration sources and determine how much influence each source is having on the overall vibration in a press section.

The following TAPPI procedures can be used to analyze press and press clothing performance [7].

(1) TIS 014-46 Press Section Monitoring: This method discusses the following parameters: fabric thickness, fabric total mass, fabric water content, sheet water to fiber ratio, degree of fabric filling/openness (permeability), running fabric length, fabric tension, fabric tradeline distortion, fabric and roll speeds, press vibration, press roll alignment, press roll crown.
(2) TIS 014-54 Paper Machine Clothing Performance Analysis: This method provides guidelines for good clothing performance analysis

3.5 Press Types

Press types have evolved over the years to achieve faster speeds, improve water removal, and influence sheet properties. The following illustrations and discussion will comment on these press types and how they relate to the degree of two-sidedness in the sheet and how they can influence press fabric performance.

Two sidedness is defined as the percent drop in brightness from top side of the sheet to the bottom side based on the K&N ink test [8]. K&N ink test measures the percent drop in brightness between the uninked and inked sample.

% Drop

$$= \frac{\left(\frac{\text{Initial}}{\text{Brightness}} - \frac{\text{Final}}{\text{Brightness}}\right) \times 100}{\text{Initial Brightness}}$$

(3.2)

Two Sidedness = % Drop Bottom

$-$ % Drop Top   (3.3)

The sheet from a fourdrinier is more dense on the top side due to the wash out of fines and fillers on the bottom side of the sheet. The surface density of the sheet is increased on the fabric side of each nip. The effect on sheet density is greater with each proceeding nip. Surface smoothness is used to indicate two sidedness. Sheet smoothness is measured with the Sheffield and Parker Print methods.

3.5.1 *Straight-Through™ Press (Figure 3.36)*

This press type covers an extremely wide grade and speed range as well as a range of nip conditions. Specific fabric requirements depend on water load, speed, sheet properties, sheet weights and nip conditions.

- used on virtually all grades
- variations in this type of press include inverted and reversed 2nd press (fabrics on the top side of the sheet)

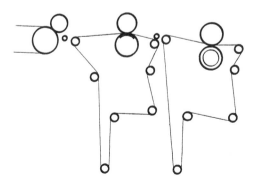

FIGURE 3.36. Straight-Through™ press.

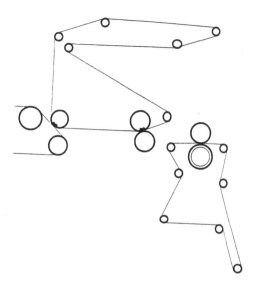

FIGURE 3.37. Inver press.

- open draws prone to sheet breaks
- prone to blowing and wad burns, if top roll not doctored
- speeds limited due to sheet breaks at open draws.
- installation a problem with multilayered fabrics

3.5.2 *Inver Press (Figures 3.37)*

This type of press is generally used to make light to medium weight fine paper grades at low to medium speeds.

- common press arrangement for fine paper grades
- prone to drop-offs with some grades
- 2nd press prone to sheet blowing
- KN: +2 to +4

3.5.3 *Twinver™ Press (Figure 3.38)*

This type press configuration was a forerunner of the no-draw cluster press. Newsprint and LWC grades at medium to high speeds are generally made on the Twinver™.

- common press arrangement for light weight, high-speed grades (news and fine)

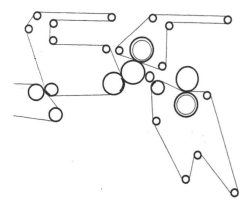

FIGURE 3.38. Twinver™ press.

- prone to sheet stealing or edge flipping by pick-up at the center roll
- 3rd press prone to vibration
- prone to drop-offs with some grades
- prone to sheet blowing in the 3rd press
- KN: +5 to +7

3.5.4 *Combi Press (Figure 3.39)*

This press has many of the characteristics of the Twinver™ press but without the disadvantage of a long inverted run which is prone to drop-offs. It is a compact arrangement that can be applied in a limited space.

- compact press arrangement
- no-draw effect through first two nips

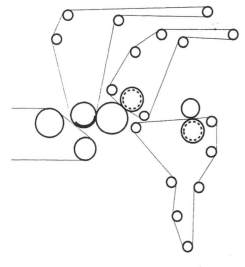

FIGURE 3.39. Combi press.

- less prone to drop-offs compared to a Twinver™ or Inver press
- pick-up prone to shadow mark with suction holes greater than 9/16 inch
- loads limited by single-felting (crush) and load differential between 1st and 2nd nip on center roll
- without 3rd press, KN: +10 to +13 (very two sided); with 3rd press KN: about +7
- difficult press fabric installations

3.5.5 *Four-Roll, No-Draw Presses*

TriNip™ is the most popular press arrangement for all high speed fine and newsprint grades (Figure 3.40). Double felted first nip allows for higher press loads without crush. A 4th press adds symmetry to sheet quality.

- popular no-draw, high-speed press
- double-felted first press prevents crushing and allows higher loads
- prone to shadow mark due to two nips on suction roll
- prone to sheet stealing by pick-up with some geometries
- 3rd press prone to vibration
- KN: +9 to +12 (very two-sided); the last two presses have fabrics on the top side.

TriVent™ is evolved from TriNip™ press arrangement (Figure 3.41). Separate 2nd press

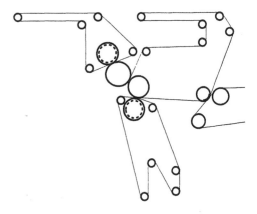

FIGURE 3.40. TriNip™.

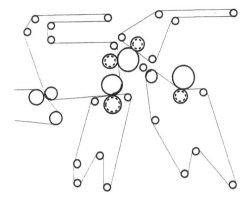

FIGURE 3.41. TriVent™.

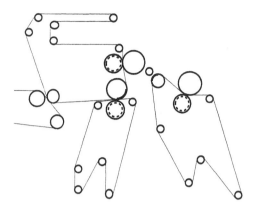

FIGURE 3.43. BiVent™.

reduces propensity for shadow mark and allows for higher loads in the 2nd and 3rd nips.

- able to obtain higher press loads in the 2nd and 3rd press nips
- prone to water rejection in the 2nd nip
- other characteristics same as TriNip™

3.5.6 *Three-Roll, No Draw Presses*

Alternative to TriNip™ arrangement, the BiNip™ press is popular on medium to heavyweight grades (Figure 3.42). It improves bottom side sheet densification.

- less two-sided due to last felt on bottom side; KN: +5 to +7
- double-felted first press prevents crushing and allows higher loads
- prone to shadow mark due to two nips on suction roll

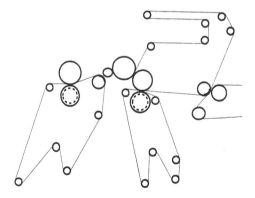

FIGURE 3.42. BiNip™.

- prone to sheet stealing by pick-up with some geometries
- 3rd press prone to blowing and wad burns

BiVent™ is evolved from the BiNip™ press arrangement (Figure 3.43). Separate 2nd press reduces propensity for shadow mark and allows for higher loads in the 2nd nip.

- prone to water rejection in the 2nd nip
- able to obtain higher press loads in the 2nd and 3rd press nips
- other characteristics same as BiNip™

3.5.7 *Shoe Presses*

Shoe presses were developed to increase the nip residence time for improved water removal and sheet densification. These presses operate up to 8500 pli with nip widths up to 10 inches. Sheet consistencies in some instances have improved 3 to 4% out of the press section. In all cases, sheet density is increased. The predominant grades made with shoe presses are linerboard and corrugating medium with some bleached board machines equipped with shoe presses. Recent innovations have seen application of shoe presses for fine paper and newsprint grades.

The original shoe press developed by Beloit consists of a special blanket or belt that runs inside the felt run and is pressed by a hydraulic shoe that forms the nip. Recent developments have eliminated the blanket run and contain

Beloit ENP   Voith FLEXONIP   Nipco INTENSA-S

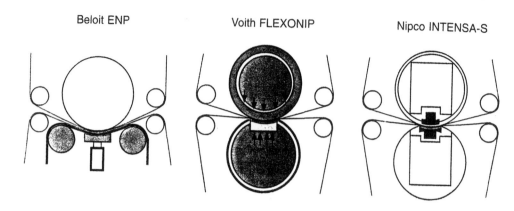

FIGURE 3.44. Shoe presses.

the shoe press inside a large diameter roll (Figure 3.44). Felt compaction and mark are potential problems with shoe presses. High water loads necessitate high void volume press fabrics.

3.6   References

1 Reese, R. A., "Pressing Operations" in *Pulp and Paper Manufacture, Volume 7: Paper Machine Operations,* Eds. B. A. Thorp and M. J. Kocurek, TAPPI Press, 1991.

2 Wicks, L. D., "Press Section Water Removal Principles and Their Applications," *Southern Pulp and Paper Manufacturer,* January 1978.

3 *Press Fabrics Training Manual,* Asten, Inc., 1995.

4 Walstrom, P. B., *Pulp and Paper Mag. Can.,* Aug. and Sept. 1960.

5 Walker, K., and Foulger, M. F., "Press Section Performance," *TAPPI Journal,* Vol. 79, No. 11, November 1996.

6 Antos, D., "Advantages of Nonwoven Press Felts," *Paper Age,* April 1993.

7 Technical Association of the Pulp and Paper Industry (TAPPI), Atlanta, GA.

8 Wicks, Laurie, "The Influence of Pressing on Sheet Two-Sidedness," *TAPPI Engineering Conference Proceedings,* 1982.

General References

Helle, T., and Forseth, T., "Influence of Felt Structure on Water Removal in a Press Nip," *TAPPI Journal,* Vol. 77, No. 6, June 1994.

Laivins, G., and Scallan, A., "Removal of Water From Pulps by Pressing," *TAPPI Journal,* Vol. 77, No. 3, March 1994.

3.7   Review Questions

1 Explain how water is removed from the sheet in the press section.

2 What are the differences between flow controlled nips and pressure-controlled nips?

3 Compare the requirements of press fabrics to those of forming fabrics. Which type requires more stringent properties?

4 Explain the differences in manufacturing of press fabrics from forming fabrics.

5 What are the reasons for press fabric treatments? Explain.

6 Define press vibration. What are the causes of it? What are the remedies?

7 What are the effects of distance- and speed-producing devices?

8 Find out how the K&N ink test is done. What is the significance of this test?

Drying

Drying is the last major process in paper manufacturing. The sheet, which is formed in the forming section and consolidated and dewatered to a certain extent in the press section, enters the dryer section for the removal of remaining extra water in its structure. The consistency of pulp at the entrance to the forming section is approximately 99% water and 1% fiber. A significant amount of this water is removed in the forming and press sections. Typical sheet consistency after the press section is 40% fiber and 60% water (in shoe presses, the ratio is 50%/50%). The remaining extra water in the sheet has to be evaporated in the dryer section. During the drying process, the loose network of fibers is consolidated into a more compact, continuous web structure. The moisture content of the sheet after drying is approximately 5%.

Although the effect is not very pronounced, the drier the sheet is coming to the press, if all other things remain constant, the drier the sheet will be going to the dryers. A dry, well compacted sheet from the forming and pressing sections will help the dryer section to run more efficiently, resulting in better machine operation. Dryer cleanliness and fewer sheet breaks are advantages but the big effect comes from better heat transfer and easier drying. Better dryer contact from a smooth, well formed sheet will result in easier drying. The drier the sheet is coming to the dryers, the less steam will be required to dry the sheet. Drying is a costly process. Therefore, every

effort should be made to remove as much water as possible from the sheet prior to drying section of the paper machine. A 1% increase in sheet dryness to the dryers will result in about a 4% increase in dryer efficiency.

Significant improvements in sheet runnability in the dryer section have been accomplished in the last fifteen years. These systems have not only increased the efficiency of existing machines, they have made possible ever increasing machine speeds and sheet quality improvements.

Dryer fabric quality and design have also had to keep up with these innovations in machine design. The dryer fabrics are becoming a more crucial element in successful machine operation.

4.1 Paper Drying Theory

Figure 4.1 shows the schematic of a typical dryer configuration. The sheet is pressed against the heated cylinder by the dryer fabric. The drying of paper is accomplished by two interdependent actions: heat transfer and mass transfer. Heating of the water in a sheet of paper is accomplished by contacting the sheet with the hot surface of a steam heated dryer. This contact results in energy (heat) transfer from the dryer to the sheet and progresses to mass transfer of the water into the surrounding air with evaporation.

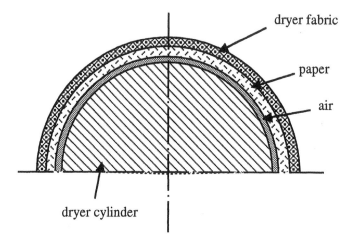

FIGURE 4.1. Drying configuration.

Heat transfer takes place from hot dryer to cooler sheet. As a result, the water in the sheet is warmed up. The evaporation of water under any drying conditions requires heat be transmitted to the water in the sheet to raise the temperature of the water to the evaporating temperature. Once the water molecules absorb enough energy that they leave the sheet to be carried away by the surrounding air, the second activity known as evaporation or mass transfer begins.

Now that the water molecules are ready to leave the sheet, they must be allowed to leave the sheet and enter the surrounding air. The temperature of vaporization (temperature necessary for evaporation to occur) is lowered in the presence of unsaturated air by increasing the velocity of the air movement. Therefore, if we provide air around the sheet that is dry enough to hold the moisture evaporating from the sheet at an adequate velocity, the air will carry the moisture away. The need to supply dry air to all areas around the sheet and to eventually exhaust the newly moisture laden air from the machine is critical.

If we did not supply dry air and exhaust the wet air, the air surrounding the sheet would become saturated. At this point, the drying slows down or stops. Pocket ventilation systems are used to feed air into the dryer pockets for the entire width of the machine. This air picks up evaporated water and carries the air and water to the sides of the machine where

it, in turn, is picked up by the exhaust fans and exhausted from the machine.

When the moisture evaporates from the sheet, this evaporation cools the sheet. The now cool sheet can then be heated by the next dryer and the process starts over again.

Many things affect both the heat transfer and the mass transfer rates. Drying will take place at almost any sheet temperature and air moisture content. Even if a sheet of paper is laid out on the floor on a very humid day, it would eventually dry. In fact, the first sheets of paper were made this way. Today's drying technology requires fast and effective drying of the sheet.

4.1.1 Drying Rate

The term *drying rate* refers to the amount of water evaporated per hour per unit area of drying surface. Since most papermakers are interested in getting maximum production which is commonly limited by drying capacity, an improvement in the drying rate without a negative impact on quality is an ever constant pursuit. If the focus of effort to improve the drying rate on a given machine is centered around the heat transfer side of drying, then production will only be increased if heat units get into the paper more rapidly.

The law of heat transfer states that the amount of heat transferred varies directly as the temperature difference between the source of heat (i.e., the steam inside the dryer) and the material being heated (i.e., the paper sheet in contact with the dryer) is increased [1]. Increasing the temperature difference will increase the amount of water evaporated or paper dried in direct proportion. The temperature difference can be increased by raising the steam's temperature (i.e., raise the steam pressure) or decreasing the drying temperature of the paper.

4.1.2 *Saturated Steam Pressure*

The first step is to heat the sheet to get more of the water molecules to leave the sheet in a shorter time. This is done using steam heated dryer cylinders. Saturated steam at pressures which generally range between 30 psi up to 150 psi is supplied into the dryer. The temperature of the steam at 30 psi is 274°F while at 150 psi the steam temperature is 365°F. The drying rate does not increase linearly with increased dryer steam pressure. Table 4.1 shows this characteristic about steam.

As steam pressure increases, the condensing saturation temperature increases and the latent heat decreases. Therefore, the heat transfer rate increases with pressure, but more steam must be condensed for a given amount of heat transferred. In other words, the rate of increased drying capacity resulting from an increase in dryer pressures decreases as the steam pressures are raised. For example, the drying rate for a dryer with 150 psi of steam is only twice that of a dryer with 25 psi of steam.

4.1.3 *Superheated Steam*

An alternative to raising the saturated steam pressure in a given dryer is to superheat the steam. Superheat in the steam raises the steam temperature at a given pressure. This would allow an increase in the temperature differential between the steam and the paper without the need for higher steam pressures. Unfortunately, it does not work that way. Superheated steam is a gas which gives up its heat slowly to the dryer shell as compared to the flow of heat from saturated steam. In addition, the amount of heat available above the saturation temperature is small compared to the latent heat.

Paper drying is not just a simple function of water removal. The use of high steam pressure is impractical if the result is an unacceptable quality paper. High steam pressure accentuates or causes picking, puckering, and cockles. In addition, high steam pressures in the first dryers may actually retard drying (by sealing the sheet) and are detrimental to sizing.

4.1.4 *Condensate*

We must maintain a given differential pressure between the steam supply and condensate siphon within the dryer to remove the condensed steam as it forms. Condensate removal must function properly or the dryer will fill up with hot water. The condensate buildup within the dryer acts as an insulator inhibiting the direct heat transfer from the steam to the dryer shell. The thinner the condensate layer, the more efficient the heat transfer. There is a minimum required condensate layer for heat

TABLE 4.1. Change of Enthalpy with Steam Pressure and Temperature.

Saturated Steam Pressures (psig)	Saturated Steam Temperatures (°F)	Enthalpy (latent heat) (Btu/lb)
0	212	971
5	227	961
10	239	953
15	250	946
20	259	940
30	274	928

transfer to take place and the thickness of this minimum layer is determined by siphon design.

4.2 The Role of a Dryer Fabric in Drying

Another requirement for heat transfer is to maintain and/or improve the sheet's contact with the dryer. This is the primary function of a dryer fabric. If a wet object is placed against a heat source, the water on the surface will heat up. In the case of a sheet of paper, the evaporation of the heated water creates a positive pressure which will tend to lift the sheet off the dryer. This reduction or loss of contact with the dryer will result in poor heat transfer.

There is also a layer of air between the sheet and the dryer that is carried by the moving sheet and dryer (Figure 4.1). The dryer fabric presses the sheet onto the dryer reducing the air layer and minimizing any sheet lifting due to heating the sheet.

It is very important to maintain adequate tension on the fabric. This tension is normally in the 8 to 12 pound per linear inch range. The general tendency is towards increased tensions. Higher fabric tensions (20 pli and above) are being utilized on some newer or rebuilt machines and grades to provide further heat transfer efficiencies.

Most of the dryer fabrics are woven textiles made of synthetic yarns. The fabrics are permeable in that air can pass through the fabric structure. This characteristic is called permeability.

Permeability is expressed as the amount of air (cubic feet per minute) that will pass through a square foot of fabric per minute at a pressure drop of 0.5 inch of water. The common value is stated as cfm. The range of permeability for today's dryer fabrics is from a low of 50 cfm up to 1000 cfm.

Fabric permeability is determined by the structure of the fabric and the sizes and types of yarns used. In order for more air to pass through a fabric, the open spaces in the fabric have to be larger. One way to accomplish this is to simply use fewer yarns. There is a lower limit for the number of yarns to be used before the fabric becomes too flimsy to run on the machine. Other factors that affect permeability are design pattern and yarn size.

Dryer fabrics carry or pump air as they run through the machine. This air carrying is a benefit to drying efficiency in machines with inadequate pocket ventilation systems. Studies in a wind tunnel showed that the amount of air carried by the dryer fabric is the highest right next to the surface of the fabric and it decreases very quickly as the distance to the fabric surface increases. The amount of boundary air carried is directly proportional to the speed of the fabric. At a given point of a specific fabric, the height of boundary layer air remains unchanged, independent of different shapes. Air permeability and material type have the most influence on the amount of air carried. The effect of surface roughness is negligible [2–4].

Similar to drag coefficients on cars, a drag coefficient, C_A was developed for dryer fabrics to characterize the air carrying capacity. This coefficient is used to compare the running performances of fabrics regarding their capacity of carrying air. The C_A value of a fabric is the quotient of the division of the amount of air carried by the fabric and the amount of air carried by the reference surface which is a glass plate (Figure 4.2). This proportion remains constant for any running conditions. The smallest possible value of C_A is 1. Typically, dryer fabrics have C_A values in the range of 2.0 to 14.7.

As an example, if Fabric A has a C_A of 4 and Fabric B has a C_A of 8 and both fabrics are running in the same paper machine position under the same conditions, then Fabric B would carry always twice as much air as Fabric A and always eight times more than an "ideal" reference surface of the glass plate.

There is a limit on how high a permeability can be utilized which is dependent upon the machine speed, paper grade and basis weight regarding heat transfer properties. A fabric can pump so much air that it creates sheet flutter and other sheet handling difficulties, depending on C_A value. Normally, machines

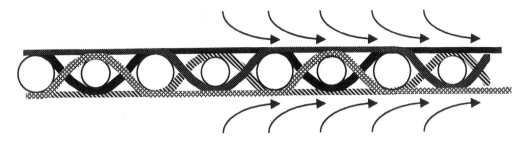

The air carrying ability of a flat ribbon fabric with symmetrical weave design is the same on bots sides: $C_A = 1.5$.

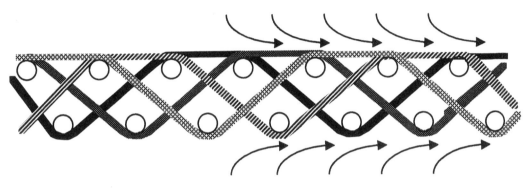

The air carrying ability of a common fabric: Paperside $C_A = 2$ Backside $C_A = 2.4$

The air carrying ability of a glassplate:

(reference value) $C_A = 1$

FIGURE 4.2. Air carrying by dryer fabrics.

run as high a fabric permeability and air carrying characteristics as possible, but below the point that they cause sheet handling problems. Pocket Ventilation Systems are designed to blow the air through the fabric. These systems are typically designed for fabrics with a maximum permeability of 150 to 400 cfm. There is no benefit to use higher permeability fabrics.

There is some evaporation that takes place through the fabric when it is pressing the sheet to the dryer but in most cases, this is not a major factor, especially with the recent paper machine constructions where the dryer diameter can be up to 1.8 m.

Sheet surface quality can also be a limiting factor in the fabric permeability. In most conventional weave patterns, the higher the permeability, the coarser the fabric surface. A coarse fabric surface can damage the sheet surface.

4.3 Effects of Drying on Sheet Properties

Although the main function of drying is to remove the remaining extra water in the sheet, several properties of the sheet are influenced by the drying process. The drying process affects the sheet structure, strength and behavior during the end use later on [5]. Various tensions are applied to the paper from entrance to the exit of the dryer section. The paper quality is affected by the draw (wet stretching or straining) in the dryer section. Increasing draw reduces bursting strength. Tight draw reduces extensibility and increases machine direction (MD) tensile strength. In-plane stretching reduces the strength in the thickness direction [6–8].

Changes in moisture content during drying change the sheet dimensions. For example, reducing moisture content will result in sheet shrinkage. If shrinkage is restricted during drying, the tensile strength of the paper will be increased and residual stretch decreased. As a result, the sheet will carry high applied loads but will have low yield and impact resistance. It was shown on handsheets which were dried freely that sheet shrinkage takes place mostly in the region between 55% and 85% dry solids content (Figure 4.3).

Restraint-free drying produces lower tensile strength paper which will be tougher due to ability to stretch. Cross-direction (CD) restraint is applied to the paper when the sheet is on the drying cylinders on multicylinder machines. Continuous sheet support provides cross-direction restraint. When the sheet is not supported in the draws, it shrinks freely in the CD. The shrinkage is not uniform and usually concentrates at the edges [9]. Excessive shrinkage negatively impacts sheet smoothness and contributes to cockle and edge curl.

Overdrying of the sheet causes brittleness

and reduces strength of the sheet. Blistering is caused by an inappropriate dryer fabric or by an excessively hot cylinder. Wrinkles in the sheet are mainly caused by sheet fluttering and edge lifting. Coarse seams in dryer fabrics may mark the sheet.

4.4 Dryer Fabrics

Figure 4.4 shows the dryer fabric design characteristics. The dryer fabrics are similar to both forming and press fabrics in many aspects. An important characteristic of a dryer fabric is the requirement for hydrolysis resistance.

4.4.1 *Yarns*

A greater variety of synthetic yarns is used in manufacturing of dryer fabrics than the forming or press fabrics. Resistance against hydrolysis, dry heat, abrasion, acids and alkalis is important for dryer materials. Other required properties include uniform density, tensile strength, elongation strength, thermal shrinkage, loop and knot resistance, affinity for resins, dirt repellence, cleanability and thermostability.

Figure 4.5 shows several yarns that are used in dryer fabrics. The four major categories of materials used in dryer fabrics are:

- staple yarns
- multifilaments
- monofilaments
- special yarns

Staple Yarns

The polymers that are used for staple yarns are polyamides, acrylic, polyester and aramids. Polyamide (nylon) has excellent tensile strength, good spinnability and good properties, which have very positive effects on the affinity of impregnation. Hydrolysis resistance is good and acid resistance is mediocre. Acrylic is 100% hydrolysis resistant and has good resistance against chemical attacks (acid,

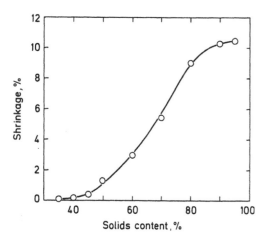

FIGURE 4.3. Effect of dry solids content on shrinkage for a handsheet made of unbleached kraft pulp [5].

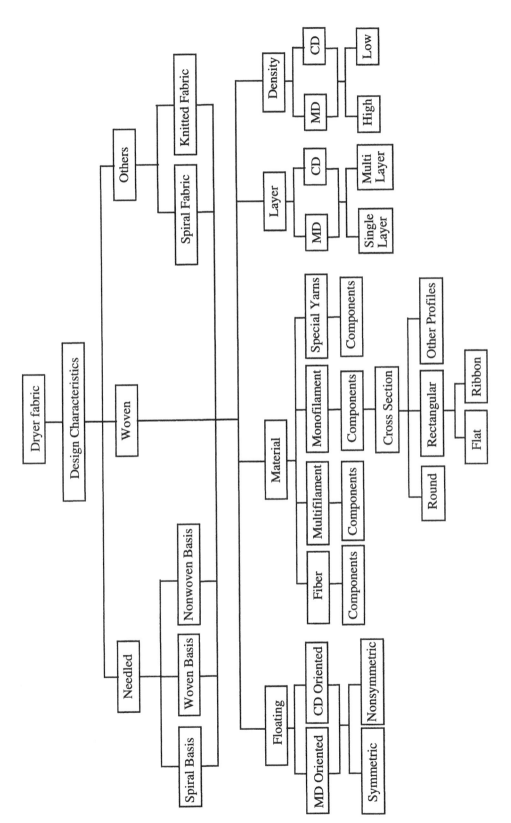

FIGURE 4.4. Dryer fabric characteristics.

197

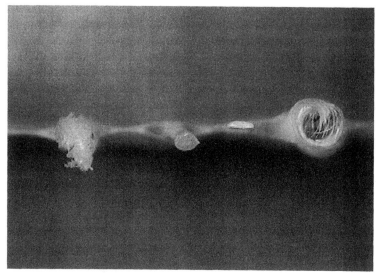

FIGURE 4.5. Major types of yarns used in dryer fabric manufacturing in addition to round monofilaments. Top: longitudinal view, bottom: cross sections. From left: staple Nomex® aramid, multifilament polyester core with nylon sheath, flat PET (PolyEthylene Terephthalate, i.e., polyester) and Nomex® aramid wrapped glass.

alkali, solvents, etc.). Its disadvantage is mediocre tensile strength. Polyester has good tensile strength and acid resistance. However, most of the time acrylic will be preferred due to some hydrolysis problems with polyester. Nomex® aramid is a high performance fiber with excellent hydrolysis, heat and chemical resistance. It also has good affinity for synthetic resins (impregnation).

Fibers from these polymers are spun either purely or as a mixture with other fibers in order to optimize their properties depending on their specific use. Polyamides and acrylics are usually mixed in order to produce yarns with a very high hydrolysis resistance and a good impregnation affinity. Depending on the density and twist, they are used as filling for multilayer fabrics in order to reach a specific air permeability and surface structure. It is for example possible to reach a smooth, homogeneous dryer fabric surface without any marking by using a bulky fiber yarn. Polyester spun yarns are used to mantle monofilament yarns for specific applications in cases where hydrolysis resistance is a less significant factor. Aramids are either spun purely or as a mixture with acrylic to obtain high performance yarns. These are used as filling yarns for fabrics requiring a high hydrolysis, heat and chemical resistance.

Multifilaments

Multifilaments are made of a multitude of very thin monofilaments. In addition to the polymers used in spun yarns, fiberglass is also used for multifilaments. Multifilaments, except glass yarn, are supplied by the manufacturer without or with very little twist. Therefore, they must be twisted in house. During the process, several multifilaments are twisted to obtain one single yarn (twisted yarn). This process is very often used to obtain a yarn with very specific properties by twisting several synthetic yarns:

- polyester with acrylic: The excellent tensile strength of polyester is combined with the hydrolysis resistance of acrylic. This compound has given excellent results both in MD and CD of multifilament fabrics.
- acrylic with Nomex® aramid: This compound has high performance properties and is used as a warp material for fabrics used in an environment requiring excellent hydrolysis, chemical, heat and mechanical resistance.
- acrylic, Nomex® aramid and polyester: Hydrolysis properties are the same as for acrylic with Nomex® aramid, but this combination has even better mechanical properties due to polyester. This MD yarn is therefore used for dryer fabrics with a small number of warp threads running in dryer sections with very strong conditions.

100% pure multifilament yarns are also used. Polyester is used in fabrics that do not require much hydrolysis resistance. It is the standard MD yarn for transport fabrics and CD yarn for polyester monofilament fabrics. Polyamide is used as a filling yarn with high impregnation affinity for multifilament fabrics. Polyacryl is suitable as a warp yarn for fabrics requiring high hydrolysis but low mechanical resistance. Glass filaments are supplied by the manufacturer in different densities. Each single filament is impregnated with a special resin in order to avoid abrasion. Glass yarns have high tensile strength, good heat resistance and stability. Their bending strength, however, is low. Glass yarn is a traditional filling material for positions with high temperatures.

Monofilaments

The cross section of monofilaments can be round, oval or rectangular. Special cross sections are mainly used as filling yarns.

Polyester is the most widely used monofilament material. Hydrolysis stabilizers are added to polymer of monofilaments to improve hydrolysis resistance. In addition, fluoropolymers are added to polyester to impart contaminant resistance properties. Several types of polyamide monofilaments can be used depending on the mechanical, physical and chemical property requirements. However, in dryer fabrics, Nylon 6.6 has given the best results. It is often used with polyester in the same fabric.

Polyphenylensulfide (PPS), e.g. Ryton®, has very good resistance against heat, chemicals and hydrolysis; however, it has mediocre tensile strength and abrasion resistance. Additives are used to improve some of these properties. PPS is relatively expensive.

Polyetheretherketone (PEEK) has excellent heat, hydrolysis and chemical resistance. However, due to its high price, it is only used in special cases in fabric form. PEEK is very often used in special parts of dryer fabrics, e.g., seam spirals, insert wires and possibly also reinforced edges.

Polypropylene (PP) has good hydrolysis

and chemical resistance and also it is not expensive. The main disadvantage is low permanent heat resistance. Therefore, PP is not used anymore in dryer fabrics.

Polytetrafluoroethylene (PTFE), e.g., Teflon®, is a pure fluorocarbon polymer, the chain of carbon atoms of which is totally protected by fluor atoms. PTFE offers the best anticontaminant properties with excellent chemical properties. However, its mechanical properties such as elongation strength, abrasion and tensile resistance are rather low. Furthermore, it is difficult to extrude and it is very expensive. PTFE is not used as a yarn material in dryer fabrics.

Small monofilament yarns (usually polyester and larger than multifilament) twisted together are called cabled yarns. They are used in certain monofilament fabrics.

Copolymers

The term *copolymer* stands for all monofilaments where two or more polymer systems are mixed. Strictly speaking, this term should only be used for polymer systems, which differ very much in their molecular structure. This is for instance the case for polyester with a fluoro polymer additive or for PPS with a polyamide additive.

PCTA is a very well known monofilament which is very often called a copolymer. In reality, it is a copolyester, i.e., a mixture of two different polyester raw materials. PCTA has very good hydrolysis resistance, but low mechanical and technical properties. Therefore, its use is limited to humid-hot groups in the paper machine.

Special Yarns

In the dryer fabric industry, some special yarns are used, which are specifically produced for this purpose.

Preimpregnated Twisted Yarns

They are generally made with multifilaments. There are also multimonofilament twisted yarns. These yarns are impregnated with resins that allow stabilization of the final product (twisted yarn). The disadvantages of these yarns are not only their high price, but also their thermoplastic impregnation: the stabilizing effect will get partly lost when in contact with heat in the paper machine.

Wrapped Yarns

These consist of a carrier material which is equipped with a cover or coating, mostly made from a different type of material. Cabled fiberglass, resin treated and wrapped with coated polyester and nylon multifilament is an example of wrapped yarns. This yarn is used as CD yarn in multifilament fabrics.

Yarns Coated during Extrusion

A prefabricated core out of multifilament, multifilament twisted yarn or monofilament is coated through extrusion. Examples:

- polyester multifilament coated with polyamide: This yarn is very effective as filling.
- polyester multifilament coated with different fluorocarbon polymers: These yarns are mainly used as warp or filling material in anticontaminant fabrics.
- polyester monofilament coated with polyester: The coating is made with polyester which has different properties than the polyester used in the core, e.g., low melting point. These yarns are specially developed to be used as filling.
- polyester monofilament coated with polyamide: The special advantage of this yarn is a slightly increased abrasion resistance compared to pure polyester. The main advantage however is that it is relatively smooth, i.e., the coating can be easily formed. It can be easily cropped and therefore has a very good form stabilizing effect, especially for fabric types with a low warp and filling number. This yarn is generally used

as a filling material.

- polyester monofilament with fluorocarbon polymers: The relatively stiff monofilament core increases the stability of the fabric.
- polyamide and/or polyester core coated with polypropylene: This relatively less expensive combination is very unusual due to the low melting point of PP.

Yarns Coated during the Spinning and/or Twisting Process

During this process, a core yarn is coated with multifilament or staple yarn. Examples:

- Glass yarn coated with multifilament gives a hard and very smooth yarn. Glass yarn with polyacrylnitrile gives a very hydrolysis resistant yarn.
- Glass yarn with staple yarn; e.g., acrylic also gives a bulky, hydrolysis resistant yarn. As it can be easily impregnated, it is an excellent filling material for stabilizing piece impregnated skip-dent fabrics.
- Polyester core with staple yarn is used in positions that are less subjected to hydrolysis. It is also an excellent stabilizing filling material for monofilament fabrics.
- PPS core with acrylic multifilament is a high performance product used to meet stringent requirements such as high hydrolysis, dry heat and chemical resistance. Its use is limited to a filling yarn for fabrics with a monofilament warp and to fabrics with a multifilament warp, which are impregnated afterwards.
- Each type of core coated with multifilament and/or staple yarn in an open-end process. During this process, core and coat are combined very tightly and permanently. It was however found that the uniformity of these yarns did not meet the requirements. Therefore, they are at best to be used as a filling material.
- Coating made during special processes.

There is at least one yarn that can be classified under this category: flock yarn. In this process, a carrier material (core) of multi- or monofilament is equipped with a bonding agent and put in an electrostatic field, where very short (0.5 to 1 mm) fibers are added and bonded onto the core. The fibers do not entangle, but are bonded evenly perpendicular to the surface of the core yarn such as in a cylinder brush. The average number of fibers per meter of yarn is 800,000. This yarn has proven good as a surface filling against different contaminations, but especially against white pitch.

Table 4.2 shows several properties of a polyester yarn used in dryer fabrics. Figures 4.6 and 4.7 show hydrolysis and dry heat resistance of various dryer fabric yarns, respectively.

4.4.2 Dryer Fabric Structures

Table 4.3 gives the general requirements of dryer fabrics. Some of these requirements are machine specific, i.e., they depend on the geometry of the machine, the position where the fabric will be used on the machine and the type of paper to be produced.

Theoretically, a dryer fabric should meet all of these requirements. Since some requirements are often in opposition to others, there is no dryer fabric that meets all possible requirements without any exception. Therefore, the dryer fabric manufacturer has to meet a maximum of general requirements and possibly some well defined strategic requirements. This results in production of so called "all-round" fabrics as well as fabrics with very special properties. Figure 4.8 shows the MD profiles of the major dryer fabric types used by the paper industry. Depending on their manufacturing process, dryer fabrics can be divided into three main groups:

(1) Woven fabrics
(2) Spiral fabrics
(3) Needled fabrics (batt on base)

TABLE 4.2. Properties of Polyester Dryer Yarn Material (diameter: 0.5 mm).

	Density (dtex)	Tenacity (cN/tex)	Breaking Load (daN)	Elongation (%)	Free Shrinkage (%) (180°C, 30 min)	Testrite (%) (180°C, 2 min)	Loop Strength daN	%
	2832	36.5	10.33	41.6	4.0	2.7	17.2	82.6
	2835	36.8	10.41	40.6	4.1	2.5	18.21	87.5
	2836	37.2	10.54	43.2	4.0	2.7	15.39	73.9
	2826	37.2	10.54	41.0	4.1	2.6	17.21	82.7
	2829	37.0	10.49	41.1	4.1	2.7	14.52	69.8
	2830	35.9	10.16	37.4	4.1	2.6	16.80	80.7
	2833	36.8	10.41	41.7	3.9	2.6	15.75	75.7
	2832	36.9	10.45	41.8	4.0	2.6	18.61	89.4
	2830	36.4	10.32	43.0	3.9	2.6	15.31	73.6
	2837	36.8	10.41	41.5	4.0	2.5	16.54	79.5
Average	2832	36.7	10.41	41.3	4.0	2.6	16.55	79.5
Minimum	2826	35.9	10.16	37.4	3.9	2.5	14.52	69.8
Maximum	2837	37.2	10.54	43.2	4.1	2.7	18.61	89.4
s	3.2	0.4	0.12	1.6	0.1	0.04	1.32	
v	0.1%	1.1%	1.1%	3.8%	1.9%	1.5%	7.9%	
Certificate		36.8		39.2	3.0			
Product tolerance:	2750 ± 150	36.0 ± 4.0	10.2 (9.5–11)	41.0 ± 6.0	3.5 ± 1.0	2.5 ± 1.0		

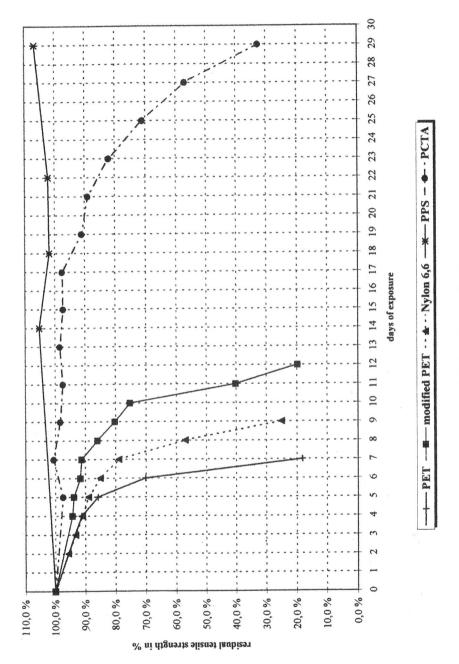

FIGURE 4.6. Hydrolysis resistance of various dryer yarns.

203

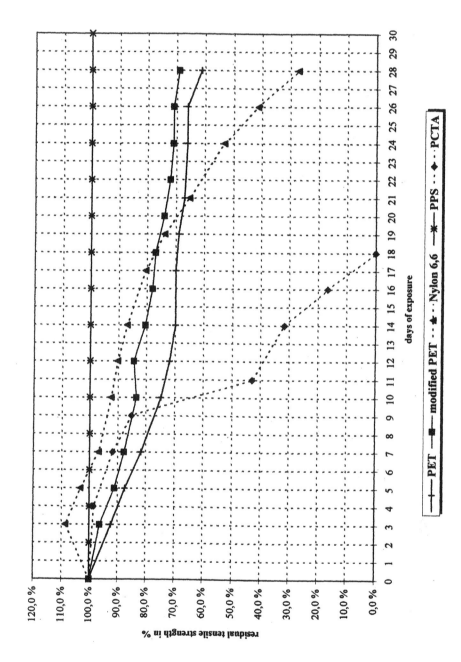

FIGURE 4.7. Dry heat resistance of various dryer yarns.

204

TABLE 4.3. General Requirements of Dryer Fabrics.

> Heat transfer
> Reconditioning
> Contact area
> Caliper
> Permeability
> Amount of carried air (C_A value)
> Surface properties, i.e., nonmarking
> Fabric stability (mechanical and thermal)
> Tensile strength and modulus
> Contaminant resistance properties
> Cleanability
> Physical and chemical resistances:
>   • Hydrolysis
>   • Dry heat
>   • Abrasion
>   • Acid and alkaline
> Sheet stretching (neutral line)
> Sheet fluttering
> Moisture profile
> Permeability profile

Woven Dryer Fabrics

The big majority of dryer fabrics are woven. This may be due to the fact that, from a historical point of view, woven fabrics have been closely linked to the development of paper production than the other fabric types. Moreover, nonwoven fabrics and fabrics manufactured with other methods do not meet the same wide range of requirements as woven fabrics do.

The main characteristics which allow the differentiation of woven fabrics and have a direct influence on the requirements are:

• raw materials
• types of yarns
• yarn crimping and floating
• density
• number of warp and filling layers

The choice of raw materials depends on the specific requirements that have to be met in a special position. Therefore, different materials may be combined in the same fabric such that the positive characteristics of one raw material make up for the negative characteristics of another.

This type of combination is frequently used for slight to medium strength hydrolysis conditions: polyester is combined with another hydrolysis resistant material such as PPS. The sequence of the raw materials in the fabric construction depends on the application of the fabric. For example, in the case of slight to medium strength hydrolysis environment, it is advisable to use all hydrolysis resistant raw materials on the fabric surface since it is the place where hydrolysis usually attacks. Such a sequence can be chosen for the whole fabric width or just for the edge areas.

Regarding the yarn types, woven dryer fabrics may be roughly subdivided into two categories, regardless of any other characteristics: monofilament and multifilament fabrics.

Monofilament Fabrics

Today, monofilament fabrics are used in almost all positions in the dryer section. Their main characteristics in comparison with multifilament woven fabrics, nonwoven fabrics and spiral fabrics are as follows:

• Most of the time, the monofilament fabric is less thick, which improves heat transfer and reduces steam consumption.
• For the same raw materials and fabric construction, monofilament fabrics have a better tensile strength which increases the lifetime of the fabric.
• One hundred percent monofilament fabrics are easier to clean. This is a major characteristic considering the increasing number of fabric cleaning installations in the paper machine due to increased use of recycled paper.

Fabrics are considered monofilaments if at least the warp yarns are made with monofilament yarns. These may be made with all possible yarn profiles. Monofilament fabrics can be divided into three types according to the shape of the yarns used:

• round yarn fabrics
• flat yarn fabrics
• ribbon yarn fabrics

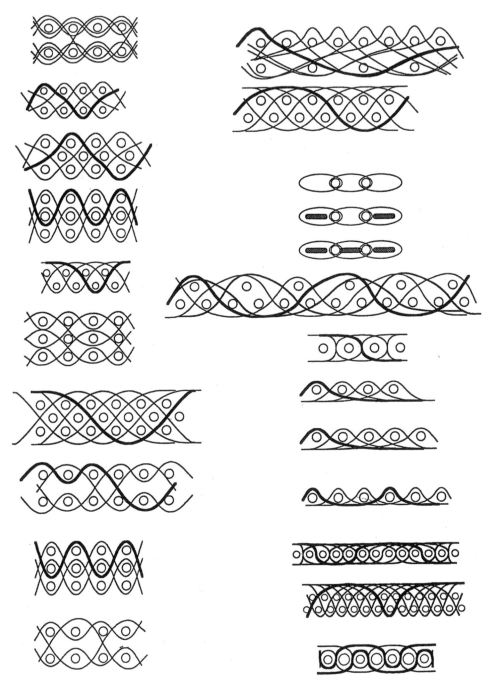

FIGURE 4.8. MD profiles of the major dryer fabric types.

Different yarn cross sections allow further construction possibilities for the manufacturing of woven dryer fabrics and may have an influence on the different technical properties such as contact area, maximum tensile strength, elongation, abrasion resistance, diagonal stability, air carrying capacity and others. From a technical application point of view, however, the advantages of a specific yarn cross section and thus of a whole type of a woven monofilament dryer fabric cannot be exclusively defined according to one single yarn, but depend on several other construction characteristics.

In monofilament woven dryer fabrics, single monofilament yarns, mostly with a round cross section and with different diameters, are also used as filling. Moreover, any other types of yarns may also be used as filling yarns (Figure 4.9). These are mainly used in order

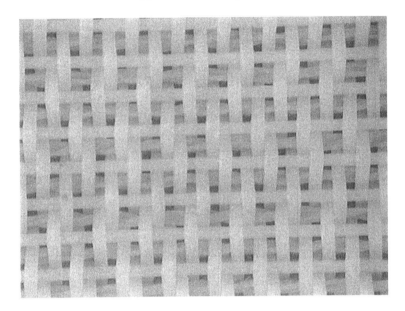

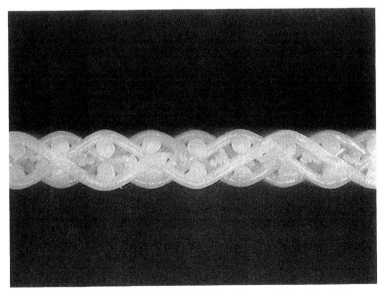

FIGURE 4.9. Top view and MD profile of a monofilament fabric with stuffer yarn. MD: flat PET, CD: round PET, stuffer: cabled PET.

to create a special fabric structure to control air permeability or to increase the stability of the fabric.

Most of the woven monofilament fabrics used today are made of round and flat yarns. The so-called flat ribbon fabrics are a new generation of dryer fabrics. The flat ribbons have a rectangular cross section where the proportion between the width and height is more significant than the traditional flat yarns.

With the use of flat ribbon yarns, it is possible to make new kinds of fabric and seam structures as shown in Figure 4.10 [10]. With the rectangular cross section, the warp yarns can be stacked on top of each other without slippage. It is also possible to realize a fabric with two separate warp layers based on one single filling layer. In traditional fabrics, two

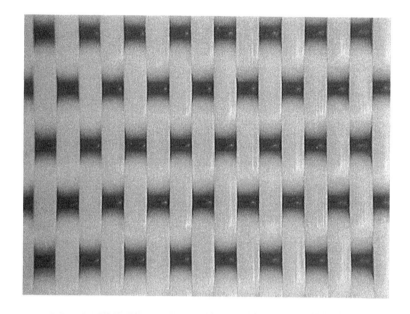

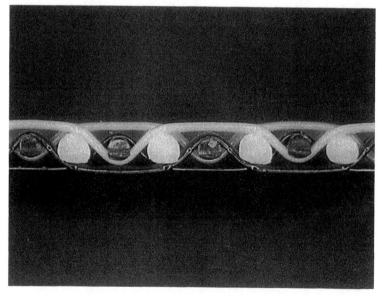

FIGURE 4.10. Top view and warp profile of Monotier® dryer fabric.

separate warp layers necessarily require three filling yarn layers that makes the fabric much thicker which is a disadvantage. The new type of construction also allows a twistless warp loop seam. This kind of fabric and seam construction is a fundamentally new development in the construction of dryer fabrics that meets a maximum number of requirements with a single dryer fabric construction:

- longer life
- smooth surface with high contact area
- wide air permeability range (50–600 cfm)
- uniform permeability profile
- low air carrying capacity
- reduced sheet stretching (Figure 4.11)
- contamination resistance and cleanability
- high tensile strength
- no seam marking

Multifilament Fabrics

Fabrics in which at least the warp yarns are made with spun and multifilament yarns are considered as multifilament fabrics (Figure 4.12). The twisted warp and filling yarns may be made with different raw materials in order to combine several special properties. The use of multifilament fabrics in the dryer section is decreasing.

Multifilament fabrics can be classified as piece-treated and pre-treated. In piece-treated multifilament fabrics, multifilament yarns are usually processed as raw yarns. Therefore, in order to be stabilized, a multifilament fabric has to be treated before being used in the paper machine. This is called piece-treatment.

Pre-treated multifilament fabrics are made with the yarns that are already pre-treated. Stability of fabrics made with pre-treated yarns cannot be altered. Moreover, the crossing points of the fabric construction do not have the same stability as in a piece-treated fabric, which is another disadvantage.

For all multifilament woven dryer fabrics, any possible type of yarn can be used as filling yarns.

Crimping and Floating

Similar to forming fabrics, crimping and floating (long knuckles) are significant for the dryer fabric construction and its subsequent properties. They also determine the construction of the fabric surface, which is very important. Floating is the case where a warp yarn is woven over several filling yarns or if a filling yarn is woven over several warp yarns. The main attention is given to the construction of the fabric side which is in contact with the paper. Thus, the crimping and floating also determine the contact area of the dryer fabric. Figure 4.13 shows contact areas of several dryer fabrics.

Referring to floating, woven dryer fabrics may be divided into two groups: fabrics with length oriented floating (warp float) and fabrics with cross machine oriented floating (weft float). Fabrics where the contact area with paper mainly consists of floating warp yarns are called warp float weave patterns (Figure 4.14). In this case one warp is floating over several filling yarns. Fabrics where the contact area with the paper mainly consists of floating weft yarns are called filling float weave patterns. In this case, one weft yarn is floating over several warp yarns. In general, a filling float weave pattern has a much higher yarn density on the upper weft layer. The covering of one fabric side with a high number of weft yarns is on purpose, e.g., in order to reach the best possible cover with yarns having special properties (such as anticontamination) or in order to protect the weft yarns against abrasion. Dryer fabrics with warp float weave patterns are used more than the fabrics with filling float weave patterns. Figure 4.15 shows a filling float weave pattern with a large spun face side, which is called "soft face" fabric. Soft face fabric has a limited market and is used primarily on slower, non-blow box UnoRun positions and marking sensitive grades.

Floating can be used to design the contact area of the fabric for a particular application. If the float weave patterns are the same on both sides of the fabric, the fabric design

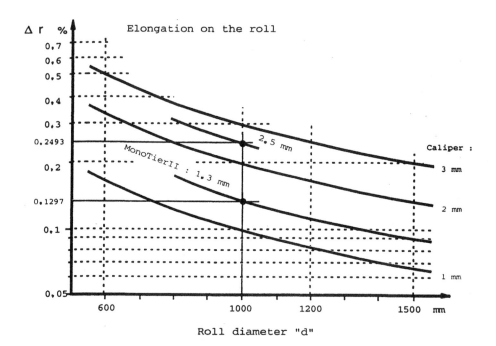

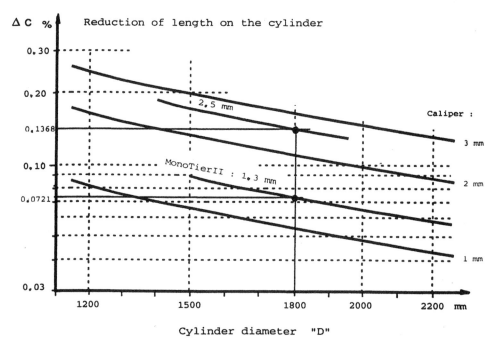

FIGURE 4.11. Sheet stretching for single fabric machines.

FIGURE 4.12. Hydrolysis resistant multifilament fabric. MD: spun Nomex® aramid, CD: acrylic wrapped glass.

is called a symmetrical weave pattern construction (notice the difference in symmetry definition of forming fabrics). Symmetrical weave patterns have the same smooth or rough embossing on both sides of the fabric. If the yarn cross sections of all warp and weft layers are the same, the neutral line (Section 4.6.3) in symmetrical weave patterns lies exactly in the middle of the fabric construction. If the float weave pattern is not the same on both sides of the fabric, this is called an asymmetrical weave pattern construction which has one smooth and one rough surface. The neutral line in an asymmetrical weave pattern is not situated in the center of the fabric construction, but more towards one side of the fabric.

FIGURE 4.13. Contact areas of various dryer fabrics.

Mesh

Similar to forming fabrics, the number of warp and filling yarns per unit length has a direct influence on many properties of dryer fabrics including:

- contact area
- fabric and seam tensile strength
- air permeability
- air carrying capacity
- stability

Warp density can exceed 100% in some designs. In certain designs, the warp yarn count is decreased on purpose to reach high air permeabilities and/or to obtain a good fabric ventilation.

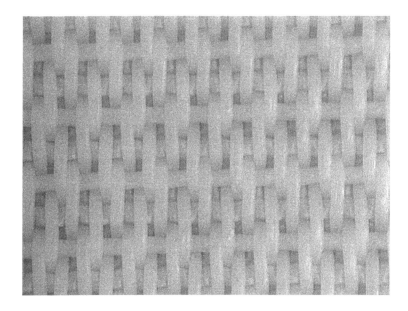

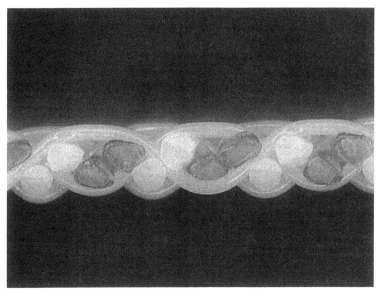

FIGURE 4.14. Top view and MD profile of a warp float weave fabric. MD: flat PET, CD: round PET and sheath-core.

Number of MD and CD Layers

Usually a woven dryer fabric has 1 to 4 MD (warp) layers. In some special cases, there may also be more layers. More than one warp layer is used to obtain a special dryer fabric surface and/or to improve the stability of the fabric. Figure 4.16 shows 1 and 4 MD layer fabrics. The disadvantage of a traditional woven fabric with several warp layers is the thickness. However, this disadvantage is minimized in recent multilayer fabric constructions [10].

Woven dryer fabrics generally have 1, 2 or 3 CD (filling) layers. Fabrics with only one CD layer are generally thinner than fabrics with 2 or more CD layers. For fabrics with

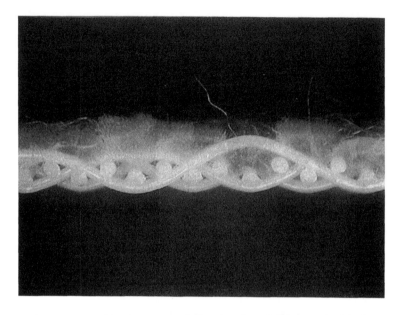

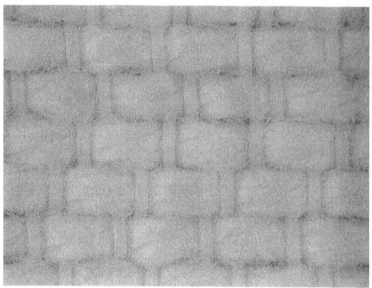

FIGURE 4.15. MD profile and top view of a soft face fabric. MD: round PET, CD: spun acrylic and round PET.

only one weft layer, it is difficult to obtain a warp loop seam which is easy to close, without marking and which meets the dynamic requirements of the paper machine. This problem can be solved with a design with only one weft layer combined with two warp layers [10].

Spiral Fabrics

Spiral fabrics are not woven but are made with monofilament yarns that are put in spiral form. The spirals are linked together by monofilament cross yarns or pintles (Figure 4.17).

FIGURE 4.16. One (left) and four MD layer fabrics.

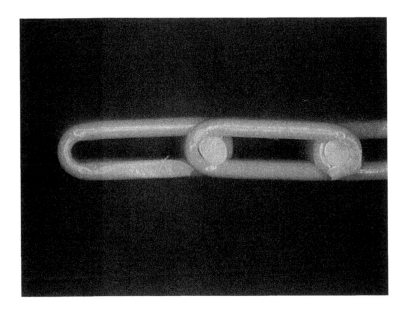

FIGURE 4.17. Spiral fabric. Coil: round PET, pintle: round PET.

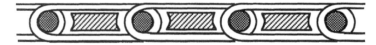

FIGURE 4.18. Schematic of filled spiral fabric.

Dryer fabrics manufactured with this process have been known since 1978 and have reached a limited market share. The main advantages of spiral fabrics are good dimensional stability and good fabric ventilation. The seam joint has the same construction as the rest of the fabric. Nevertheless, from a technical application point of view, woven dryer fabrics are preferred over spiral fabrics. Disadvantages of spiral fabrics are a relatively high fabric thickness and a limited contact area. High air carrying capacity also limits the use of spiral fabrics. The high air permeability of spiral fabrics (800–1000 cfm or 14632–18290 $m^3/m^2 \cdot h$) can be reduced by filling the spirals with different filling materials (Figure 4.18). Spiral fabrics are mainly made with polyester monofilament yarns.

Needled Fabrics (Batt on Base)

Needled dryer fabrics are made up of a base fabric with a nylon web layer needled to face on sheet side. This web layer is fixed onto the base fabric in a special process called needling (Section 3.2.5). The structure of the base fabric may be a woven monofilament or multifilament structure or a spiral fabric.

Needled dryer fabrics are especially used in positions or for paper types liable to marking. Air permeability is limited to about 50–200 cfm (915–3658 $m^3/m^2 \cdot h$). A disadvantage of these designs is the relatively high thickness of the fabric.

4.4.3 *Seam Styles*

Seams are often the most sensitive part of a dryer fabric. In general, dryer fabrics are woven flat and seamed later to form an endless belt. The seam consists of formed loops or inserted coils at the two ends of the fabrics which later on are connected together. To join the seam, a stainless steel wire is used as a leader. The real pintle wire (yarn) is connected to the end of the steel wire. This pintle wire is either a monofilament yarn or a braided multifilament with a monofilament wire. Today, use of one pintle wire is common. The diameter of the pintle wire depends on the channel size formed by the loops and coils. The form, the size, the amount and arrangement of the spirals/coils as well as the dimension and structure of pintle wire determines if a seam is easy and quick to join together.

Besides easy installation, the major requirements of seams are good tensile strength, wear resistance and nonmarking surface. The seam tensile strength, which is usually lower than that of the fabric, can be increased by proper selection of yarn materials, yarn profiles and seam construction. Press-marking can be avoided if the thickness of the seam is made identical to the fabric. Wet-marking can be avoided if the permeability of the seam and loops/coils are identical with the fabric permeability. The nonmarking will be reached by using a suitable yarn profile and seam construction.

Dryer fabric manufacturers have developed various devices for joining seams. Mainly these devices remove the tension of the seam so that loops/coils can build a channel for an easy inserting of the pintle wire.

Pin Seams

There are different variations of pin seams. The standard pin seam is nonmarking and designed for wear resistance (Figure 4.19). Standard pin seam may mark in the bolt area in some grades. In some designs, caliper of the weave back may be slightly higher than the caliper of the body. Potential wear areas are weave back, loop and tie back loop.

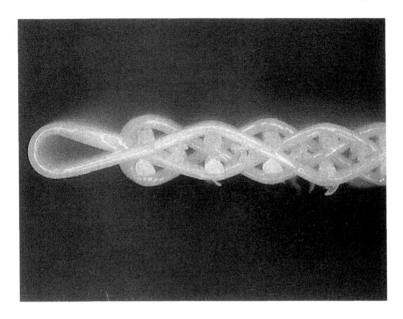

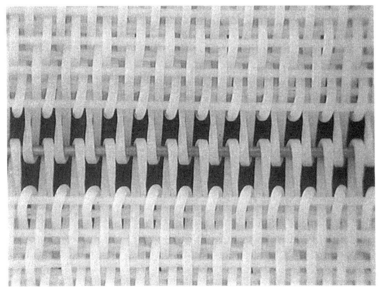

FIGURE 4.19. Standard pin seam.

The air permeability of the weaveback area is often not identical to the permeability of the fabric itself. Air permeability in the loop area is significantly higher which may result in wet-markings on the sheet. With flat ribbon fabrics, the same permeability in weave back area and fabric is obtained. In addition, the permeability of the loop area can be altered by using additional filling yarns in order to avoid wet-markings.

Coil Seam

In this seam type, coils are wrapped by all MD yarns which results in a strong seam (Figure 4.20). Half of the CD yarns are removed in fold back area. It is recommended for less mark sensitive grades or positions not requiring a thin line coil seam. There is a potential for thermal mark on some grades by the sewing lines or coil. The seam type relies

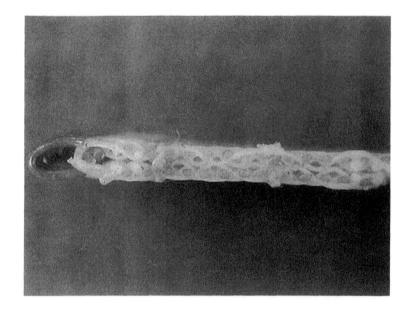

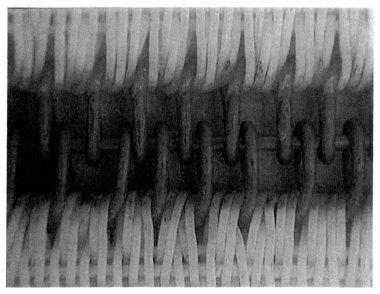

FIGURE 4.20. Coil seam.

on the sewing lines for seam strength. There is a difference in air permeability of the seam area (joint and fold over) and the body of the fabric. Potential areas for wear on the sheet side are sewing lines, tie back and coil (very rare). Potential nonsheet side wear areas include fold back shoulder, sewing lines, tie back and coil (very rare).

Thin Line Coil (TLC) Seam

Thin line coil seam is nonmarking and uniform with the sheet side and back side of the body of the fabric (Figure 4.21). It is very easy to join. TLC has good seam wear resistance.

TLC seam is recommended for less mark sensitive grades or positions that do not

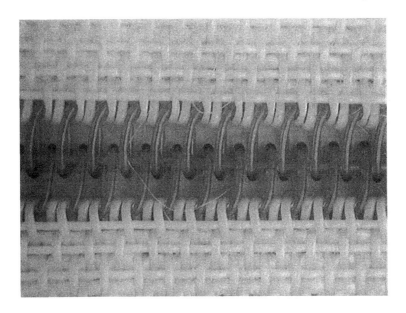

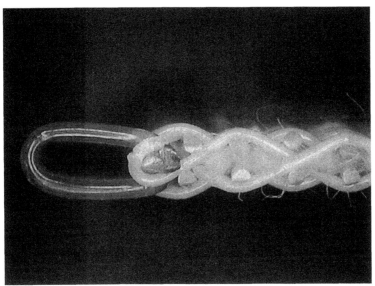

FIGURE 4.21. Thin line coil seam.

require a pin seam. It has potential to thermal mark on some grades. There is a potential for side wear on the coils with excessive distortion of the seam. Possible wear areas are tie back, coil and weave back. Air permeability of the seam joint is usually higher than that of the body of the fabric in some designs.

King Pin Seam

Figure 4.22 shows the schematic of the King Pin Seam. Seam strength is a function of the sewing, coil, MD warp and weave back. It has excellent seam wear resistance because the sewing lines are embedded and the shoulder is ramped. Weave back area

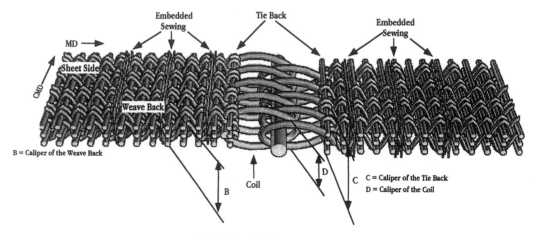

FIGURE 4.22. King pin seam.

provides improved seam strength and a ramp reducing the shoulder wear seen in coil seams. Potential wear areas are sheet side and nonsheet side tie back and nonsheet side shoulder ramp. There is a permeability difference between the weave back and seam area to the body. Mono King Pin Seam is a variation of this seam style.

Double Wire Double Offset Clipper Seam (DWDOCS)

Metal hooks are clipped to a tape which is sewn to the fabric (Figure 4.23). This seam is easy to manufacture. Seam strength relies totally on the sewing. Seam mark potential is

high. Potential wear areas are side wear on hooks, nonsheet side flap and nonsheet side sewing on flap. There is a permeability difference between the weave back and seam area to the body.

4.4.4 Dryer Fabric Manufacturing

The dryer fabric manufacturing is similar to forming and press fabric manufacturing. The common manufacturing steps are warping, weaving, heatsetting, seaming and finishing. However, some of the manufacturing materials and process parameters are different as explained below.

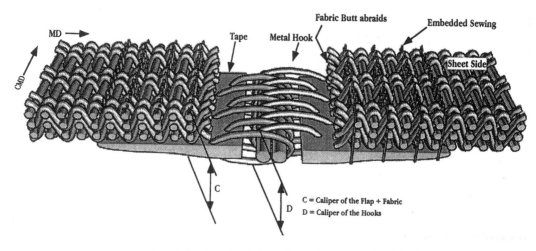

FIGURE 4.23. Double wire double offset clipper seam (DWDOCS).

Warping and Weaving

The major formation method for dryer fabrics is weaving which is explained in Section 2.2.3. The machine direction (warp) yarns are wound onto drums containing an exact amount of the specified yarns which is called warping (Figure 4.24). These drums are loaded onto the loom and provide the machine direction yarns for the specific fabrics scheduled to weave.

A wide variety of cross machine direction yarns is used as explained earlier. The looms can be equipped with shuttle, projectile or rapier as the transportation mean for CD yarn insertion. These systems maintain a supply of the CD yarn at each fabric edge. The carrier grips the filling yarn and carries it across the loom. The shuttle is usually pirnless (i.e., without quill) which allows the looms to run nearly continuously without having to shut down to repenish the yarn carried by the shuttle in the old systems (Figure 4.25).

Wide and heavy duty looms are used for dryer fabric weaving (Figure 4.26). Spiral fabric manufacturing, on the other hand, does not require a traditional weaving method. The yarns are made into spirals which are assembled together in groups to form the fabric (Figure 4.27).

Once the fabric has been woven to the required length, it is removed from the loom and the next fabric is begun. The fabric is next examined completely on both sides in burling.

Heatsetting

Temperature, moisture, and tension in the paper machine affect the stability of a dryer fabric because of the thermoplastic properties of the different raw materials used. Besides, the yarns woven into the fabric have not fully conformed to the weave pattern. For stability of the fabric, they must permanently conform to the weave by developing crimp. For these reasons, to achieve a sufficient dimensional stability, a dryer fabric has to be heatset. An unheatset (raw) fabric that is

FIGURE 4.24. Canister warping.

FIGURE 4.25. Pirnless shuttle.

FIGURE 4.26. Dryer fabric weaving.

FIGURE 4.27. Spiral fabric assembly.

put onto a paper machine would go through heatsetting while heating up the machine but in an incomplete and insufficient manner. The uncontrolled changes in length and width of the fabric would quickly lead to problems like guiding the fabric or tension limits.

The heatsetting is done with the fabric under tension on either a hot oil cylinder or a hot air oven (Figure 4.28). The fabric is put on the calender in an endless manner and is rotated several times (called "passes") at the right heatsetting temperature. Tension and high temperature during heatsetting do not only affect the length and width of the fabric, but also its permeability and thickness.

Heatsetting using hot oil cylinder calender with a surface temperature of 425°F generates a fabric surface temperature of 360°F. In general, the face side of the fabric is towards the cylinder during heatsetting. Usually, a hood is used with the cylinder to blow hot air on the other side of the fabric to reduce the difference in temperature on both sides of the fabric. Heatsetting using the hot air calender is done at a typical temperature of 420°F. It should

be noted that these temperatures may vary slightly depending on the fabric thickness.

Heatsetting of common dryer fabrics is done around 8 pli of tension which is considered to be the optimum tension to avoid slack edges while getting length stability. Ideally, heatsetting of a fabric should be done at a tension equal to or slightly greater than the paper machine running tension.

Seaming

Dependent upon fabric style and seam type, the flat woven fabric is seamed either before heatsetting, after a partial heatset with full heatset following seaming, or after full heatset with the seam "leveled" on the calender. Figure 4.29 shows pin seaming of a dryer fabric.

Resin Treatment (Impregnation)

All multifilament fabrics which are not manufactured with preimpregnated yarns are piece impregnated after seaming to obtain stability. For these fabrics, generally thermoset

FIGURE 4.28. Heatsetting of dryer fabrics.

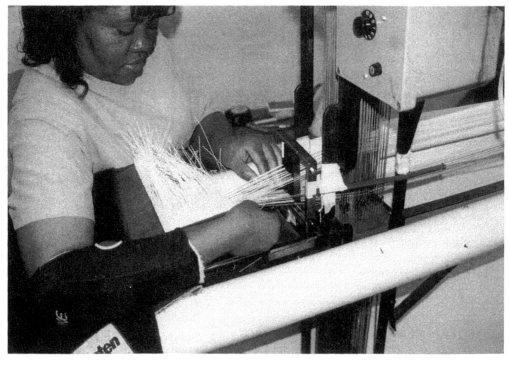

FIGURE 4.29. Pin seaming of a dryer fabric.

resins are used. These resins keep their stiffness under heat in the paper machine. However, in some cases thermoplastic resins are also used. The use of thermosetting or thermoplastic resins and/or a mixture of them depends on the type of fabric and/or the conditions in the paper machine. The recipe (concentration of resin components) is prepared according to the application conditions.

For fabrics with a smooth fiber yarn surface thermosetting resins are used in low concentration. For open fabrics, stabilization is a significant factor. In this case, thermosetting resins are used in high concentration. Fabrics with a hard resin and high stiffness tend to develop folds. In this case, a mixture of thermosetting and thermoplastic resins gives good results. Depending on the type of resins used and their concentration, there are many possibilities to vary the grade of stiffness. For fabrics with high elasticity warp yarns, the use of a highly stabilizing mixture of thermosetting and thermoplastic resins gives good results.

Special Treatments

There are other treatments that are used to impart different characteristics to the dryer fabrics. These treatments are not necessarily used for stabilization nor are they specific to multifilament fabrics.

In some cases, it is necessary to have a low air permeability on the edges of a dryer fabric (Figure 4.30). A thermoplastic, smooth resin is used for this purpose to obtain a similar stiffness in the edge areas as in the other, nonimpregnated areas.

Anticontamination impregnation is also applied to dryer fabrics. Considering that the paper material is an aqueous suspension with dirt particles in it, a fluorocarbon resin is applied to the fabric to reduce contamination. As a result, monofilament or multifilament fabrics develop hydrophobic properties. The resin layer is very thin which makes it susceptible to abrasion on the yarn surface. However, hydrophobia is maintained longer in the fabric gaps which helps to slow down

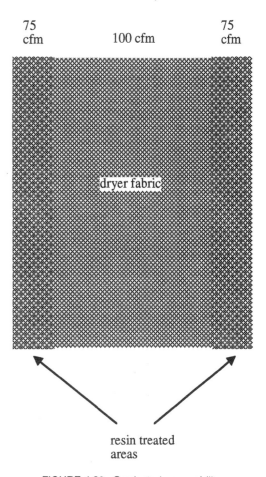

FIGURE 4.30. Graduated permeability.

the dirt accumulation on the fabric and keep the air permeability high.

Edge Treatments

Due to the conditions in the paper machine, the edges of dryer fabrics must be protected against mechanical damage. Mechanical damage generally occurs due to a permanent or temporary contact with the pallets. An edge protection treatment must meet the following requirements which are specific to the conditions on the paper machine:

- protection against edge raveling
- elasticity and flexibility adapted to the fabric

- edge protection should not attack the fabric or further enhance the existing damage
- sufficient temperature, hydrolysis and abrasion resistance
- sufficient acid and alkali resistance
- have no or minor influence on the thickness of the fabric

Fabrics are trimmed to width and edges are treated on a calender. Basically there are three methods for edge treatment:

- welding
- bonding
- sealing

Welding simply consists of welding the edges of the fabric. The warp and filling yarns are welded to each other on their crossing points with heat. The elasticity, flexibility and other properties of the fabric edge should not be altered significantly during the welding process. During bonding and sealing, a sealant is put onto the fabric edge. Warp and filling yarns are combined through the action of bonding. Sealing does not result in an adhesion between warp and filling yarns; it encapsulates the warp and filling yarns in the edge area (Figure 4.31).

The efficiency of an edge protection treatment depends on the type of welding or the properties of the respective bonding or sealing products. The edge protection treatment can also be improved by combining different processes.

Finishing

Auxiliary items like Zip or Velcro seam assist (Figure 4.32), leader skirt, hook protector, trade line, seam alignment arrows, etc., are installed in the finishing process. After finishing, fabrics are packed in a tube or box container.

Zip tape device is a stapled-on zipper. When closed, the zipper holds the seam in exact position for insertion of the joining wire. Once the seam is joined, two strips are pulled to remove the assembly. Zip tape is installed

FIGURE 4.31. Sealant application to fabric edges.

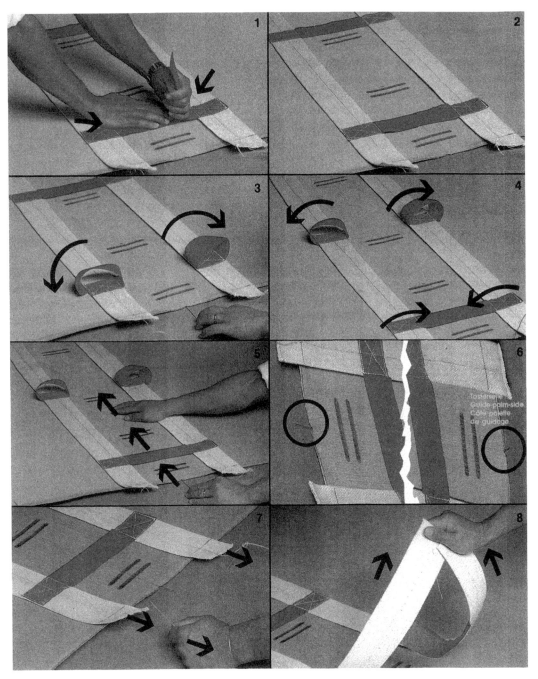

FIGURE 4.32. Velcro seam assistant device.

on the side opposite from which the seam is joined. Velcro assist systems consist of a strip of self-adhesive Velcro across the width of the fabric on each end (just behind the seam). Upon installation, the two ends of the fabric are brought together and the strips applied across the seam. This holds both ends together while seaming to avoid tension from the fabric.

All fabrics are equipped with a leader skirt attached to the seam end on the outside of the roll and a hook protector on the seam end at the core.

4.4.5 *Dryer Fabric Testing*

The following tests are done on dryer fabrics:

- abrasion resistance
- dry heat resistance
- distortion resistance (shear force)
- weight
- tensile strength

- percent elongation
- modulus
- seam tensile strength
- air permeability
- air carrying capacity
- heat transfer
- thickness (caliper)
- coefficient of friction
- diagonal stability
- hydrolysis resistance
- stiffness
- acid/caustic resistance
- contact area
- edge binder tensile strength
- thermal stability (dimensional)
- dynamic behavior
- neutral line
- contamination tendency and cleanability

Hydrolysis tests are done in autoclaves (Figure 4.33). One type of test is done at 250°F and 1.5 lb gauge pressure which may last between 1–30 days. In a faster test, which is called Parr Bomb test, the sample is left in

FIGURE 4.33. Autoclaves for hydrolysis testing.

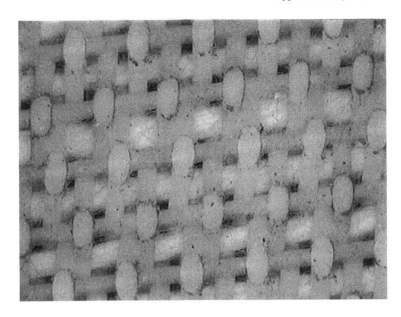

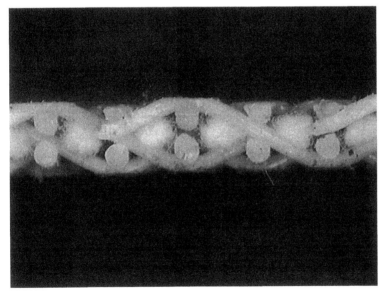

FIGURE 4.34. Photomicrographs of a used dryer fabric (which is the fabric shown in Figure 4.9).

the pressure cooker for 5.5 hours at 325°F. After the tests, the residual tensile strength is measured.

Some of these tests are done on returned fabrics from the field. Figure 4.34 shows the photomicrograph of a used dryer fabric. Figure 4.35 shows the comparison of original and residual tensile strength and air permeability values of a dryer fabric that is run 362 days making uncoated free sheet.

4.5 Applications of Dryer Fabrics

4.5.1 *Installation*

Proper installation of a dryer fabric is the first requirement for a trouble-free performance. Any maintenance work to correct possible misalignment on the machine should be done prior to installation [11].

A careful examination will identify the face

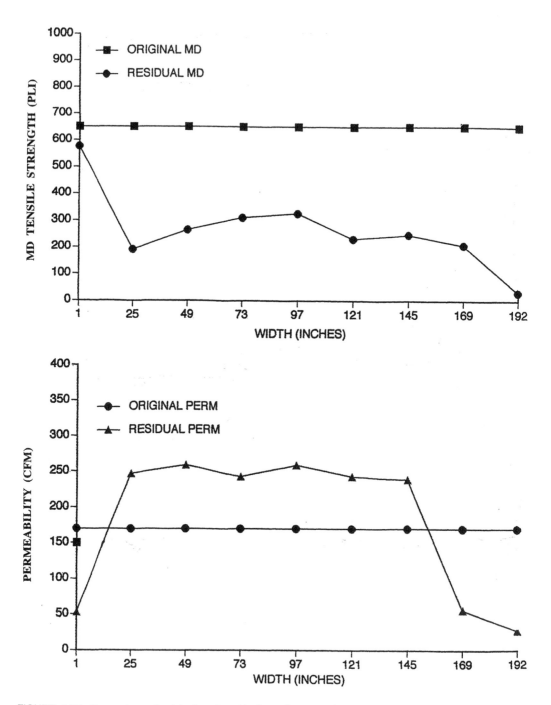

FIGURE 4.35. Comparison of original and residual tensile strength and air permeability values of a typical dryer fabric.

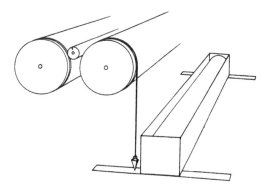

FIGURE 4.36. Use of plumb bob for proper placement of box.

the box onto the machine, proper placement of the box is critical. It must be level, perpendicular to the machine center line, and placed to provide adequate clearance from the machine frame and other obstructions. Figure 4.36 shows a method of using a plumb bob to find the proper placement of the fabric. Once this position has been accurately determined, lines should be painted on the floor to save time on subsequent installations. Once properly in place, the box should not be moved during the rest of the installation.

It is important to use the crawl speed assistance to pull the fabric as far as possible in the section. Therefore, normal running direction has to be chosen and the entry should be at the first "back way" roll (Figures 4.37 and 4.38).

Mostly, old fabrics are used to pull the new fabric onto the position. In this case, the old fabric should be stopped with the seam spotted near the point where the new fabric will be tied on. This serves a twofold purpose:

- It allows observation of seam bow or misalignment so that the proper corrective action can be taken.
- The seam area provides a tougher and

and back sides of the fabric as well as the running direction. If necessary, the fabric is lifted and placed into its saddles so that it can turn freely during installation. When these steps are accomplished, the fabric is ready to be placed into the machine.

Some machines have permanent saddles bolted to the machine frame to hold the fabric during installation. Fabrics for these positions may be packaged in tubes. The machine frame saddles minimize the problem of potential misalignment as the fabric is pulled onto the machine.

If the fabric is to be pulled directly from

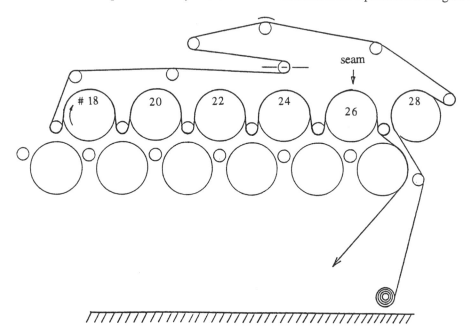

FIGURE 4.37. Example of a top fabric in conventional geometry.

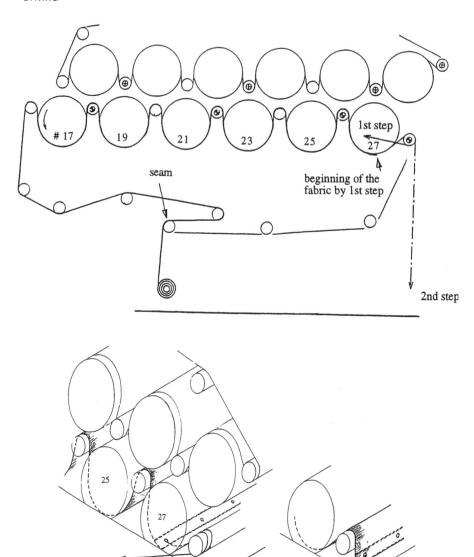

FIGURE 4.38. Example of a bottom fabric with roll, driven apron.

stronger portion of the fabric for the tie-in.

The old fabric should be cut just behind the seam area, and a series of holes should be punched just ahead of the seam area. This allows a series of ropes to be used for the tie-in as shown in Figure 4.39. If the old fabric is bowed, the lengths of the ropes are varied so that the new fabric will be pulled on straight and true. This is most easily accomplished by

tying a set of long lengths of rope onto the old fabric and then tying off each one at the length required to correct for the seam bow.

If the bridle arrangement is to be used, a rope is tied into the center of the old fabric as it is cut. The old fabric is then run off into the basement as it threads the rope through the section. If there is no old fabric on the position, the rope must be passed through by hand. In either case, it is essential that the rope run straight along the center line of the

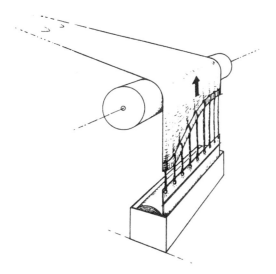

FIGURE 4.39. Tying new fabric to the old fabric.

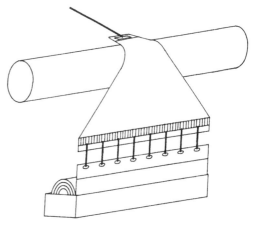

FIGURE 4.41. Lead-on harness arrangement.

machine. Figure 4.40 illustrates a ring bridle arrangement. In using this rig, it is important to make sure that the fabric is pulled onto the machine properly aligned. Each of the ropes attached to the fabric must carry an equal portion of the load.

Some of the potential problems with the ring bridle can be solved by using a lead-on harness as shown in Figure 4.41. The harness consists of a triangular-shaped leader which connects to the fabric leader and has a fringed

area across its width to which the ropes can be tied.

The usual procedure in cases where the fabric box seems to be in the wrong position is to go ahead and pull the fabric completely onto the machine without moving the box and to join the seam even if the fabric is not in the right position and alignment.

During these operations the guide palm has to be turned and tied off the swing arm at the midpoint of its range as shown in Figure 4.42. This will place the guide roll in the center of its range, and the fabric will go on straight. When seaming is completed, the guide palm is turned back to its normal position and untied before the fabric is started up. A newly installed and seamed fabric should be turned slowly (crawl speed) until it has time to completely align itself before any tension is applied ("hanging felt"). This will often eliminate or minimize any errors made during installation and ensure that the fabric will track true and correct upon start-up.

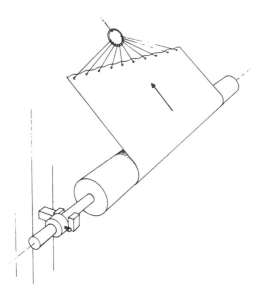

FIGURE 4.40. Ring bridle arrangement.

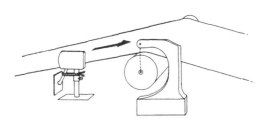

FIGURE 4.42. Centering guide roll during installation.

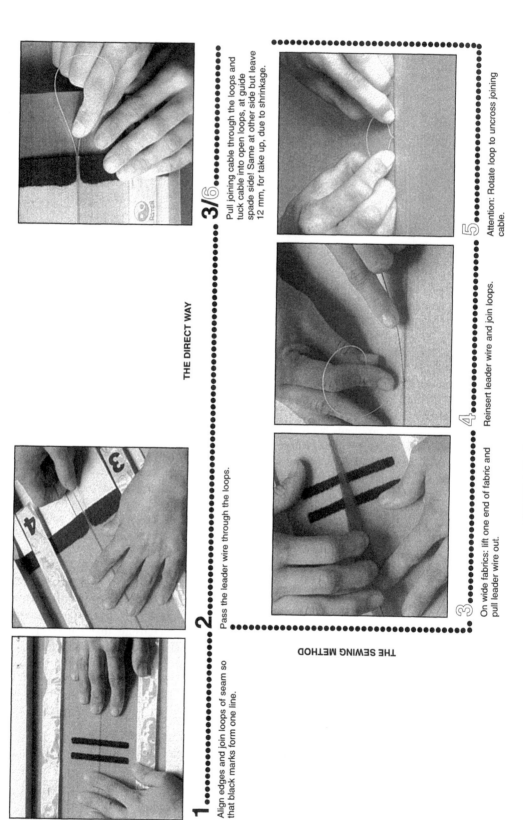

THE DIRECT WAY

1 Align edges and join loops of seam so that black marks form one line.

2 Pass the leader wire through the loops.

3/6 Pull joining cable through the loops and tuck cable into open loops, at guide spade side! Same at other side but leave 12 mm, for take up, due to shrinkage.

THE SEWING METHOD

3 On wide fabrics: lift one end of fabric and pull leader wire out.

4 Reinsert leader wire and join loops.

5 Attention: Rotate loop to uncross joining cable.

FIGURE 4.43. Seaming instructions.

Seaming

Joining the ends of the fabric and installation of the pintle wire in the seam should be done on a good, flat surface if possible. This might be a catwalk or a portable platform which can be inserted in the machine frame at the point where the seam is to be joined.

A flat surface is a better work area, and it makes the joining operation easier. If it is not practical to provide a flat surface, a dryer drum surface is the next most preferable location. A felt roll is usually the least desirable location for joining. As the amount of curvature at the joining point increases, the operation becomes more difficult.

As the pintle wire is inserted some distance into the seam, it may become difficult to push. This difficulty may be overcome by using the "sewing method" (Figure 4.43). This technique consists of pointing the needle out of the seam when progress becomes difficult. The full length of the cable is then pulled into the seamed part of the fabric and a large loop is made with the pintle wire. The needle is then reinserted and progress across the seam continues until seaming is completed or another sewing operation is required.

A zip tape can be used to minimize the problems encountered with seam joining (Figure 4.44). The two ends of the fabric are joined without effort and zipped together over the entire width of the fabric. This removes the stress and strain from the fabric to make the seaming operation easier and faster. Use of a zip tape can reduce the manpower required to join a seam, since it eliminates the need for a group of people trying to hold the seam together across its width as the wire is inserted.

A properly installed seam should have the edges flush and no mismatched hooks. After joining the seams at the factory, the edges are sealed and trimmed to assure matched edges during installation on the paper machine. Observation of mismatched or "left-over" hooks at the back edge is an indication that hooks have been skipped or mismatched as the wire was inserted and this condition should be corrected.

The recommended procedure is to tuck both

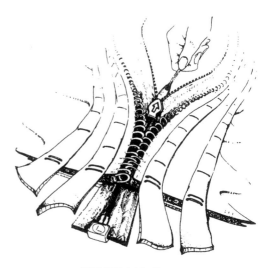

FIGURE 4.44. Zip tape.

edges back into the seam so that it is flush with the front edge where the fabric contacts the guide palm. When tension is applied to the fabric, the joining wire will be held in place securely (Figures 4.45 and 4.46).

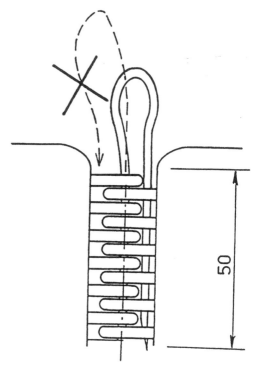

FIGURE 4.45. Step 1, tie back of pintle wire (dimension in mm).

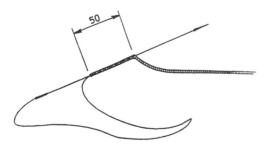

FIGURE 4.46. Step 2, tie back of needle wire (dimension in mm).

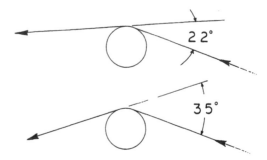

FIGURE 4.47. Increasing fabric contact with the increase of angle of wrap.

4.5.2 *Guiding*

When dryer fabric guiding problems are encountered, the first step should be to examine the seam. The seam usually could provide a key to the underlying causes of guiding problems.

Close attention should be given to front-to-back misalignment. If one edge of the seam is running ahead of the other edge, then there is a high likelihood of misalignment somewhere in the section. Misaligned roll(s) can affect guiding of the fabric. Before any other changes are attempted, the section should be realigned and front-to-back seam misalignment should be minimized. In this case, the lateral displacement of the fabric is cyclical with the running length and this problem is commonly called "oscillation" (independent of guiding system). A symmetrical seam bow, with the edges roughly even and the middle leading, is evidence of other problems which do not affect guiding. If there are oscillations and the seam remains straight, the origin of the problem may be "too early tightened felt" after mounting (not yet aligned before tightening and crashed near the tension roll). If the lateral displacement is not cyclical but happens during a time much longer than one revolution of the fabric, then there must be a guiding problem.

Guiding of dryer fabrics is dependent on friction between the guide roll and the fabric. Anything that increases the coefficient of friction between fabric and roll will improve the guiding action of the roll.

Increasing the diameter of the roll, or covering it with rubber or fiberglass, will increase the friction coefficient and guiding effectiveness. However, these solutions require capital outlays and downtime, which may make them impractical.

Increased fabric tension provides more friction and can help guiding, but there is a limit. Fabric tensions above 10–15 pli can cause other problems.

Friction is directly proportional to the amount of fabric surface in contact with the guide roll. As shown in Figure 4.47, increasing the angle of wrap increases friction and guiding effectiveness. The minimum acceptable angle of wrap is 30°. This wrap should be equally divided on each side of the roll, i.e., the moving direction of the roll must be perpendicular to the bisection of the wrap angle. Experience has shown that 17° into the guide roll and 17° on the outgoing side, as shown in Figure 4.48, will give significantly better performance.

```
angle "IN" = angle "OUT"
```

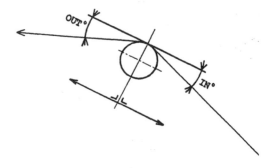

FIGURE 4.48. Splitting the angle of wrap evenly.

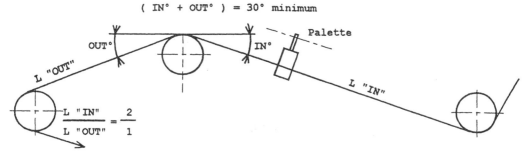

FIGURE 4.49. The 2:1 ratio.

The ratio of lead-in and lead-out distances between the guide roll and rolls just ahead and just following also have a dramatic effect. These distances must be correctly interrelated with angle of wrap for good guiding.

There is a common rule of thumb among papermakers: "the 2:1 ratio," i.e., the lead-in distance should be twice the lead-out distance. This longer lead-in distance allows the guide roll to work with the fabric and steer it. The lead-out roll is located closer to the guide roll to take hold of the correction and maintain it (Figure 4.49).

A second point to consider, although not as important as the 2:1 ratio, is the lead-in distance from the roll just ahead to the guide roll should be between one-third and one-half of the width of the fabric. Machine width is also a factor. On narrow machines the lead-in distance should be one-half the width of the fabric. As machine width increases, the relative lead-in distance may be decreased, and one-third of the fabric width is adequate on the widest machines.

Today's dryer fabrics are constructed to provide a lot of internal resistance to bowing and narrowing. This internal resistance causes the fabric to try to return to its natural path of travel before it was guided. A single lead-out roll may be incapable of holding the correction if it is not wrapped enough. To overcome this problem, a third roll is located after the lead-out roll and can be chosen to be the tension roll. The 2:1:1 configuration is shown in Figure 4.50.

The first lead-out roll should have at least 70° to 90° wrap (then, no third roll is needed). The roll eventually following the lead-out roll

should have as much wrap as practical; 180° wrap on this third roll has proven to be the most effective.

Both oscillation and guiding problems could appear simultaneously. In that case, they have to be solved separately.

4.5.3 *Bowing, Skewing, Narrowing and Distortion*

If all conditions on a paper machine were perfect, the dryer fabric seam would run straight throughout its life. In reality, this seldom happens because of a number of factors. Analysis of the cause and effect of seam bow or skew is a key step to getting the most from a dryer fabric and the paper machine on which it runs. Schematics of a bow and skew are shown in Figure 2.68.

Seam bows and skews have a variety of effects on dryer fabric and paper machine performance, and most of these effects are undesirable. The most notable results are:

(1) Loss in width of the fabric: A severe bow or skew can cause so much width loss that the edge of the sheet will be unfelted. This in turn, leads to a host of paper machine problems and also to premature fabric removal. Table 4.4 shows the amount of width loss for bowed fabrics. For example, a 240 inch wide fabric with a 30 inch long bow will regain 9.8 inches of width when the fabric seam is straightened.

(2) Variations in fabric tension and drying are also caused by seam distortion of the fabric. The tension profile and amount of

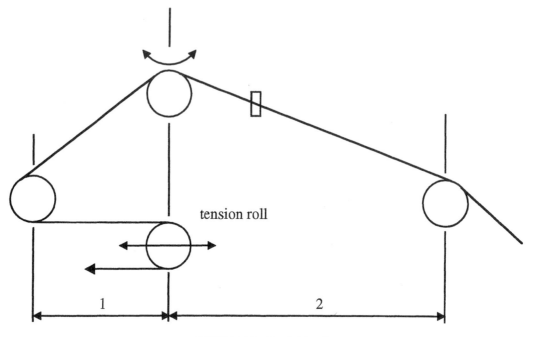

tension roll

1

2

FIGURE 4.50. The 2:1:1 ratio.

variation change with the type of length of seam bow and skew. A fabric with a symmetrical bow runs tight on the edges and relatively slack in the middle. This tension profile tends to add to overdrying of the edges of the sheet and high moisture in the middle. Moreover, this condition places unduly high stress on the edges of the seam. The edges of the seam carry most of the tension load of the entire fabric. If left uncorrected, a symmetrical seam bow can contribute to early seam failure at the fabric edge. A fabric with a front-to-back seam skew has lower tension on the edge that is running ahead. The greatest amount of stress is being applied to the edge of the seam that trails. A seam running with a front-to-back skew leads to uneven moisture profiles in the front and back edges of the sheet and lower overall permeability.

(3) Variations in permeability are characteristics of bowed or skewed fabrics. A distorted fabric will not pass air at the same rate as an undistorted fabric. When woven, a dryer fabric is "square" and the openings in it are uniform in size. This produces a fabric with uniform permeability. If this fabric is distorted as it runs on the machine, the openings change shape and may restrict the passage of air. These variations in permeability can cause nonuniform drying and moisture profile problems.

The major causes of bows and skews and the possible solutions are as follows:

(1) Symmetrical bows are usually caused by roll deflection. The most common causes of roll deflection are high fabric tension (depending on the paper machine construction) and/or old rolls that merely sag or deflect. For new paper machines, the accessories are calculated and constructed for tension up to 20 pli or more. High tensions always improve drying ratio as well as fabric guiding. High tension in old paper machines can cause seam failure, high fabric wear rate, increased likelihood of accidental damage and bearing failure, in addition to roll deflection. Roll diameter differences, due to buildup of stock, rust, or other foreign materials on

TABLE 4.4. Width Regained by Straightening Bowed Fabrics (inch).

Bow Length (inch)	Fabric Width (Measured in Bowed Condition, inch)								
	80	120	160	200	240	280	320	360	40
3	0.2	0.1	0.1	0.1	0.1	0.0	0.0	0.0	0.0
6	1.1	0.7	0.5	0.4	0.3	0.3	0.2	0.2	0.2
9	2.6	1.7	1.3	1.0	0.8	0.7	0.6	0.5	0.5
12	4.7	3.1	2.3	1.9	1.5	1.3	1.1	1.0	0.9
15	7.3	4.9	3.7	2.9	2.4	2.1	1.8	1.6	1.4
18	10.3	7.0	5.3	4.2	3.5	3.0	2.6	2.3	2.1
21	13.9	9.5	7.2	5.8	4.8	4.1	3.6	3.2	2.9
24	17.9	12.4	9.4	7.5	6.3	5.4	4.7	4.2	3.8
27	22.4	15.5	11.8	9.5	8.0	6.8	6.0	5.3	4.8
30		19.0	14.6	11.7	9.8	8.4	7.4	6.6	5.9
33		22.8	17.5	14.2	11.9	10.2	8.9	8.0	7.2
36		26.9	20.7	16.8	14.1	12.1	10.6	9.5	8.5
39		31.3	24.2	19.7	16.5	14.2	12.5	11.1	10.0
42		35.9	27.9	22.7	18.9	16.5	14.5	12.9	11.6
45		40.8	31.8	25.9	21.4	18.9	16.6	14.8	13.3
48		45.9	35.9	29.4	24.1	21.4	18.8	16.8	15.1
51			40.3	33.0	26.9	24.1	21.2	18.9	17.1
54			44.8	36.8	29.9	26.9	23.7	21.2	19.1
57			49.5	40.8	33.1	29.9	26.4	23.6	21.3
60			54.5	44.9		33.1	29.2	26.1	23.5
63							32.1	28.7	25.9
66							35.1	31.4	28.4
69							38.3	34.2	31.0
72							41.5	37.2	33.7

the roll face, can also be a cause of a symmetrical bow.

Replacement of deflecting rolls with either stronger rolls or roll of larger diameter will normally solve a symmetrical bow problem. However, this can be a costly operation. Old rolls can also be removed from the machine and fiberglass or rubber covered to add enough stability to prevent sagging or deflection.

It may be more practical to install concave rolls. The installation of one or two concave rolls will usually eliminate even an extreme symmetrical bow, but it increases the fabric tension at the edges. In a concave roll, the center has a smaller diameter than the edges (Figure 3.32). At any given rotational speed (rpm), the surface speed of any point on the roll varies with the circumference at that point. Thus, the middle has the lowest speed and the edges higher speeds. This causes the edges of the fabric to run ahead to correct the symmetrical bow. Proper placement of a concave roll and the amount of negative crown required are dependent on a complex set of factors. Generally a high wrapped roll gives the best results.

(2) Skews result from rolls that are out of line and/or differences in roll diameter. All the rolls in a dryer fabric run, including all the return rolls and all the pocket rolls, should be parallel to one another and to the dryer cylinders. If any roll is out of line, the seam will react by showing a front-to-back misalignment. The magnitude of the effect of a single out-of-line roll is dependent on the amount of wrap

on that roll. Thus, pocket, hitch, corner and stretch rolls are potential problem points. When a fabric encounters a misaligned roll, the edge that has the shortest distance to travel will begin to move ahead, causing a misalignment skew. Changes in roll diameter can also be the culprit. Rolls may be reduced in diameter by wear or abrasion. An increase in diameter can result from buildup of stock, rust or other foreign materials. If one side decreases or increases diameter more than the other, the result will be a front-to-back misalignment skew.

If buildup of stock or contaminants on roll surfaces is causing misalignment, the rolls should be scraped or cleaned as required. If roll diameter differences are caused by wear or abrasion, roll replacement or recovering may be necessary. If all rolls in the section are in good condition, they may need to be realigned. Theoretically, the ideal solution would be optical alignment which is time consuming, expensive and not always possible. Other more practical and common methods are available and can be used.

Tension force calculations of dryer fabrics are given in Appendix D.

4.5.4 *Yarn Degradation*

All yarns that are used in modern dryer fabrics are polymers, some of whose chemical bonds can be damaged by various agents or combination of agents. Damage to these bonds results primarily in a loss of tensile strength since the yarns become weak and brittle.

Hydrolysis

The most common of these agents is a combination of heat and moisture which is called hydrolysis. Generally, steam pressures greater than 60 psi are required but machines with poor pocket ventilation can experience hydrolysis at lower steam pressures.

Conversely, machines with good pocket ventilation can run higher steam pressures with no serious hydrolysis. This is due to the lower moisture levels in the air. Poor pocket ventilation results in high moisture content of the air.

The most common form of hydrolysis occurs on the fabric edges where the fabric extends out beyond the sheet. Severe hydrolysis, however, can occur across the entire fabric width.

Dry Heat

Dry heat can also degrade yarns. Yarns either can have an additive to provide dry heat resistance or they can be made inherently resistant. The only commonly used yarn with poor dry heat resistance is PCTA.

Chemicals

Caustic degrades polyester and acid degrades nylon. Chlorine degrades both nylon and polyester but nylon is affected the most. Diluted chemicals, such as cleaning agents, can be left on the fabrics for a short time (half hour) provided that the chemical is thoroughly rinsed off the fabric following this exposure. The small amount of fabric damage due to the chemical is far outweighed by the benefits of removing contaminants.

Hydrolysis and Chemical Resistance

Some polymers, due to their composition, are more chemically inert. As a result, yarns made from these polymers are not affected by chemicals, dry heat or hydrolysis. These yarns include Ryton®, Nomex® aramid or PEEK. Acrylic, which is generally used as CD yarn, is in between polyester and these yarns in its chemical resistance. However, these yarns are much more expensive than polyester. Extensive research is being done by dryer fabric suppliers to find a lower cost alternative yarn.

4.5.5 *Fabric Contamination and Cleaning*

Due to intensive use of recycled fibers, dryer fabric contamination has become a major issue in the manufacturing of paper. Production problems such as breaks, holes, sheet flutter, wrinkles and capacity losses are often the result of contamination.

Contamination is mainly composed of adhesives (polyvinylacetate), bitumen, wax, latex and resin. These components are deposited on the drying cylinders, outer rolls and fabrics in the dryer section. Once they reach a certain dimension, debris falls off and is carried on by the dryer fabrics and causes production problems in size presses and coaters.

As contamination increases, the air permeability decreases. In UnoRun positions, the efficiency of stabilizers (UnoRun boxes) is then reduced and the adherence of the web to the dryer fabric decreases. As a result, sheet flutter may occur which may lead to wrinkles and breaks.

Another result of low air permeability is the reduction in drying capacity. Since most paperboard machines run at the drying capacity limit, this has a direct impact on the productivity of paper machine.

Contamination can be reduced in two ways. The first way is to use anticontaminant materials in the fabrics. Teflon® fluorocarbon is an excellent dirt repellent material (Section 4.4.1). The surface of the traditional monofilament polyester fabrics can also be coated with a dirt repellent solution. The second way to control contamination is to use cleaning installations built into the paper machine. The fabric structure should allow easy cleaning.

There are several cleaning devices installed on paper machines. Brushes and doctor blades are mainly used for monofilament fabrics. The surface of a multifilament fabric would be worn out by these types of devices.

Brushes

Brushes can be fixed or rotary. The position and material of the brushes are important. Steel-Teflon® brushes are very efficient but synthetic hair brushes may also be used. When using a steel brush, the fabric should be regularly checked for abrasion.

Brushes have to be installed on the paper side at the height of an inner roll and in the first third of the fabric return, if possible. In top positions, a collecting basin (dirt collector) is additionally required. In lower positions such a collector is not necessarily required since the dirt falls directly onto the floor.

This method can be applied to all paper machines and is not very expensive.

Doctor Blades (Scraper)

Use of doctors is not recommended for paper machines running at a speed over 800 m/min (2625 ft/min). However, the speed of the paper machine can be reduced during the cleaning process.

Doctors can be used continuously or discontinuously. For continuous running, the doctor should be installed in the first third of the fabric return, perpendicular to the fabric, about 40 cm before an inner roll. The scraper blade may be out of steel or synthetic material. Although abrasion is generally not a problem, the fabric, especially the seam should be regularly checked. A dirt collector is required in top positions.

For discontinuous application of doctors, the position should be the same as for continuous application. The angle with the fabric however should not be 90° but only 30°. The cleaning is usually performed once or twice a day and if necessary, more often. The doctor is put in contact with the fabric at four to five rotations and debris created by dirt accumulation are scraped off. In top positions dirt should be collected in dirt basins.

Installations Run by Water/Air

Since fabric contamination has become a major issue in recent years, machine builders have developed cleaning installations operated with air and water.

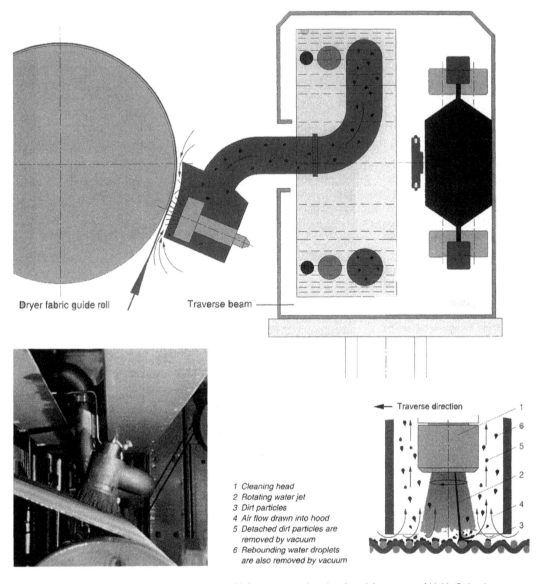

1 Cleaning head
2 Rotating water jet
3 Dirt particles
4 Air flow drawn into hood
5 Detached dirt particles are
   removed by vacuum
6 Rebounding water droplets
   are also removed by vacuum

Dryer fabric guide roll

Traverse beam

Traverse direction

FIGURE 4.51. Continuously traversing high pressure cleaning head (courtesy of Voith Sulzer).

Traversing Nozzle Head

Examples of these systems are EMA-Jet®, Stamm-Jet®, Scan-Jet®, etc. These systems work with air and water nozzles crossing the whole fabric width. They can also be used only on some areas of the fabric. Figure 4.51 shows the schematic of a continuously traversing high pressure cleaning head. Figure 4.52 shows the position of these showers on paper machines. Typical running characteristics of these systems are:

- water pressure: 30 to 60 bar
- air pressure: 6–8 bar
- nozzles: needle nozzles
- number of nozzles: 1 to 9 depending on the manufacturer

To remove or blow out accumulated water,

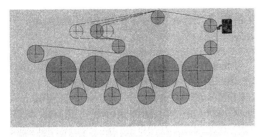

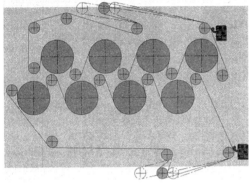

FIGURE 4.52. Position of high pressure cleaning head on paper machines: single tier (top) and double tier configurations (courtesy of Voith Sulzer).

the water nozzles are followed by an air nozzle, the pressure of which corresponds to the paper machine pressure. On the fabric side opposite to the nozzle head, a gutter should collect the accumulated dirt/water mixture.

It should be noted that this kind of cleaning is especially applied for slight contamination. The air nozzle generally does not remove all accumulated water and humidity marks may appear.

Shower-Pipe

These pipes oscillate and are equipped with needle or fan nozzles. The following measures can be considered as standard:

- needle nozzle
- nozzle distance: 100 mm
- distance between nozzle and fabric: 150–200 mm
- water pressure: 40–60 bar

These installations are mostly used together with a chemical treatment which is applied by means of an additional shower-pipe or by

hand. After a soaking time of about 15 minutes, the fabric is sprayed out with shower pipe. Some manufacturers also offer systems with subsequent air nozzles that blow the rest of the water out of the fabric. Accumulated dirt components and water are collected in a basin on the opposite fabric side and are then evacuated.

A disadvantage of this system is the use of large water amounts. Furthermore, steel components of the paper machine may oxidize, which can, however, be avoided to some extent by heating the paper machine during the cleaning process.

Manual Cleaning

The fabric is sprayed out by means of a high pressure water jet when running. Here again, chemical auxiliary means may be used, either by prior soaking or during spraying.

Cleaning Outside of the Paper Machine

The fabric is taken out of the paper machine and cleaned either in the paper mill or at the manufacturer. The dirt is removed from the fabric by use of cleaning agents and is sprayed out under high pressure. The main disadvantage of this method is the high cost and time consumption. However, after such treatment, the fabric may be considered as almost new.

As to the use of alkali cleansing agents, it must be mentioned that polyester is not alkali resistant. High temperatures also have a negative effect. It is therefore of vital importance to thoroughly rinse the fabric after treatment by such agents. When using chemical agents, there are often bad smells within the paper machine hood and adequate measures should be taken. Since cleansing agents mostly have a high alkali content, necessary safety measures should be observed.

A new method of cleaning fabrics uses deep-frozen CO_2. This method is presently being developed and can be applied outside or inside of the paper machine but requires a relatively long time. It takes about 8 hours to clean a fabric of 250 m^2 (2691 sq. ft).

4.6   Dryer Section Configurations

Major developments have been made in dryer section design and equipment to improve sheet runnability and sheet quality. Paper machine speeds with 5,000 feet per minute (1524 meters per minute) and more are now possible through the use of UnoRun and other fabric configurations as well as other supplementary equipment. Specially designed dryer fabrics are required to fully utilize this equipment.

The forming and press sections of modern paper machines had evolved to the point that the dryer section was becoming the major obstacle to attaining higher machine speeds. The areas of concern have been the open draw between the last press and the dryer section, the draws between the top and bottom dryers in the early sections and the open draws in the transfer between dryer sections. Significant improvements have been made in the dryer sections of paper machines in recent years.

4.6.1   *Conventional Top and Bottom Dryer Configuration*

Sheet flutter and billowing in the early part of the dryer section was one of the major obstacles to achieving higher speeds with this configuration which is shown in Figure 4.53. This flutter occurs in the open draw between dryer cylinders and since the sheet is the most fragile at this point in its drying, frequent breaks and edge cracks would result.

4.6.2   *UnoRun*

The disadvantage of conventional dryer configuration has been significantly reduced through the use of UnoRun or serpentine fabric configuration in the first dryer section. The UnoRun configuration was successfully applied in the late 1970s and today it has become the standard. The UnoRun uses a single fabric arranged so that the sheet is completely supported by the fabric throughout the dryer section (Figure 4.54). This

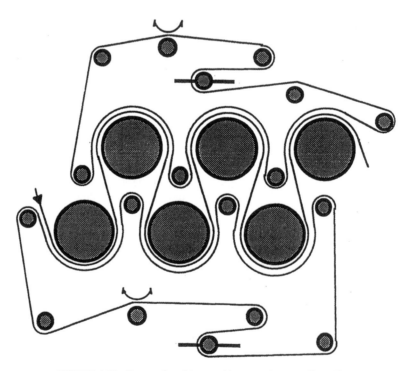

FIGURE 4.53. Conventional top and bottom dryer configuration.

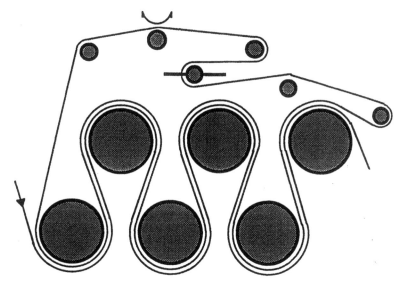

FIGURE 4.54. UnoRun configuration.

configuration has the sheet in contact with the top dryers while the fabric is between the sheet and the dryers on the bottom dryers.

This configuration significantly reduced sheet breaks, improved the sheet edge quality and increased the sheet width. These quality and width increases were due to the sheet being restrained by the fabric thereby reducing the sheet shrinkage that would normally occur.

Some loss in drying capacity results from the use of UnoRun configuration. The amount of fabric wrap is increased on the top dryers but the fabric is between the dryers and the sheet in the bottom position. The insignificant heat transfer from the bottom dryers to the paper through the fabric cannot compensate the loss of direct contact of the sheet with a cylinder, even with higher surface temperatures. In any case, it is not recommended because of the possibility of other mechanical problems. Usually, if possible, these bottom dryers are closed.

The inverse UnoRun configuration (Figure 4.55) was also tried but this brings the same problems added to very hard removal of paper breaks.

Some machines that were configured with the first dryer as a bottom dryer experienced problems with proper adherence of the sheet to fabric. This was due to the sheet not being "nipped" until it passed over the first top dryer. This problem was reduced by installing a "baby" dryer ahead of the first bottom dryer (Figure 4.56). The "baby" dryer provides the nip point for the sheet and the fabric.

Fabric Requirements

The most successful fabric type for the UnoRun is a smooth faced monofilament with a permeability in the 50 to 80 cfm (915 to 1463 $m^3/m^2 \cdot h$) range. The smooth fabric face provides better initial adherence of the sheet to the fabric. The fabric permeability had to be low and very uniform to overcome several forces that were in effect, trying to remove the sheet from the fabric as they traveled together through the dryer section as shown in Figure 4.57.

Figure 4.58 shows the forces acting on the sheet. The first force is an adhesion force between the humid sheet and the polished cylinder surface. The second force is the negative pressure which is caused by the divergence of the dryer from the sheet and fabric (Point A in Figure 4.57). The next force is the positive pressure created at the incoming nip on the bottom dryer (Point C). The positive pressure results from the convergence of the dryer with

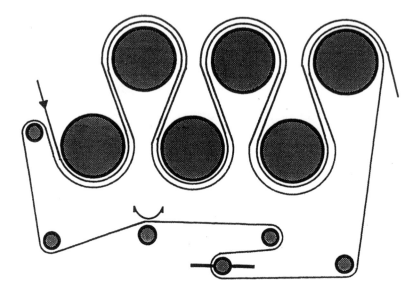

FIGURE 4.55. Inverted UnoRun configuration.

the fabric. A low fabric permeability as well as a low C_A value prevent too much air from passing through the fabric and blowing the sheet off the fabric (Point B).

The last problem force is the positive pressure at the incoming nip of the top dryer (Point D). If the sheet had lifted at any time since leaving the preceding top dryer, this positive pressure will force the sheet to the fabric which would form a bubble of air between the sheet and fabric resulting in sheet wrinkles. Avoiding this problem requires a high fabric permeability to allow the air to pass through the fabric.

We now have two contradictory fabric permeability requirements. Fabrics in the 50 to

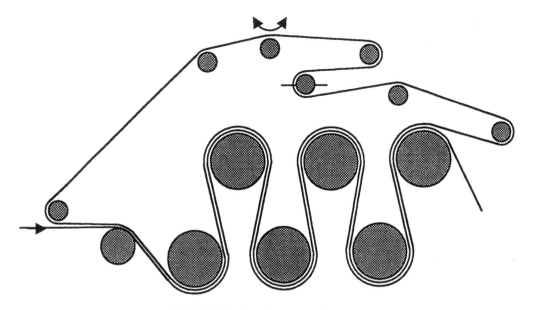

FIGURE 4.56. "Baby" dryer configuration.

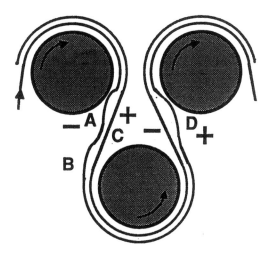

FIGURE 4.57. UnoRun sheet instabilities.

80 cfm range normally will run successfully at speeds up to 2500–2800 feet per minute (762–853 m/min) dependent upon the paper grade. At speeds higher than this there is no fabric permeability that allows a UnoRun to run without either edge lifting on the bottom dryer or bubbling on the top dryer. This problem was eliminated by a device called a blow box.

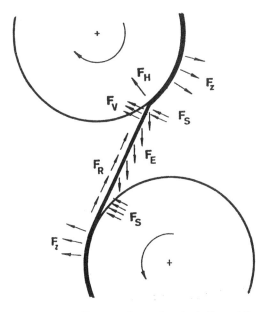

FIGURE 4.58. Forces acting on the sheet. F_z: centrifugal force, F_E: gravitational force, F_H: adhesion force, F_V: vacuum force, F_R: frictional force, F_S: jamming force.

UnoRun with Blow Boxes

UnoRun or serpentine fabric configurations were developed to support the sheet on the dryer fabric during the early phase of its drying. This is where the sheet is the most fragile and when most sheet breaks occur.

This concept worked well until machine speeds increased to the point that the sheet would not stay on the fabric and it would drop off as the sheet went around the bottom dryers. This drop-off was caused by the air pressure that developed at the incoming nip of the bottom dryer. This air pressure developed due to the air carried by the fabric and dryer surface. This air pressure in the nip would blow the sheet away from the fabric. In addition, the adherence at the dryer and the loss of vacuum in the nip may cause lifting of sheet edges. This sheet/fabric separation would cause wrinkles on the top cylinder and edge cracks.

A device called a blow box or sheet stabilizer was developed to block and remove the air being carried into the nip (Figure 4.59). These devices today are full machine width and cover the entire distance from where sheet and fabric leave the top dryer to the nip point of the bottom dryer. These devices direct air to block the air carried by the fabric and dryer. In addition, they create a very slight vacuum on the back side of the fabric.

The placement and air flows of the boxes are critical to the proper operation. Although they are custom made for each machine, in general, the boxes require a minimum of 15 cfm per inch of width air flow with a static pressure of seven inches of water.

The placement should be checked frequently since a paper wad or fabric installation can cause the boxes to move too close and they will rub the fabric causing wear and premature removal. If they are too far away, they will not keep the sheet on the fabric. The air slots should also be checked periodically since these slots can become plugged with airborne debris. Blow boxes are very good runnability assistance devices but they do require periodic maintenance.

The blow boxes have now evolved to provide a negative pressure on the back side of

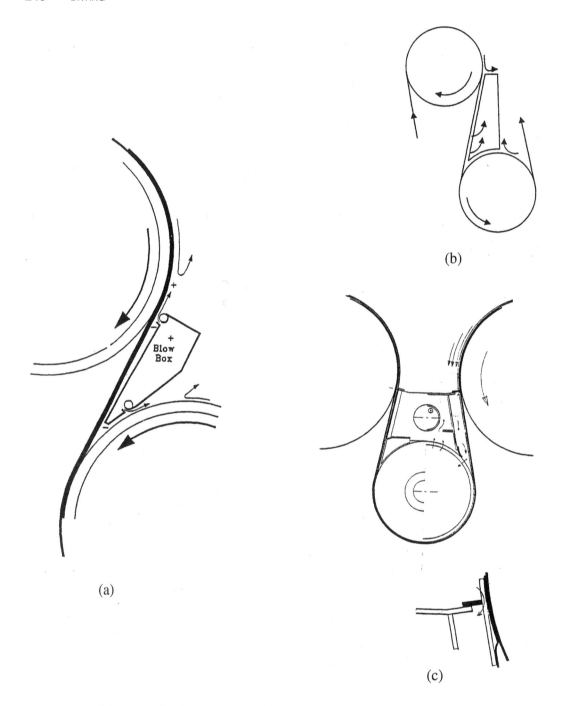

FIGURE 4.59. Schematics of various blow boxes: (a) Valmet system, (b) Andritz system and (c) Voith system.

the fabric for the full length of the draw from the top dryer to the bottom dryer. This full draw box diminishes the effect of the negative pressure at the outgoing nip of the top dryer and the positive pressure at the incoming nip.

These devices use air blowing away from the incoming nip and air blowing as an air doctor to purge air carried by the fabric at the top of the box. The boxes are also positioned closer to the fabric at the incoming (top) end than

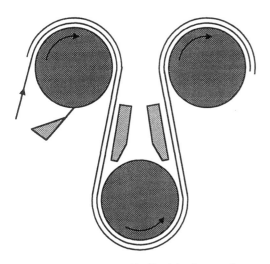

FIGURE 4.60. A modern UnoRun blow box configuration with a second box.

at the nip point at the bottom dryer. This positioning creates a negative pressure. The recent trend is to use a second box opposite to the first one for better sheet stabilization (Figure 4.60).

The use of the blow boxes allows the use of a higher permeability fabric. Fabric permeabilities of 120 to as high as 180 cfm depending upon C_A value, grade and machine, are now being run. The fabric construction is dependent upon machine configuration, speed and the grade being produced. The close placement of these boxes requires a very low bulk seam, normally a pin seam.

UnoRuns are now common and normally included in all new fine and newsprint machines. This configuration is being run on nearly all grades not only for sheet control but also for sheet edge quality improvement. Many mills have seen a reduction in the amount of sheet edge graininess or cockle since the sheet is held by the fabric which retards the sheet edge from shrinking during its drying in these sections.

UnoRun with Press Transfer

The next step in the first section UnoRun development was to extend the dryer fabric

toward the last press roll to reduce the length of the open draw between the last press and the dryer section. If the sheet can be made to adhere to the dryer fabric the sheet will be supported by the fabric through most of this distance (Figure 4.61). The problem had always been to develop a means of maintaining a low vacuum on the back side of the fabric to keep the sheet adhered to the fabric. If any part of the sheet dropped away from the fabric, this separation would result in sheet wrinkles as the sheet was pressed between the fabric and the first dryer.

Any type of a "contact" vacuum box where the fabric came into contact with the box would result in very short fabric or seam life. This short life is due to the heat and wear generated by the fabric dragging over the stationary box.

Many machines are now running with press transfer boxes, developing vacuum using the same basic principles as the full draw blow box. These units blow air away from the fabric and are positioned so that the leading end is closer to the fabric than the trailing end. This positioning creates the negative pressure by the divergence of the moving fabric away from the stationary box.

Sometimes, depending on the paper grade, a short bottom fabric may be used to form a "sandwich" to convey the sheet to the dryers as shown in Figure 4.62.

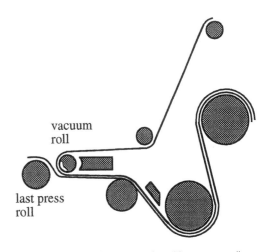

vacuum roll

last press roll

FIGURE 4.61. Press transfer with vacuum roll.

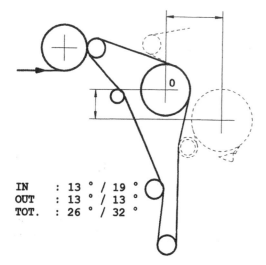

IN   : 13 ° / 19 °
OUT : 13 ° / 13 °
TOT. : 26 ° / 32 °

FIGURE 4.62. Use of a short bottom fabric to convey the sheet.

Multisection UnoRun (Figure 4.63)

Many machines are running multiple sections utilizing the UnoRun configuration. On some machines, the percentage of dryers using the UnoRun is as high as 74%. If the machine has the drying capacity it is only logical to extend the advantages of this configuration as far down the machine as drying capacity will allow.

Another advantage of multiple UnoRun sections is that this allows the use of lick transfers between sections. This configuration eliminates the open draws between sections.

4.6.3 Single Tier

Two recent developments combine many of the previously discussed sheet runnability elements. These designs are the single-tier arrangements. Both of the designs are essentially UnoRun configurations that eliminate the costly and marginally effective bottom dryers.

One design utilizes vacuum rolls in place of the bottom dryers as shown in Figure 4.64. The fabric and sheet leave the dryer and travel around the vacuum roll and back up to the next dryer. The sheet to fabric contact is maintained by a relatively low negative pressure in the vacuum rolls.

The other configuration substitutes large, grooved rolls, 48 inch in diameter for the bottom dryers as shown in Figure 4.65. This configuration uses blow boxes. Recent installations also have holes drilled in the bottom of the grooves, and a small negative pressure is maintained inside the roll.

Both of these concepts were initially utilized in the early dryer sections, replacing what would have been conventional UnoRun positions. These installations proved so successful that this concept was expanded to include entire main dryer sections. These designs heat the sheet only from one side, and the concern for developing sheet two sidedness resulted in a design that alternates the dryer layout as the sheet heating occurs on both sides of the sheet (Figure 4.66).

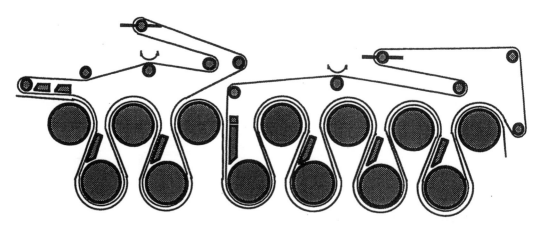

FIGURE 4.63. Multisection UnoRun.

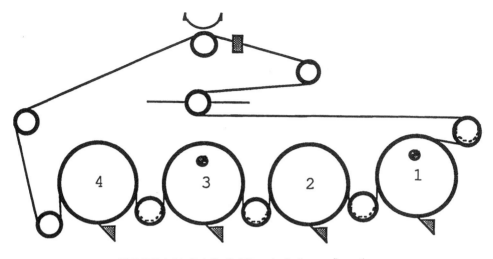

FIGURE 4.64. Beloit's Bel Run single-tier configuration.

These designs also allow no draw transfers between sections. The vacuum roll arrangement utilizes a fabric "sandwich" to vacuum roll transfer (Figure 4.67). The grooved roll configuration utilizes a "lick down" transfer (Figure 4.68).

These configurations provide full length sheet support by the dryer fabrics and have virtually eliminated the positive speed difference previously required through the dryer section. This positive speed difference was necessary to keep the sheet tight in the transfers between sections. Some machines are now operating without any ropes since the fabric accomplishes the sheet transfer even when threading in a tail.

Sheet width shrinkage and edge cockling have been reduced with these designs since the sheet is supported and restrained by the dryer fabric during the drying process.

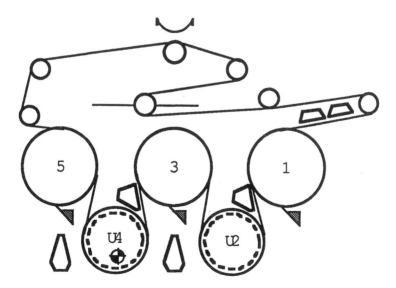

FIGURE 4.65. Valmet's Uno Roll single-tier configuration.

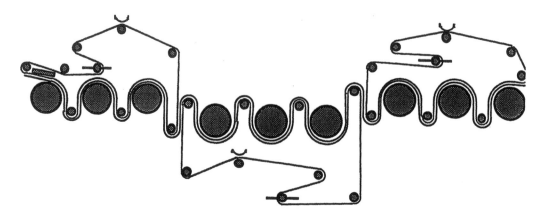

FIGURE 4.66. Beloit's Bel Champ alternating configuration.

Single-Tier Dryer Fabric Requirements

These designs, plus other new or rebuilt conventional fabric configured sections, typically do not drive all the dryers. The dryer fabric acts as a drive belt for those undriven dryers. Therefore, a dryer fabric is required just to run the section.

The sheet and dryer fabrics are in constant contact throughout the drying process. There is the possibility that a dryer fabric or seam could mark the sheet surface resulting in degraded sheet surface quality. A very smooth, nonmarking fabric and seam are essential.

Some very sensitive grades have noted a sheet "welting" or drying difference in the area in contact with the dryer fabric seam on some of the vacuum roll machines. It is now generally accepted that this condition is caused by a higher air permeability in the seam area, primarily in the joint, than the rest of the fabric body. Therefore, seam permeability must be close to body permeability. Fabric permeabilities are typically in the 120 to 180 cfm range and must be very uniform.

Fabric caliper is also a concern since the sheet and fabric are traveling together, and they have to make the transition from bending around the dryer to the opposite bend around the bottom roll and back to the dryer many times. Some machines have noted sheet contact area wear on the dryer fabrics and seams, particularly in the later sections.

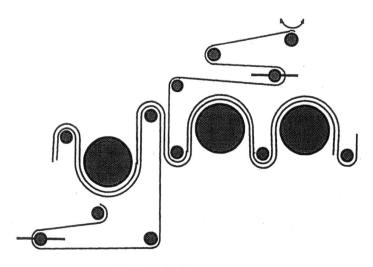

FIGURE 4.67. No-draw transfer.

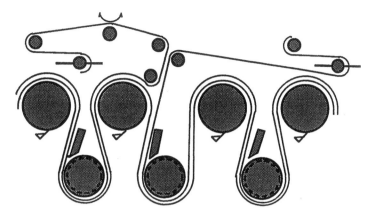

FIGURE 4.68. "Lick down" transfer.

The shrinkage of the sheet in MD which is particularly high during the overdrying time of the paper, reduces its elasticity. The length difference between bending around the dryer and that around the bottom rolls could cause paper breaks. Low fabric caliper and/or asymmetric neutral lines minimize these problems.

Neutral Line

In the positions "single tier" and "Uno-Run," the fabric thickness influences a possible shrinkage or elongation of the paper web, because the paper web has to follow once the shorter distance (contact with the cylinder) and once the longer distance (contact with the fabric surface) as shown in Figure 4.69. Thus, speed differences are the result of the fabric thickness. The point where the speed differences are equal to zero is defined as the Neutral Line.

The position of the Neutral Line is the distance between the back side of the fabric and the point having the same speed in the fabric; the resulting percentage refers to the total fabric thickness. The position of the neutral line can be influenced in all directions by the fabric construction (design, material used).

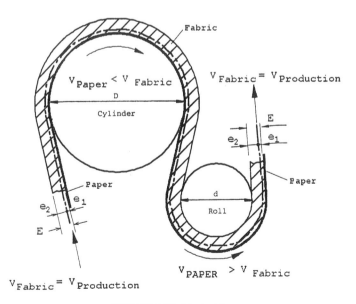

FIGURE 4.69. Neutral line.

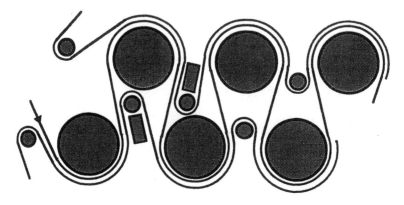

FIGURE 4.70. Short draw transfer between sections.

From the figure

$$E = e_1 + e_2 \qquad (4.1)$$

Percent paper web stretch, Δ_{paper}:

$$\Delta_{roller} = 2e_1/d + 2e_2, \text{ and}$$

$$\Delta_{cylinder} = 2e_1/D + 2e_1$$

$$\Delta_{paper}\% = (\Delta_{roller} + \Delta_{cylinder}) \times 100 \quad (4.2)$$

4.6.4 *Sheet Flutter within a Section— Non-UnoRun*

The UnoRun configuration has been very successful in eliminating sheet flutter and breaks. However, some machines cannot afford the loss of drying or may not run a UnoRun configuration. To eliminate sheet flutter in these machines, trials are now being run using the short draw concept where felt rolls and fabrics runs are modified to shorten or eliminate the open draws between top and bottom dryers within a section [12,13].

One approach is the same as the short draw transfer between sections (Figure 4.70) with the blow box modified to provide pocket ventilation air as well.

Another concept is to utilize two pocket rolls per dryer as shown in Figure 4.71. These rolls are placed so that the preceding felt roll overlaps the succeeding roll thereby providing a closed fabric to fabric transfer.

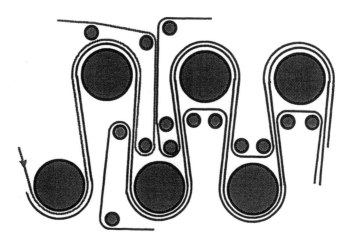

FIGURE 4.71. Two pocket rolls per dryer.

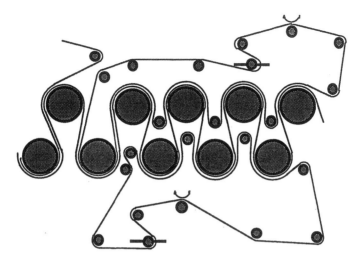

FIGURE 4.72. Pistol grip lick transfer.

4.6.5 *Section to Section Transfer*

"Pistol Grip" Lick Transfer

It is possible to accomplish a lick transfer without having the entire section utilize a Uno-Run. Many machines are running a "pistol grip" UnoRun where only the first bottom dryer is included in a UnoRun (Figure 4.72). The top fabric reverts back to a conventional configuration. The bottom fabric enters the section by wrapping the second bottom dryer. A deflection or "chimney roll" should be used to prevent this bottom fabric pumping too much air into this pocket and causing sheet flutter.

Problems with a slack sheet in the draw between the first top dryer and the second bottom dryer have been experienced on machines running over 2500 feet per minute. This arrangement should be considered only for machines running less than 2500 feet per minute.

Lower permeability fabrics of up to 200 cfm (3658 m^3/m^2·h) are normally required. A blow box at the pistol grip dryer has proven to be beneficial.

Short Draw Transfers

One method to significantly reduce the unsupported length of the sheet in the transfers is the short draw (Figure 4.73). The preceding section lead-out roll is moved down and toward the wet end. The next section lead-in roll is moved up and toward the dry end. The lead-in roll is placed slightly beyond the tangent line toward the dry end to create a slight "S" sheet path.

The rolls are placed three to four inches apart, with a blow box in the top run. The sheet is carried by the top fabric to the short gap where it is transferred to the next section bottom fabric. This has proven to be very effective in eliminating sheet flutter and breaks in these transfers.

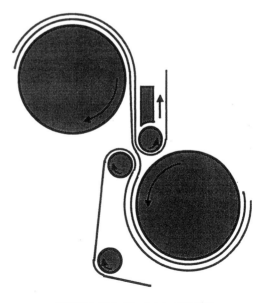

FIGURE 4.73. Short draw transfer.

4.7 References

1 Treece, R., and Deschene, P.F., *Dryer Field Service Training Manual,* Asten, Inc., 1995.

2 Siquet, M., and Stoppelmann, G., "Bestimmung der Mitgenommenen Luft an Trockengeweben in der Papiermaschine," *Wochenblatt fur Papierfabrikation,* 114, 1986, Nr. 23/24 (German).

3 Siquet, M., and Stoppelman, G., "La Valeur C_A, Nouveau Critere de Selection des Toiles de Secherie Determinant Leur Aptitude a un Certain Entrainement D'air," Papier, Carton et Cellulose, October 1986 (French).

4 Siquet, M., Stoppelmann, G., Fonck, R., and Oury, F., "Ein Neues Beurteilungskriterium fur Trockengewebe: Der C_A—Wert zur Bestimmung der Mitgeschleppten Luft," *Wochenblatt fur Papierfabrikation,* 116, 1988, Nr. 23/24 (German).

5 Htun, M., "The Control of Mechanical Properties by Drying Restraints" in *Paper Structure and Properties,* Eds. J. A. Bristow and P. Kolseth, Marcel Dekker, Inc., 1986.

6 Steenberg, B., "Behavior of Paper under Stress and Strain," *Pulp Pap. Mag. Can.,* 50(3), 1949.

7 Schulz, J. H., "The Effect of Straining during Drying on the Mechanical and Viscoelastic Behavior of Paper," *TAPPI,* 44(10), 1961.

8 Parsons, S.R., "Effect of Drying Restraint on Handsheet Properties," *TAPPI,* 55(10), 1972.

9 Ring, J. F. G., Introduction to Pulp and Paper Properties and Technology, Course Notes, The Paper Science Department, University of Wisconsin-Stevens Point, 1991.

10 "Papermakers Fabric with Stacked Machine Direction Yarns," U.S. Patent Nos 5,199,467, 5,343,896, 5,449,026, 5,167,261, 5,230,371, 5,411,062, 5,238,027.

11 Butler, T.A., *Getting the Most Out of Dryer Fabrics,* Asten, Inc., 1996.

12 Edgar, C., "Sheet Flutter Can Be Reduced through Use of Single Felting," *Paper Trade Journal,* January 1977.

13 Sahay, A., "Single Fabric in First Dryer Section Ends Sheet Flutter, Lower Breaks," *Canadian Pulp & Paper,* March 1977.

General References

Barnet, A. J., and Harvey, D. M., "Wet Web Characteristics and Relation to Wet End Draws," *Pulp and Paper Canada,* 81(11), 1980.

Lindem, P. E., "An Instrument to Assess the Biaxial Stress during Paper Drying," *TAPPI Journal,* Vol. 77, No. 5, May 1994.

Palazzolo, S., "Positive and Negative Aspects of Serpentine Felt Drying—Benefits Outweigh Detriments," *Paper Trade Journal,* May 1978.

Palazzolo, S., and Wedel, G., "Advances in Dryer Section Runnability," *TAPPI,* 70 (9): 65, 1987.

Paper Machine Felts and Fabrics, Albany International Corp., Albany, NY, 1976.

Smith, T. M., "Heat Transfer Dynamics," *TAPPI Journal,* Vol. 77, No. 8, August 1994.

Widlund, O.N.G. et al., "Aerodynamics of High-Speed Paper Machines," *TAPPI Journal,* April 1997.

Wilhelmsson, B. et al., "An Experimental Study of Contact Coefficients in Paper Drying," *TAPPI Journal,* Vol. 77, No. 5, May 1994.

4.8 Review Questions

1 What is the cause of yarn degradation due to hydrolysis?

2 Why are some multifilament fabrics resin treated?

3 What causes the sheet to cool between dryers?

4 Where does most of the drying take place?

5 Is a slight contact of the fabric with a blow box or press run box acceptable? Why?

6 Which of the following are the fabric requirements of a UnoRun dryer fabric for a machine running 3600 fpm with blow boxes?
   (a) Smooth surface monofilament
   (b) 90 to 120 cfm
   (c) Soft face fabric
   (d) 50 to 65 cfm

7 Define the following terms
   • monofilament yarn
   • multifilament yarn
   • spun yarn
   • cabled yarn

8 What type of treatments are done on dryer fabrics? Explain.

9 How can dryer fabric air permeability be controlled in all monofilament and all multifilament fabric designs?

10 How do heat and mass transfer take place in paper drying? Explain.

11 Superheated steam at a given steam pressure:
   (a) transfers heat more quickly than saturated steam at a given pressure
   (b) contains significantly more energy than saturated steam at its corresponding temperature

(c) all of the above

(d) none of the above

12 The term "drying rate" is the amount of paper dried per hour per unit area of drying surface. The "drying rate" will be

  (a) negatively impacted by an increase in steam pressure

  (b) positively impacted by an increase in steam temperature

  (c) will be negatively impacted by a reduction in the drying temperature of the paper

  (d) will be positively impacted by a reduction in the drying temperature of the paper

  (e) a and c

  (f) a and d

  (g) b and c

  (h) b and d

13 Which sheet quality defects do high steam pressures accentuate or cause?

14 What are the advantages of UnoRun configuration?

15 Answer the following questions as True or False:

  (a) The most common area of fabric hydrolysis is where the sheet contacts the fabric.

  (b) Caustic degrades polyester.

  (c) Caustic degrades nylon.

  (d) Acid degrades nylon.

  (e) Chlorine degrades polyester.

  (f) Chlorine degrades nylon.

  (g) It is essential that the sheet adhere to the fabric throughout a UnoRun section.

  (h) A pressure is created in the incoming nip of the bottom dryer and fabric in a UnoRun. This pressure tends to force the sheet away from the fabric.

  (i) Typical blow box systems use a high volume vacuum pump.

  (j) Slight contact of the fabric with a blow box or press run box is acceptable.

  (k) Seams are always installed after final heatset.

  (l) Polyester has poor dry heat resistance.

  (m) The temperature at which water evaporates is not affected by air velocity.

5

Paper Machine Auditing

Many paper machine clothing manufacturers offer services to paper companies related to operation and performance of fabrics. The main purposes of a paper machine audit are to gather and analyze operational data that may be used by the paper company to benchmark operations, troubleshoot operational problems and find opportunities to improve machine performance. A typical paper machine survey may take a few days and usually consists of several stages as explained below (Figure 5.1). Examples from actual measurements are given throughout the chapter to demonstrate the type of information that can be gathered from paper machine auditing. It should be emphasized that paper machine auditing is machine condition specific and therefore each audit may be different than the other. Grade structure may also affect the audit results.

5.1 Stock Approach and Pulsation Studies

A pulsation study is performed to evaluate the stability and performance of stock approach piping and rotating elements. This evaluation is then compared to the on-line basis weight signal to determine if any approach piping instability is detrimental to the overall basis weight.

The more accurate and precise the information provided, the more accurate the pulsation study will be. To insure accurate results, several key pressure tap locations must be provided. These locations must be available before a detailed pulsation study can be performed.

- before and after cleaner supply pump
- before and after headbox supply pump
- before and after pressure screens
- before and after attenuator
- headbox backside and front side
- headbox recirculation
- couch vacuum

It is possible to measure two locations from one tap, such as after headbox supply pump and before pressure screens.

Figure 5.2 shows pressure tap locations. Taps should be located at least one pipe diameter away from any elbows, pump, screens, valves or other pipe line obstructions if possible and should have a minimum of 12–18 inches of clearance beyond the end of the valve. The preferable tap size is 1 inch NPT; however 1/2 or 3/4 inch is also acceptable (Figure 5.3). Larger diameter taps can also be utilized if reduced to 1 inch.

Pump and screen speeds as well as any gearbox ratios will be needed during the analysis. The number of vanes on the pump impellers and the number of foils on the screen rotors will also be needed. It would also be helpful to know if either the pump or screen design is specified as low pulse.

FIGURE 5.1. A service engineer conducting wet-end survey.

Access to the raw analog basis weight signal is required during the study. This signal is acquired before the system computer and will involve no proprietary information. The signal is interfaced with an instrument that has several gig ohm input impedance and will not affect the measurement signal.

Using a real time Fast Fourier Transform (FFT) analyzer and other special instrumentation, the following are analyzed during stock approach and pulsation survey.

- high frequency analysis: 1–200 cycles per second
- low frequency analysis: DC—2 cycles per second
- thin stock consistency variation
- cross correlation between stock approach pulsation/vibration and consistency versus basis weight are checked for both frequency ranges
- fan pump and pressure screen speed stability

- fan pump and pressure screen mechanical performance evaluation
- head box vibration
- accepted piping practice recommendations
- attenuator performance evaluation (where applicable)
- stream flow valve position control (where applicable)

Pulsation studies are applicable to any paper grade on a particular paper machine. However, it is recommended that a critical paper grade be selected for the study on that machine.

Figure 5.4 shows the low frequency signal correlation between the basis weight backside, fan pump in, screen #1 out, screen #2 out, screen inlet and fan pump out locations of a typical stock approach system shown in the figure. These signals are conditioned through a 0.25 Hz low pass filter and recorded over an eight minute running time. The information

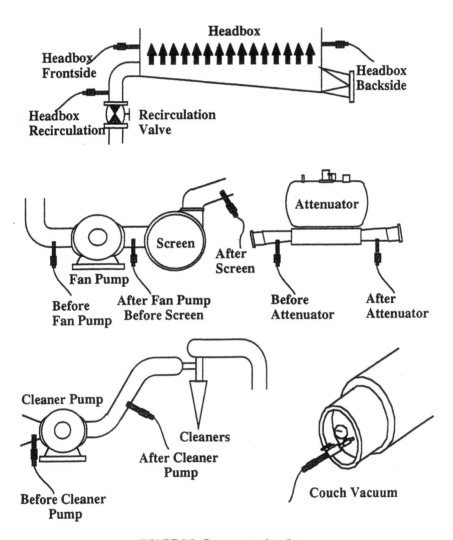

FIGURE 5.2. Pressure tap locations.

on the left side of the graphs presents the mean value and engineering units and the information on the right side of the graphs presents the peak to peak information in the same engineering units. All of the pressure signal scales are presented on the same scale of ± 7 inches of H_2O with the exception of the fan pump out. The scale on the fan pump out is ± 15 inches of H_2O.

Figure 5.5 shows the raw basis weight front side spectra and the synchronous time averaged basis weight front side spectra averaged against the pressure screens and fan pump for the system shown in Figure 5.4. Table 5.1 lists the measured pressure levels at different locations for the same stock approach system.

5.2 Forming Section Analysis (Wet-End Performance)

The purpose of this audit is to evaluate wet-end performance to highlight any conditions that may be detrimental to sheet formation, water removal or clothing performance. Headbox delivery, table activity, trim squirt efficiency and couch sheet release are among the items that are inspected.

Utilizing a Doppler laser sensing device, it is possible to measure the headbox jet velocity readings across the profile of the slice to highlight variations that may be contributing to fiber polar angle anomalies and hence promote curl, formation and profile concerns.

Using a gamma gauge, data are collected to determine table drainage profile and wet-end water balance. Other tests may be conducted as needed to inspect fabric drive loads and fabric wear.

High speed video camera and/or still photography are used to photograph the following:

- headbox/slice delivery to check for bubbles or flow irregularities

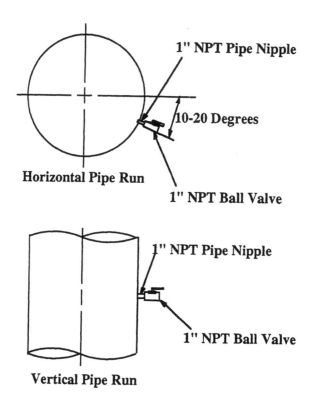

FIGURE 5.3. Specifications for pulsation pressure taps.

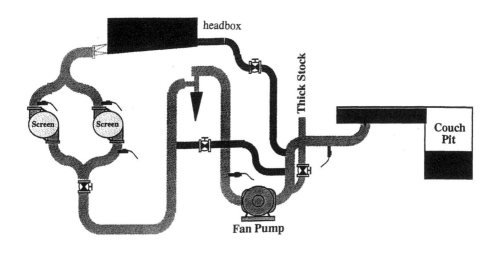

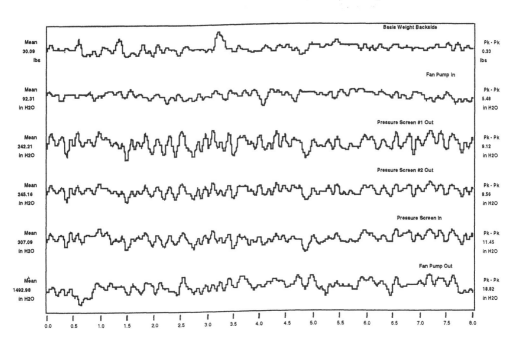

FIGURE 5.4. Low frequency signal correlation of a typical stock approach system.

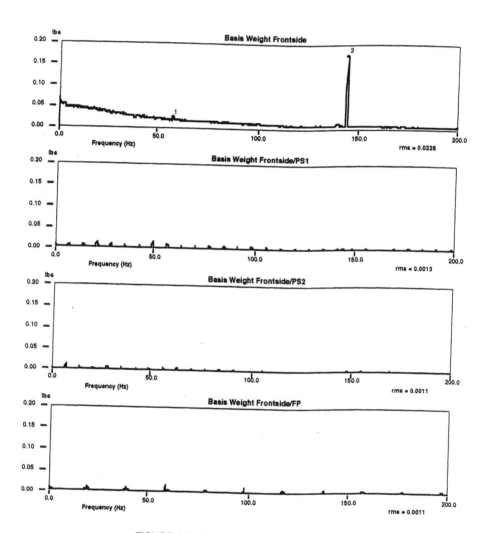

FIGURE 5.5. Basis weight front side spectra.

TABLE 5.1. Stock Approach Pressures of the System Shown in Figure 5.4.

Tap Location	Pressure (psi)	Pressure (H$_2$O)
Fan pump in	3.3	92
#1 Screen out	8.7	242
#2 Screen out	8.8	245
Screen in	11.1	307
Fan pump out	53.86	493

- elemental table activity looking for stock jump or dead areas
- trim squirts to determine cleanness of cut
- sheet release from couch

In addition to these tests, white water consistency tests and first pass fiber and filler retention tests may also be required. It is preferred that a critical paper grade is selected for the wet-end performance test of any paper machine.

Figures 5.6–5.8 show measurements of percent headbox flow drained, stock flow versus drainage and percent consistency for a fourdrinier machine, respectively.

TAPPI Technical Information Sheet, TIS 0502-12 Drainage Evaluation by Mass Measurement, describes the method of consistency measurement and calculations needed to develop an accurate drainage profile.

5.3 Press Section Evaluation

In press section, vibration and press operation analysis are done for any grade of paper.

5.3.1 Vibration Analysis

Vibration analysis is done to determine press section roll and fabric operational health to highlight potential operating concerns. Using a real time FFT analyzer, the following data are collected:

- time signature of each press section roll

- frequency spectrum of each press section roll
- time signature of each wet felt

Velocity sensors measuring speed (in inch/second) and Linear Voltage Displacement Transducers (LVDT) measuring displacement (in mils, 1 mil = 1/1000 inch) along with a real time FFT analyzer are combined to gather data. The Synchronous Time Averaging (STA) technique utilizes once per revolution trigger signals to isolate the contributions to vibration caused by individual components. Photo-optic sensors are used as triggers on the fabrics and the LVDTs are used to measure actual nip displacement. These tools can aid in isolating roll problems such as corrugation, cover anomalies, misalignment, bent shafts, looseness and unbalance. Correcting these problems can assist in gaining better fabric life, improved paper quality, uniformity and improved operating efficiency.

From the data collected, the following analysis is done:

- critical bearing frequencies (potential failures)
- roll alignment and balance
- roll surface conditions
- fabric related vibration components

The following information from the paper mill is essential for this evaluation:

- roll diameter and bearing information
- bearing listing (bearing type, complete manufacturer code, location of bearings)
- list of driven rolls and their drive train architecture (two gears, four gears, etc.)
- dates of most recent equipment and supply changes on presses (rolls, fabrics, etc.)
- reflective tape on critical roll journals to use as a "trigger"

The nomenclature of Table 5.2 is used in the discussion of the vibration data. An industry accepted practice of 2.0 mils or 0.3 inch/second peak-peak vibration at a specific frequency is considered significant.

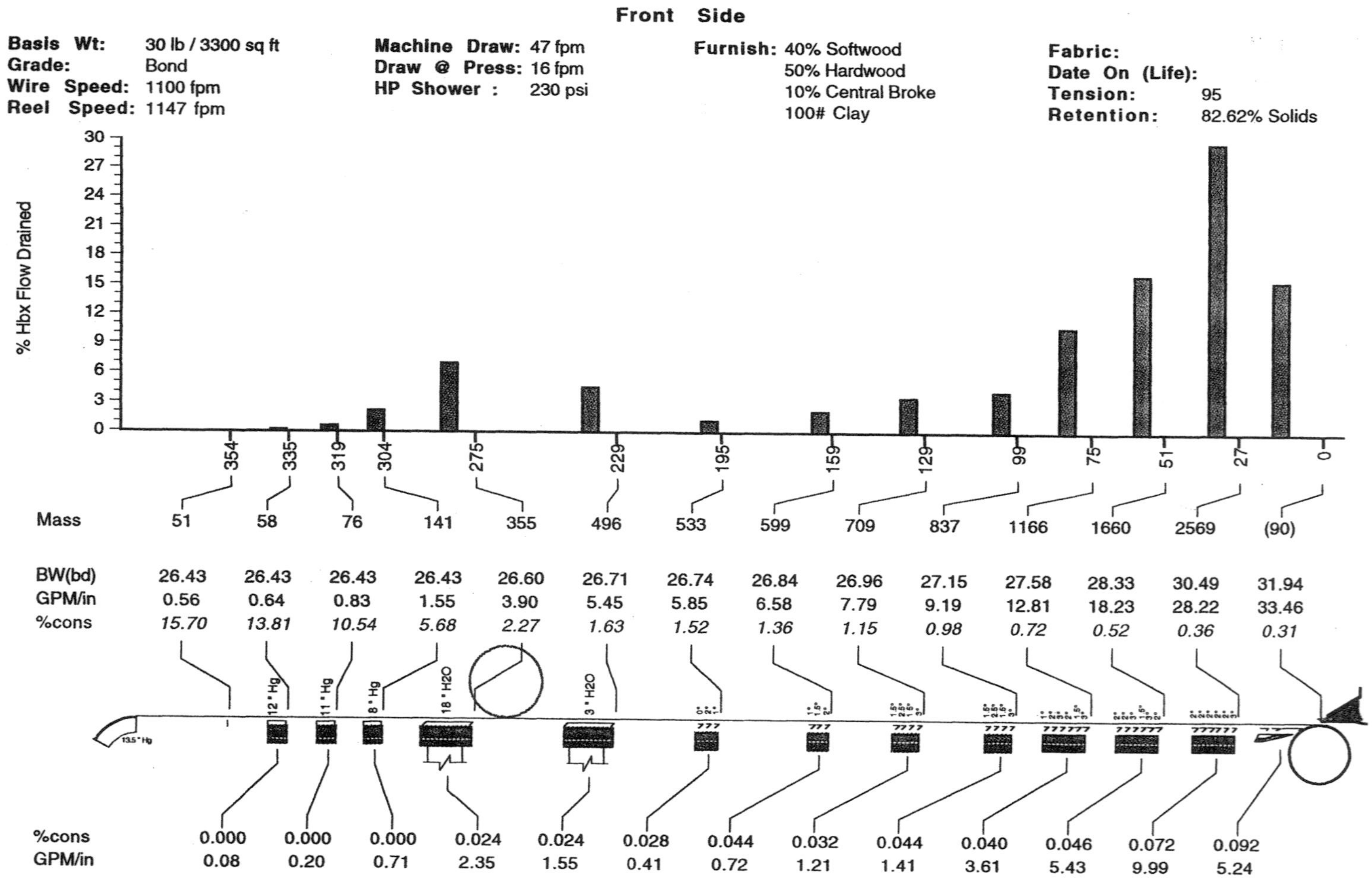

FIGURE 5.6. Percent headbox flow drained on a fourdrinier machine.

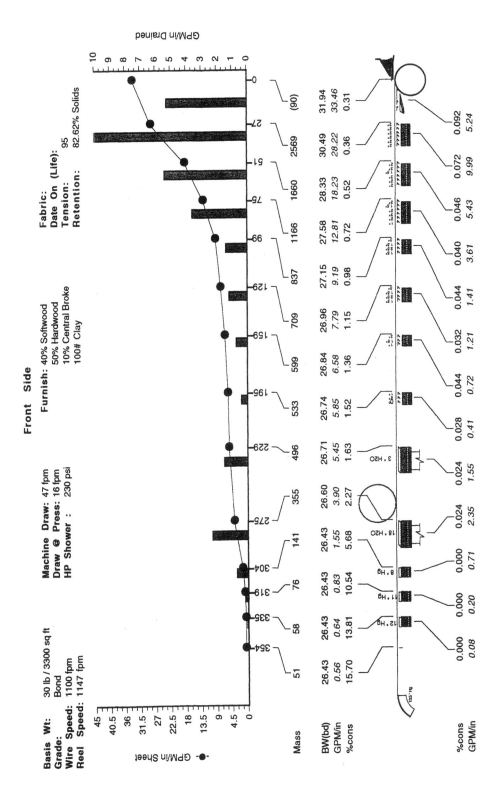

FIGURE 5.7. Stock flow versus drainage on a fourdrinier machine.

267

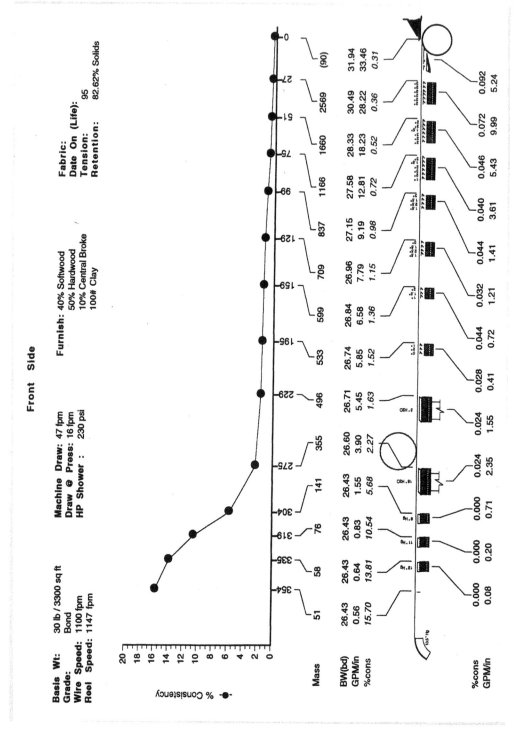

FIGURE 5.8. Percent consistency on a fourdrinier machine.

268

TABLE 5.2. Vibration Nomenclature.

Severity	Explanation	Amplitude
Acceptable		Less than 0.25 mils
Minor	No action needed	0.25–0.75 mils
Moderate	Should be monitored for degradation in condition	0.75–2.0 mils
Significant	Should be removed or repaired during next down	2.0–4.0 mils
Excessive	Should be removed or repaired at earliest opportunity	over 4.0 mils

Press Roll Trigger Placement

During the vibration study, physical movement is collected from sensors mounted on the roll bearing housings and press frame and converted to computer signals for analysis. For the analysis to be meaningful, the specific vibration contribution of each roll and felt in a nip must be known. The process for isolating these signals relies on the ability to sense each revolution of the roll or fabric (known as triggering). Currently, light-based sensors on the fabrics are used to pick up the tradeline. For a roll, reflective tape needs to be placed on the head for the sensors to effectively trigger.

Table 5.3 lists the acceptable locations for the placement of reflective tape. It should be noted that extra sections (coaters and calenders especially) take extra time and should be scheduled only if the mill has a specific vibration problem in that area. For example, tracing a vibration problem in a coater usually takes longer than completing an entire press vibration survey.

Figure 5.9 shows the acceptable locations

TABLE 5.3. Locations for the Placement of Reflective Tape.

1. Press vibration survey
   - major press rolls
2. In conjunction with machine audit
   - major press rolls
   - wire turning, lumpbreaker and couch rolls
   - breaker stack and size press rolls
3. Other areas (per mill request)
   - coater backing roll and paper rolls in coater section
   - calender stack rolls

for placement of reflective tape. The tape needs to be placed in only *one* of these locations. This tape needs to be only about 1/2 inch square and should be placed in a location that is away from bolt heads and dirty areas as much as possible. The tape needs to be stuck on with SuperGlue. Experience has shown that anything less than SuperGlue will not hold up and even SuperGlue will only stay on the machine for a couple of weeks.

Figures 5.10–5.12 show the importance of triggering roll revolutions. These graphs show a vibration signal plotted against time for press rolls in a nip. The rolls are rotating close to the same speeds (210 rpm versus 213 rpm) which makes the resulting signal in Figure 5.10 very confusing. Figure 5.11 contains the same data triggered against the top roll and shows that this roll is in "poor" condition. Figure 5.12 contains the data triggered against the bottom roll and shows that this roll has moderate problems without much significance.

5.3.2 *Press Operations*

This analysis is done to characterize fabric performance and general conditions of the press section. Using Scan Pro, high speed photography, hot wire anemometer, caliper gauge and gamma gauge, the following data are collected:

- felt water load
- felt moisture profile
- felt caliper
- uhle box air flows
- relative sheet consistency into and out of press

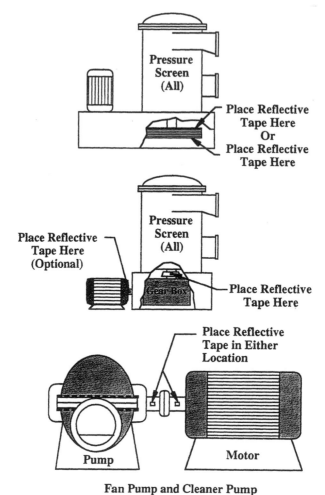

FIGURE 5.9. Reflective tape placement.

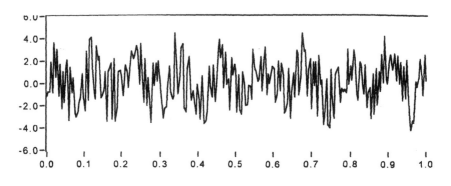

FIGURE 5.10. Raw signal.

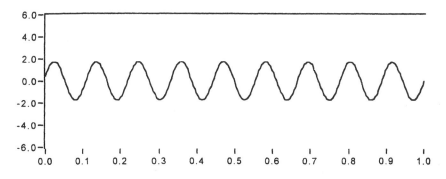

FIGURE 5.11. Top roll signal.

• proper positioning and penetration of high pressure showers.

Felt water removal and press nip anomalies such as skewed press loadings, crown/load concerns and section performance trends are evaluated using a Scan Pro. Vacuum system survey is performed to evaluate the performance of pumps and uhle boxes. Press water balance is constructed using a gamma gauge.

Figure 5.13 shows the typical moisture profiles of the first and second presses of a paper machine. Table 5.4 shows the press section water balance analysis. Figure 5.14 shows the total mass profile after the second press for the same machine.

5.4 Dry-End Performance

Dry-end analysis generally consists of four major parts.

5.4.1 Determining Balance of the Hood and the Effect of Exhaust Loading and Air Filtration

Using a vane anemometer and relative humidity analyzers, mass flows of hood exhaust and supply plenoms are measured. In addition, all equipment conditions are inspected.

Air temperature, relative humidity and barometric pressure measurements are recorded for the hood exhaust fans. A calibrated, digital readout pitot tube assembly is used to measure static and dynamic air pressures in all exhaust plenums. These static and dynamic pressure readings are then converted to air velocity readings. The data are calculated to determine the actual air permeability (cubic feet per minute, cfm), mass of dry air per minute, air density, absolute humidity, grains, dew point and mass of water vapor per minute. The dew point is the temperature at which water will condense from the air. Grains is a

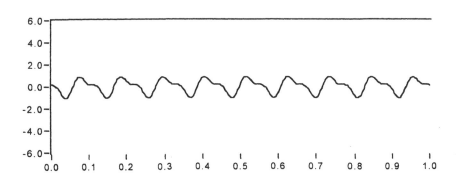

FIGURE 5.12. Bottom roll signal.

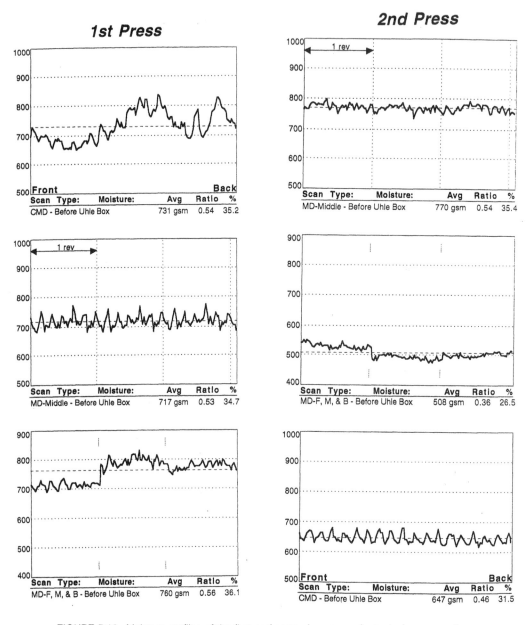

FIGURE 5.13. Moisture profiles of the first and second presses of a typical press section.

TABLE 5.4. Press Section Water Balance Analysis.

NDC Gamma Gauge

Mill: Location:

Machine: Date:

Grade:	30# Bond
Wire Speed:	1100
1st Press	1116
Reel Speed:	1147

			Vacuum - In. Hg	
Press Loading			Roll	Uhle
1st Press:	410	410	16	10
2nd Press:	400	440		9

| Basis Wt. | Size Addition | Moisture | Dry Weight |
| 29.7 | 6.00% | 5.30% | 26.44 |

	Mass Readings #/1000 Sq.Ft.		Mass Change #/1000	% Solid
Dry weight of sheet	8.01			
Weight On Wire	51	Before Couch (Drainage Study)		15.7%
		Water Removed at Couch	0	
1st felt Before Press	331			
1st w/Sheet Before	382	Sheet & Water	51	15.7%
1st Felt after Press	350	Water Gained by 1st	19	
		Water Removed From Sheet	22	
Sheet After 1st	29			27.6%
2nd Felt Before	317			
2nd Felt After	325	Water Gained by 2nd	8	
		Sheet Water loss	8	
Sheet After 2nd	21			38.2%

Results:

	Consistency	Water Removed		Press Water Flow		
			Gal/Min			
Before Couch	15.7%	at Couch	0.0		Gal/Min	
Off Couch	15.7%			Felt	28.8	63.3%
		at 1st Press	33.4	Rolls	4.6	10.0%
Out Of 1st	27.6%		73%	Felt	12.1	26.7%
		at 2nd Press	12.1	Rolls	0.0	0.0%
Out of 2nd	38.2%		27%			
		Press Total	45.50			

measurement of moisture in the air. There are 7000 grains per pound of water. The grain loading is calculated by multiplying the absolute humidity by 7000. ACFM is the actual cubic feet per minute of air in each duct. The air flow rate (in acfm) is calculated from air velocity readings and the duct size corrected for the air density at a given barometric pressure. This value is used to calculate the hood balance. Table 5.5 shows an example of typical hood performance results.

5.4.2 Determining Dryer Section Performance and Efficiency

Using a platinum RTD probe and process readings the following is established:

- TAPPI drying rate
- U-factor
- drawdowns
- steam consumption
- dryer temperature

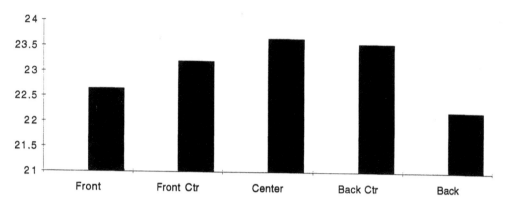

FIGURE 5.14. Total mass profile after the second press (%).

The dryer can surface temperature can be measured with a noncontact pyrometer. Figure 5.15 shows an example of the difference between the dryer surface temperature and saturated steam temperature (ΔT) at a given steam pressure in the dryer can. This difference should not be greater than 50–60°F. Although $\Delta T = 60°$ is not uncommon throughout the industry, there are dryers running at the 40–50°F ΔT levels. It is recommended that dryers with ΔTs greater than 60° be investigated. High temperature differences point out that there may be a problem with the heat transfer from the dryer can to the sheet. Some of these problems may be condensate removal, excessive blow through (i.e., too high a differential pressure), buildup on dryer cans, internal scale buildup, siphon clearance or broken siphons, or too low a differential pressure.

From the humidity surveys and fabric condition surveys, it has been shown that there is an increased potential for dryer fabric hydrolysis when the conditions of the exhaust from a dryer hood possess air temperatures of at least 180°F and an absolute humidity level of 0.12 lb H_2O/lb dry air or higher. Figure 5.16 shows a hypothetical potential hydrolysis graph for a dryer section in which the air temperature and absolute humidity of the exhaust are well below the critical levels.

Steam and condensate studies can be performed to analyze the condensing rates of various steam sections to compare them to theoretical and standard values by grade. This is useful to highlight the drying inefficiencies related to the physical condition of drying

elements such as scale buildup, siphon size/clearance inadequacies and undersized steam lines.

5.4.3 Determining Drying Impacts of Pocket Humidities and Air Velocities

With pocket humidity equipment, the relative humidity, temperature and air velocity of each dryer pocket are measured. The air temperature and percent relative humidity are usually converted into absolute humidity results for each location. The absolute humidities are used as the basis for the evaluation of the drying conditions in each of the pockets.

Dryer pocket humidity measurements can be used to evaluate evaporation from the sheet, as well as the performance of the pocket ventilation system and the effect of the dryer fabric permeability. The humidity results are an indirect measure of the drying rate in a section and are a direct measure of the pocket ventilation efficiency.

It is generally accepted that the absolute humidities in the dryer pockets should be kept at a level below 0.20 lb of water per lb of dry air. If the absolute humidity rises above this critical level, the efficiency of the drying process is reduced. The pocket humidity levels are affected by several factors including the basis weight, steam pressure, dryer fabric tension, ventilation in the pockets, machine speed and air permeability of the dryer fabrics. Figure 5.17 shows an example of humidity measurements in a typical dryer section. As can

TABLE 5.5. Example of Hood Performance Results.

Main Hood

Duct ID	Air Temp. (°F)	Rel. Hum. (%)	Abs. Hum.	Humidity (Grains)	Dew Point Temp. (°F)	Air Velocity (fpm)	Duct X-Sect. Area (Sq.Ft.)	Air Flow Volume (acfm)	Mix Density (lb/cu ft)	Lbs. D.A./min	Lbs. W.V./min
Exhaust #1	125.00	63.20	0.058	409.42	108.65	3096.50	21.229	65,736	0.063	3,948	230
Exhaust #2	127.00	49.10	0.047	330.14	101.92	2776.80	21.229	58,949	0.063	3,588	169
Averages	126.00	56.15	0.053	370	105.3			62,343	0.063		

After Hood

Exhaust #3	118.00	51.00	0.037	264.22	95.00	2744.80	18.189	49,928	0.065	3,132	188

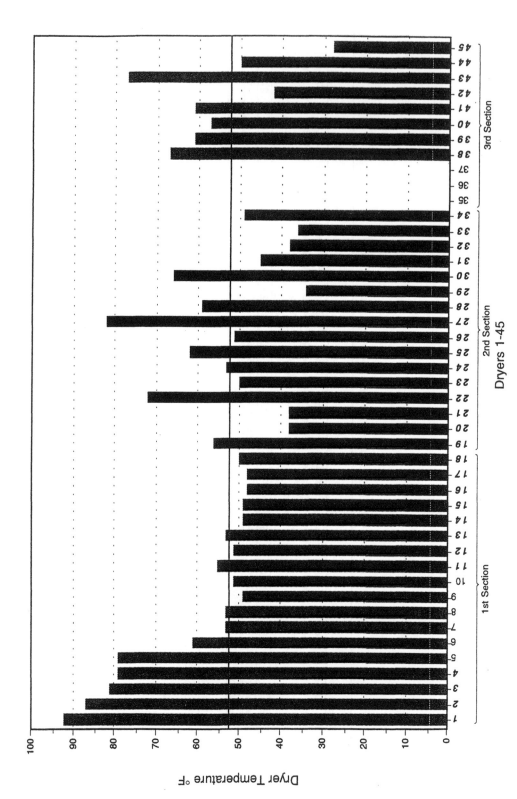

FIGURE 5.15. Example of the difference between dryer temperature and steam temperature (ΔT).

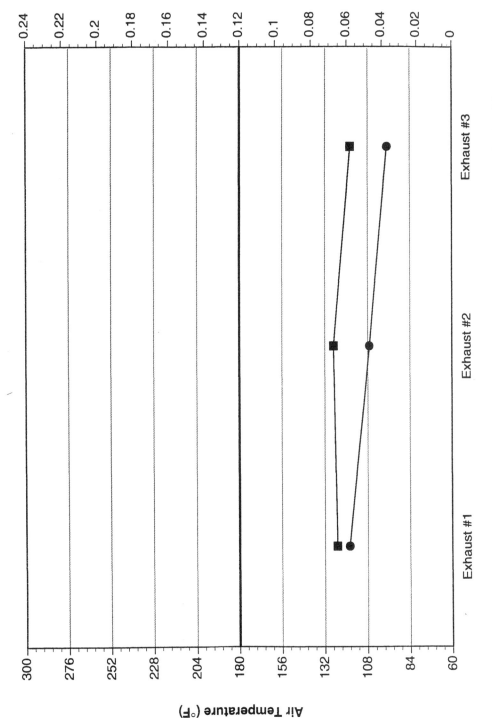

FIGURE 5.16. Example of potential hydrolysis conditions: exhaust air temperature versus absolute humidity.

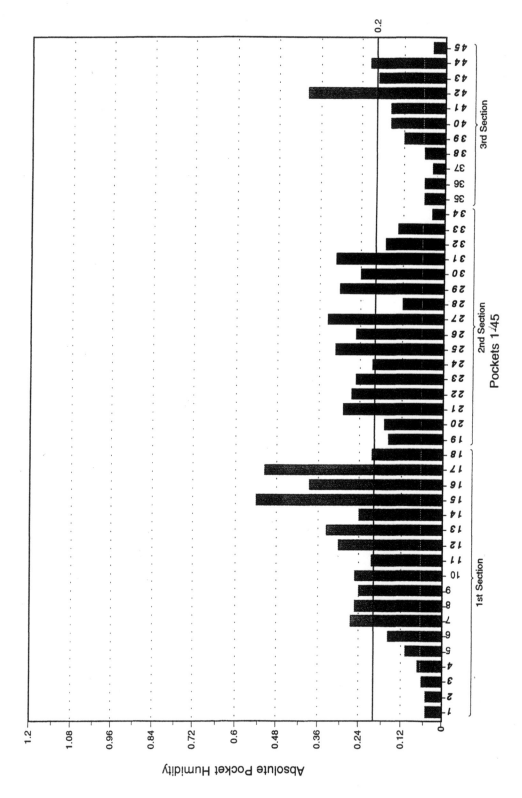

FIGURE 5.17. Example of a typical pocket absolute humidity.

278

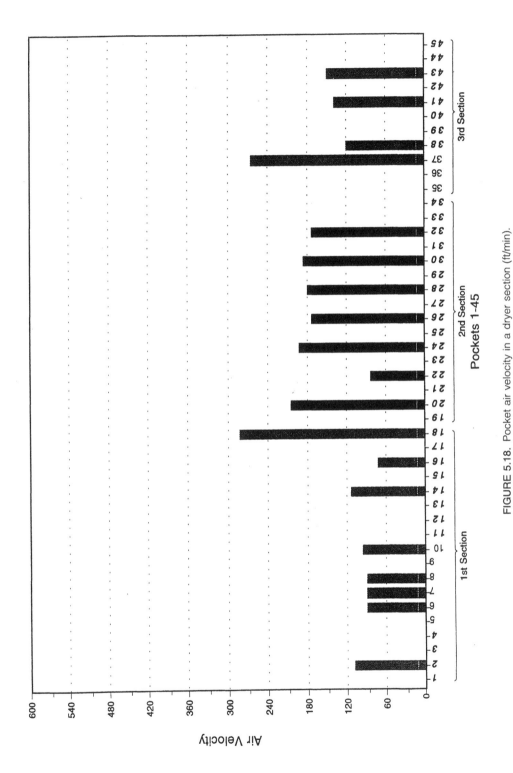

FIGURE 5.18. Pocket air velocity in a dryer section (ft/min).

279

be seen from the graph, the pocket humidities in the first section rise above the critical humidity levels starting at dryer number 7. The humidity levels remain above the critical humidity level of 0.20 lb water per lb of dry air throughout the remainder of the section until dryer number 31. The only exception is at the section transfer and at dryer number 28.

A closer review of the pocket humidity data reveals a trend that is commonly seen on conventionally clothed machines. The pockets formed above the bottom dryers nearly all had higher humidity levels than the pockets below the top dryers due to less air movement.

In order to ensure evacuation of the moisture filled air from the centers of the dryer pockets, it is generally necessary to have air velocities of 250 fpm or more exiting the machine. In areas of especially high pocket humidities, the air velocity should be kept as high as possible without affecting runnability due to edge flutter. Figure 5.18 shows the pocket air velocity distribution in a dryer section.

5.4.4 *Infrared Thermography*

Infrared cameras are used in the paper industry for trouble shooting. Characterization of any cross machine sheet or dryer can temperature differences that may highlight a machine moisture problem is possible with this technique.

Using an infrared camera, the sheet is photographed at reel, section breaks and before and after the press. Conditions of dryer fabrics and dryer cylinders are examined. Moisture variations in the paper can be determined. Infrared thermography is also used to examine paper structure [4].

5.5   References

1 *Process Study*, Asten Inc., 1996.

2 Carlson, D., "From Headbox to Couch—Consistency Measurement under the Wire Provides Benefits to the Papermaker," *PaperAge,* June 1996.

3 *TAPPI Test Methods*, 1996-1997, TAPPI Press, Atlanta, GA.

4 Kiiskinen, H.T., et al., "Infrared Thermography Examination of Paper Structure," *TAPPI Journal*, April 1997.

5.6   Review Questions

1 What is the purpose of paper machine auditing? Explain how the audit results can be used to improve the papermaking process.

2 What is FFT? How is it used in paper machine auditing?

3 How does vibration affect the press section operation? What can be done to reduce/eliminate vibration?

4 Explain the following:
   • pulsation
   • synchronous time averaging
   • pocket air velocity and humidity
   • U-factor

5 Find out the working principle of the following devices:
   • gamma gauge
   • noncontact pyrometer
   • LVDT
   • hot wire anemometer
   • pitot tube
   • platinum RTD probe
   • infrared camera

6

Paper Structure, Properties, and Testing

Since the purpose of paper machine clothing is to help make paper and paperboard, it is proper to overview the structure and properties of paper and paperboard in this chapter. Better understanding of paper structure and properties will help the textile engineer with the design and manufacturing of forming, press and dryer fabrics.

Paper has a fibrous structure. The basic building block of paper is the individual fiber. Various fillers are added to this structure to obtain certain properties such as improved strength, printability, liquid absorption, etc., depending on the end use. Although paper's structure is continuous, it is in fact a heterogeneous network of these discrete fibrous particles and fillers. The fibers and fillers, which are the main components of the paper structure, have different distributions, shapes and chemical compositions. Nevertheless, the paper is expected to behave in a homogeneous manner. In today's technology, this is achieved by evening out the variations in the structure to a certain extent during the manufacturing process. Scientific attempts to relate the paper characteristics and structure to final properties began in the 1960s [1]. The product quality in paper machine clothing is usually measured by the performance and life of the fabric. For paper, quality may require fulfillment of several requirements which sometime may conflict. The papermaker, converter and consumer have various analytical tests available to judge the quality and performance of

the paper. The uniformity of quality may be as important as the quality itself. The uniformity of quality depends on uniformity of structure.

In laboratories and research centers, paper is usually made using handsheet molds for testing and analysis purposes. It is reported that strength of handmade paper is generally better than comparable papers made on commercial paper machines [2].

6.1 Fibers Used in Papermaking

Fiber can be defined as "a thread-like body or filament, many times longer than its diameter" [3].

Both natural and synthetic fibers are used for manufacturing of paper and paper related products. Plants, which are the most abundant organic materials on earth, are the major source of natural fibers for papermaking. Although different plant fibers can be used for papermaking, wood pulp fibers are the most widely used because of their cellulose content. The use of synthetic fibers in paper and related products is increasing due to exceptional strength properties they provide. Sometimes, animal fibers such as wool or silk are also used for special types of papers. Table 6.1 shows the fibers that are used in papermaking [2,4].

Important properties of papermaking fibers include length, coarseness, intrinsic strength,

TABLE 6.1. Classification of Major Fibers Used in Papermaking.

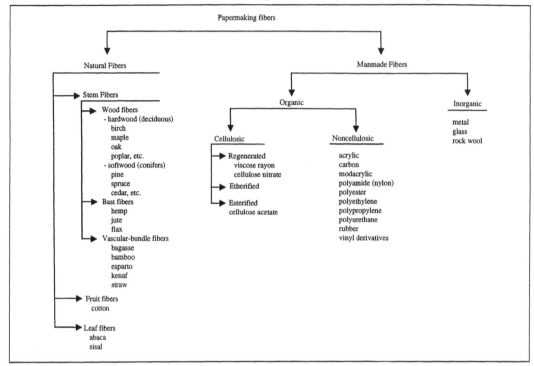

cohesiveness and flexibility. Initial and maximum fiber length, coarseness and fiber strength are determined by the genetic code of the tree. Simplified structure of a typical wood fiber is shown in Figure 6.1. The fiber has an outer primary wall and three layers of inner secondary wall. The inner layers which are rich in cellulose constitute the main body of the fiber. In the center of the fiber, there is an empty space called lumen. The individual fibers are held together with lignin. Typical wood fiber lengths vary from 1 mm to 6 mm. The fiber's cross-sectional area is approximately 1/100th of its length. Processes during papermaking usually reduce the length, coarseness and strength of fibers but increase the cohesiveness and flexibility.

6.1.1 Wood Fibers

The main ingredient of wood fiber is cellulose material. Although cellulose fibers can be obtained from different plants, wood is the most widely used material in papermaking because it is abundant and relatively inexpensive. Figure 6.2 shows the structure of cellulose. The main ingredient of paper is cellulose which is a white, hygroscopic material. It has good water absorption which affects its dimensional stability. Cellulose can form hydrogen bonds. Tensile strength and flexibility of

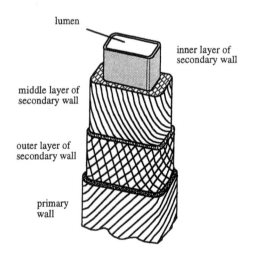

FIGURE 6.1. Schematic structure of a typical wood fiber.

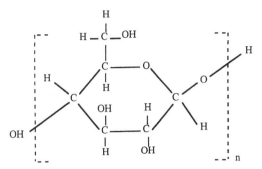

FIGURE 6.2. Structure of cellulose.

cellulose are high. It is also combustible. Since paper is made of cellulosic materials, it has all the properties that cellulose has. In addition, the properties of paper can be modified by different pulping and papermaking processes [2].

Trees are classified as hardwoods and softwoods. They have different amount of lignin in them. Another difference is that softwood fibers are longer than hardwood fibers. The length of softwood fibers is 3-6 mm while the length of hardwood fibers is 1-2 mm. A combination of hardwood and softwood is used in most papers. Short hardwood fibers provide bulk and long softwood fibers provide strength.

Hardwood Trees

Hardwood trees have broad leaves and are usually deciduous (lose their leaves annually). Examples of hardwood trees are aspen, birch, maple, oak, eucalyptus and poplar. They have short, slender, medium-walled fibers. Eucalyptus (blue gum) has good softness properties for tissue and smoothness properties for fine paper. Hardwood stem is composed of about 50% cellulose and 20% lignin. Figures 6.3 and 6.4 show photomicrographs of different hardwood fibers.

Softwood Trees

Softwood trees are also called conifers or evergreens. They have needle-like or scale-like leaves. Examples of softwood trees are

FIGURE 6.3. Southern hardwood bleached kraft (sweet gum) fibers. Magnification: 50×. Circular shapes are air bubbles (courtesy of Integrated Paper Services).

FIGURE 6.4. Northern hardwood bleached kraft (aspen) fibers. Magnification: 50× (courtesy of Integrated Paper Services).

pine, spruce, and cedar. Softwood stem consists of 50% cellulose and 30% lignin. Figures 6.5–6.8 show photomicrographs of various softwood fibers.

Southern pine, which includes Loblolly pine, shortleaf, longleaf and yellow pine, is

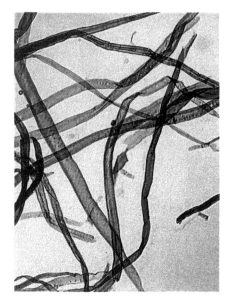

FIGURE 6.5. Southern yellow pine softwood bleached kraft fibers. Magnification: 50× (courtesy of Integrated Paper Services).

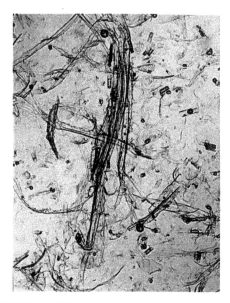

FIGURE 6.7. Northern softwood stone groundwood fibers (spruce and fir with trace of hardwood). Magnification: 50× (courtesy of Integrated Paper Services).

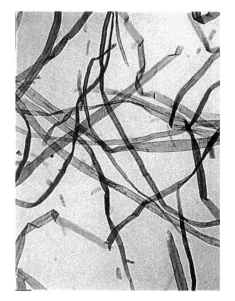

FIGURE 6.6. Northern softwood bleached kraft fibers (mostly spruce but mixed with some pine and others). Magnification: 50× (courtesy of Integrated Paper Services).

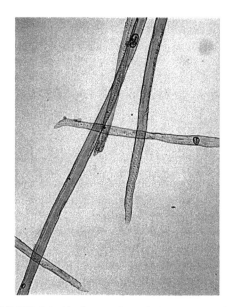

FIGURE 6.8. Northern softwood thermomechanical pulp fibers (spruce). Magnification: 50× (courtesy of Integrated Paper Services).

widely used in papermaking. They are very fast growing trees with long coarse, thick walled fibers. These fibers give excellent tear properties in the paper [2]. Another coniferous or softwood tree is spruce which has a long, medium slender, medium walled fiber. It is widely used for newsprint/groundwood and sulfite pulping due to its brightness. Western red cedar, another softwood tree, has long, medium slender, thin-walled fibers.

6.1.2 Nonwood Cellulosic Fibers

Nonwood plants are also suitable to make paper since they contain cellulose. As a result, different types of plants may be used for different types of papers. An example is cotton plant. Cotton fibers are longer (10–30 mm) and have more cellulose (90%) than wood fibers. However, cotton is more expensive than wood fibers therefore it is used in special business papers alone or in combination with wood fiber.

Other plants that are used to make paper include sugar cane, hemp and flax, barley, rye, oat, bamboo, kenaf, abaca, sisal and rice straw. It should be emphasized that wood, by far, is the major source of papermaking due to proper fiber length, ease of processability, availability and cost.

6.1.3 Synthetic Fibers

Synthetic fibers such as polypropylene, polyester, glass, etc., are mostly used in the nonwoven manufacturing (Section 2.6.3).

6.2 Properties of Single Pulp Fibers

The important mechanical properties of single wood pulp fibers are determined in tension, bending or torsion with destructive or non-destructive test methods. The tensile elastic properties are the most important properties which are measured in the range of mN (milliNewton) and μm (micrometer) [5–7].

6.2.1 Fiber Length

Fiber length is an important property that affects several sheet properties. It has been reported that for a relatively unbeaten pulp, coarseness of the fibers being equal, fiber length affects the tensile strength, modulus, burst strength, fold and tear strength of the sheet. For heavy papers with good bonding between the fibers, effect of fiber length on modulus and tensile strength is small. However, for lightweight papers, fiber length has more effect on modulus and tensile strength [8,9].

The fibers in a paper have a certain length distribution which may be due to the type of fiber. Recycling, method of pulping and degree of beating also affect fiber length distribution. TAPPI test method T-233 is designed to measure the weighted average fiber length of a pulp which is given as [10]:

$$L = \frac{\sum wl}{\sum w} \qquad (6.1)$$

where

L = weighted average fiber length (mm)
l = fiber length (mm)
w = fiber weight (mg)

Fibers that are less than 0.2 mm are considered to be fillers, debris or fines. Fibers that are narrower than 1 mm are considered to be too weak as strength bearing members. Other techniques to measure fiber length are fiber length of pulp by projection (TAPPI T-232, Fiber Length of Pulp by Projection), electronic sequential fiber length analysis (Kajanni, developed by Finnish Central Laboratory) and image analysis.

6.2.2 Fiber Coarseness

Coarseness is defined as the weight (mg) per unit length (100 m) of fibers which is called a decigrex (dg). It is 11.1 times larger than denier which is widely used to indicate the density of textile fibers (Section 2.2.3). Fibers that have a coarseness value of less than 10 dg are considered fine fibers. Above 20 dg indicates a coarse fiber. Coarseness affects several paper properties. For example, a finer pulp results in a stronger, smoother paper with better folding ability. Sheet density also increases with fine fibers. Paper made of coarse fibers has more porosity. Methods to measure coarseness include microscopic method, image analysis and Kajanni fiber analyzer [11].

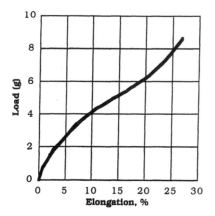

FIGURE 6.9. Load-elongation curve of a bleached-kraft, southern pine fiber [2].

6.2.3 *Intrinsic Fiber Strength*

Intrinsic strength of a fiber is, first of all, determined by the type of tree. For example, an unbeaten, unbleached, sulfate sweet gum fiber has a breaking strength of 53.6 kg/mm². A birch fiber from the same type of pulp has 97.5 kg/mm² breaking strength. Fiber treatment also has an effect on intrinsic fiber strength. Fiber modulus is increased by beating due to increased crystallite orientation. Zero-span tensile test may be used to measure intrinsic fiber strength. A typical load-elongation curve of a bleached-kraft, southern pine fiber is shown in Figure 6.9.

6.2.4 *Fiber Flexibility (Wet Fiber Compactability)*

Flexibility is the ability of fiber to conform when pressed against other wet fibers and to keep its form during drying due to the cohesiveness of the surface. Fiber diameter and wall thickness, level of fibrillation during refining and fiber cross-sectional shape are the main factors that affect fiber flexibility. Fiber flexibility can be measured in various ways such as single fiber methods and wet and dry specific volume procedures.

6.3 Paper Structure

Paper consists of a three-dimensional pore system and a skeletal fiber system. There are approximately one million wood fibers in one gram of a typical paper made by chemical pulp [12,13]. In a paper structure, a fraction of the volume is occupied by fibers and fillers and the rest is void volume.

$$V = V_f + V_v \qquad (6.2)$$

where

V = total volume
V_f = fiber volume
V_v = void volume

Table 6.2 shows typical void volume of different paper grades. The strength and stiffness of the paper structure comes from the network of fibers. This network can be considered to be of infinite dimensions in the plane of the sheet. The dimensions of the network in the direction perpendicular to the sheet plane are limited [14].

Due to the void volume in the structure, paper is a compressible material. For this reason, converting processes consolidate the sheet even further and give a more compact surface.

6.3.1 *Fiber Orientation and Directionality in Paper*

The travel direction of fabric and sheet on the paper machine is called machine direction (MD). Cross-machine direction (CD) is the direction at right angle to the MD. Machine direction is also called the grain direction. Because of the moving paper web during manufacturing, the orientation of the fibers in the plane of paper is not totally random. Due to moving direction of the slurry and forming fabric, the fibers tend to orient more in the

TABLE 6.2. *Typical Air Volume of Some Paper Grades [2].*

Paper Type	% Air Volume
Groundwood	63.1
News	53.1
Greaseproof	43.2
Bond	34.2
Glassine	13.0

machine direction during forming. As a result of this orientation preference, paper strength is usually greater in MD direction than the CD direction. Another reason for the strength difference is the higher tension exerted in the MD direction during drying. The ratio of MD/CD strength is 1.5-2.0 for fourdrinier machines and can be higher for cylinder machines. For this reason, test direction should be indicated in physical and optical tests. Since MD/CD fiber orientation may cause curl in the paper, it is desirable for most paper grades that this ratio is close to 1 [2].

The degree of fiber orientation is defined as the angle that fibers make relative to a particular direction. The fiber orientation distribution $f(\theta)$ is given by:

$$f(\theta) = \frac{1}{\pi} \sum_{i=0}^{\infty} a_i \cos 2i\theta \qquad (6.3)$$

where $a_i = 1$ and θ is the direction of a particular fiber relative to the machine direction. Other functions are also used to describe fiber orientation of fibers in paper such as the relative deviation from 45° of the average angle that the fibers make with the MD [15].

Fiber orientation has a big influence on paper tensile properties. It was reported that changing the degree of fiber orientation from 0.5 to 2.5 increased the strength three times. Increasing fiber orientation in MD increases the strength and elasticity in that direction and decreases them correspondingly in CD. It was shown that the tensile strength MD/CD ratio represents a good correlation with the degree of fiber orientation. As a result, it was suggested that the tensile strength MD/CD ratio in paper be used as a measure of the degree of fiber orientation for a given pulp furnish. Htun and Fellers proposed that the tensile strength MD/CD ratio for sheets dried under biaxial restraint be used as an index of fiber orientation, with the term fiber orientation ratio being adapted for this ratio. Figure 6.10 shows the MD/CD ratio for the specific elastic modulus, the tensile index and the compression index of sheets made of bleached sulfate pulp and dried under biaxial restraint against the fiber orientation ratio dried. It was also

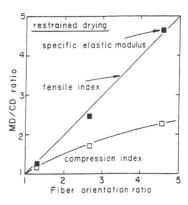

FIGURE 6.10. The MD/CD ratio for the specific elastic modulus, the tensile index and the compression index of sheets made of bleached sulfate pulp and dried under biaxial restraint against the fiber orientation ratio [19].

reported that the elastic properties are directly related to shrinkage during drying. The ratios of elastic modulus MD/CD and the tensile index MD/CD increases if the paper is freely dried. The draws (speed differentials) by the various parts of the machine may also lead to an MD stretching of the sheet. Uneven fiber distribution results in uneven printing [16–18].

The techniques that have been used to quantify the degree of fiber orientation includes averaging fiber segment orientations, X-ray diffraction, light scattering, regular and zero-span tensile strength and Fourier transform methods [20,21]. Figure 6.11 shows a typical fiber orientation distribution.

Paper has an anisotropic structure: its mechanical properties depend on the direction. Fiber orientation and shrinkage during drying

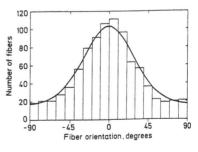

FIGURE 6.11. Typical fiber orientation distribution for a sheet of 80 g/m² [22].

(drying restraints) are the main causes for anisotropy. The anisotropic structure of sheet leads to mechanical properties that are different than those of homogeneous isotropic materials. However, in some respects and on a macroscale, paper can be considered as an orthotropic structure in which there are three mutually perpendicular directions of material symmetry. The material properties are different in each direction. For paper, these three directions are machine direction (MD), cross machine direction (CD) and the thickness direction (Z) as shown in Figure 6.12 [23–25]. The majority of fibers are aligned approximately parallel to the plane of the sheet mostly in machine direction. This alignment depends on the machine configuration, pulp and velocity difference between the jet speed and forming fabric. The biased fiber orientation in MD makes the paper stiffer and stronger in MD than in CD. Moreover, the sheet is subjected to restraining forces in MD in pressing and drying. The paper dries with little restraining force in CD. The strength and stiffness in the Z direction is the weakest: generally 1/10–1/50 of the inplane properties. In general, the in-plane and out-of-plane properties are not related. Some paper products are manufactured in layers. Each layer may have different properties [19].

The variation of paper properties is generally more abrupt in the CD compared to MD. In other words, variations occur more slowly in the MD. The reasons for more sudden variations of properties in the CD direction are possibility of uneven flow of stock from the headbox, uneven wet pressing, uneven tension during drying and edge effects [2].

The directionality of paper can be determined in various ways in addition to tensile

FIGURE 6.13. Hydrogen bonding.

test. If one side of the sheet is wetted, the expansion of the wetted side causes a curl. The expansion is more in the CD and the axis of the curl is almost always parallel to the machine direction. Stiffness in MD is greater than CD. During the bursting test, MD has less stretch and rupture line will be at right angle to MD.

6.3.2 *Fiber Bonding and Cohesiveness*

The main bonding of fibers in a sheet structure is provided by hydrogen bonds between fibers. Figure 6.13 shows schematic of hydrogen bonding. Cohesion is the hydrogen bonding force that holds the fibers together in the sheet structure which gives the paper its strength and integrity. Cohesion increases the frictional sliding resistance between adjacent fibers against a shearing action as shown in Figure 6.14. Fiber cohesiveness depends on the surface chemistry and area as well as sheet density.

There are two types of cohesion: inplane cohesion (the force is in the plane of the sheet) and transverse cohesion (the force is perpendicular to the sheet plane). It is neither practical nor meaningful to measure individual fiber cohesiveness; therefore, cohesiveness is measured empirically as a sheet property. Methods to measure cohesiveness include percent cohesion test, Clark shear test, split cohesion test and peel cohesion method.

It is a well known fact that wet sheet after pressing does not have the strength of a dry

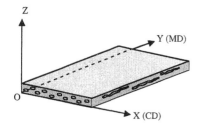

FIGURE 6.12. Three major directions of paper.

FIGURE 6.14. Cohesiveness is the resistance to shear provided by the hydrogen bonding of fibers during drying.

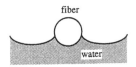

FIGURE 6.15. Surface tension on a fiber in forming section.

paper. This fact indicates that fiber bond in paper develops during drying. It is also worthy to note that if an organic solvent instead of water is used in papermaking, the resulting sheet would have very little strength after drying. This fact proves that hydrogen bonding is the major force that binds the fibers together in sheet structure. For example, glass fibers can not produce a strong sheet (although glass fibers are a lot stronger than wood fibers) due to lack of hydrogen bonding between glass fibers after water removal (for this reason, bonding agents are used to improve fiber bonding in glass sheets).

It should also be noted that unlike textile structures, there is no interlacing (as in weaving and braiding), interlooping (as in knitting) or forced entanglement (as in needlepunching) of wood fibers during paper sheet formation. Instead, fibers are brought together close enough to produce hydrogen bonds which is called the Campbell effect. At a solid content of around 8% during formation on the forming fabric (i.e. approximately where the wet line is on a fourdrinier), the surface tension force acting on each fiber is given by:

$$F = \gamma l \qquad (6.4)$$

where γ is the water surface tension and l is the fiber length (Figure 6.15).

In the press section, more water is removed from the web such that water remains only between two fibers. Since there are two menisci between the fibers (Figure 6.16), the surface tension acting on the fibers becomes:

FIGURE 6.16. Surface tension between two fibers in press section.

$$F = 2\gamma l \qquad (6.5)$$

This force pulls the fibers together. After drying, more water is removed from the sheet; however, the menisci remain continuous and pull the fibers even closer together. Around 25% solid content, the compacting pressure between the fibers becomes very high:

$$\Delta P = \frac{2\gamma}{x} \qquad (6.6)$$

where x is the distance between two fibers. The pressure at this stage can reach to 100–200 atmospheres. Highly flexible fibers require lower pressures for bonding [2].

Important factors affecting the bond strength in a sheet structure are fiber areas in contact, number of bonds in contact area and strength of each bond. The level of bonding can be determined by light scattering techniques, conductivity tests and optical techniques.

6.3.3 *Fiber Curl*

In general, the fibers or fiber segments are not straight in a typical paper even if the shrinkage during drying is prevented. The fibers may be kinked or curled in or out of the plane of the paper which causes a decrease in paper stiffness. There are two types of curl: fiber curl (which is the curl of the entire fiber) and segment curl. In addition, fibers may partially be wrapped around other fibers. Some twisting of fibers also takes place [22,26–29]. Fiber curl affects the elastic properties of the sheet. Fiber curl can be measured with image analysis techniques.

6.3.4 *Internal Stresses*

Various stresses develop in the sheet internally during manufacturing. Swollen fibers, which are molded into a sheet in forming and pressing sections, are dried at different degrees of tension in the dryer section. Various parameters during manufacture of paper influence the level of internal stresses in the paper. Beating of pulp, wet pressing, variation in fiber orientation and drying conditions all affect internal stresses in the paper. Increasing the degree of beating increases the internal stresses. This is because beating increases the density which increases the potential to shrink during drying. Wet pressing also increases internal stresses. Because of nonuniform drying rate in the thickness direction and in-plane tension forces, internal stresses develop in paper during drying. Restrained drying increases internal stresses. Internal stresses in paper can also be caused by deformation. Internal stresses can be partly released by humidity cycling and cyclic mechanical straining. Internal stresses affect the mechanical properties of paper [30,31].

6.3.5 *"Wire" and Felt Sides of Paper*

On a fourdrinier table, the side of the sheet next to the forming fabric is called fabric side (also called 'wire side' from the metal wire days). The other side is usually referred to as "felt side". A paper formed on a fourdrinier table may have the two sides different in structure. This phenomenon is called "two sidedness". Structural differences may include sizing, pigments, fines and fibrous composition. Optical differences include finish and reflectance. The fabric topography may cause marks on the sheet which is commonly known as "wire mark". Wire mark can penetrate up to 60% of the total sheet thickness [2].

Paper is coarser and more porous on the wire side than the felt side. Usually there is a high percentage of long fibers on the wire side. This is because fines are washed from the wire side before the larger fibers are laid down. A typical bond paper has 18% fines on the felt size and 11% on the wire side. Uniformity of fiber length decreases two-sidedness. Differences in two sides of a paper can be viewed best under a microscope. These differences may dictate some rules during application or end-use of the paper. For example, print quality is better on the felt side which is smoother. Laminated boards are glued together on the coarse sides. Stamps are printed on the fabric side and gummed on the felt side.

6.4 Paper Wet-End Chemistry

Various materials and additives are added to the pulp at the wet end to help with formation and improve final paper properties. Some of these materials become an integral part of the paper structure [2]. A brief summary of these materials and processes is given below.

6.4.1 *Sizing*

Sizing can be done externally (surface sizing or coating) and internally. External sizing is done after sheet is made by filling in pores on surface of paper with glutinous material such as starch, glue, PVA, wax and fluorochemicals. Internal sizing is done by additives at the wet end. Sizing makes the paper more hydrophobic by reducing the spread of fluids, thereby improving "holdout" in printing and coating.

Sizing systems on papermachines producing acid sheets normally utilize alum chemistry which can reduce paper life as a result of residual acid-forming salts. Sizing systems on alkaline sheets can minimize or eliminate alum, thereby enhancing sheet life while reducing the corrosive effects of low pH on the machine itself.

6.4.2 *Dry and Wet Strength Agents*

Dry strength agents such as starches, natural gums and emulsified latex are used to give sheets more tensile strength by improving bonding between fibers.

Wet strength agents provide paper strength in wet conditions. They prevent disruption of fiber to fiber bonds in water and therefore retain integrity of paper. Wet strength additives include Kymene, formaldehyde resins, and polyamines, which are all hydrophobic.

Dry and wet strength agents are added to the pulp before the headbox.

6.4.3 *Flocculants and Deflocculants*

Flocculants, such as cationic polymers, are the materials that bring particles together by neutralizing negative surface charge. They enhance retention during formation of sheet. The primary flocculant used in papermaking is alum.

Deflocculants, such as anionic polymers, repel negative surface charges. They improve formation.

6.4.4 *Fillers*

Fillers are used to change the optical and strength properties of paper. There are various types of fillers including clays, calcium carbonate, titanium dioxide, hydrated alumina, talc or amorphous silica. Opacity, density, brightness and whiteness of a sheet are increased by addition of fillers. Fillers are sometimes used as low cost fiber substitutes since they close pores and provide smooth surface.

6.4.5 *Drainage and Retention Aids*

Drainage aids are used to prevent fines and small fibers from plugging the forming fabric which may slow down the speed. They attach fillers, fines and small fibers to large fibers. Polymers and additives that are used as drainage and retention aids include polyamines, polyamides, gums and cationic starches.

6.4.6 *Formation and Dispersion Aids*

These are additives that keep fibers from forming clots or flocs. Examples of these aids are gums and starches. Some of these materials make fiber surface repulsive or slippery to break up the flocs.

6.4.7 *Other Agents*

Several other materials are also used in the wet-end to help with formation and improve final paper properties. These include dimensional stability agents, biocides, pitch control agents and defoamers. Pitch comes from fatty acids and resin acids in the wood. Soaps, solvents and talc are used to control pitch.

6.5 Paper Properties and Testing

Paper testing can be done on-line or off-line. On-line measurements are done on the paper machine while the paper is being produced. These measurements are automated and computerized continuous measurements. They are in general indirect measurements which means that a particular property is determined based on the measurement of another related property. Examples of on-line testing are strength tests, caliper, grammage, moisture content, color, brightness, opacity and gloss [1,2].

Off-line measurements are more direct measurements which are usually done in the lab. Several institutions have established test methods for the paper industry. Main sources of test methods in the US are TAPPI (Technical Association of the Pulp and Paper Industry), ASTM (American Society for Testing and Materials) and ANSI (American National Standards Institute). TAPPI methods are the major source of test methods for the pulp and paper industry [32]. Various institutions in other countries also have established test methods such as CPPA (Canadian Pulp and Paper Association), SCAN (Scandinavian Pulp and Paper Association) and DIN (German Standards).

6.5.1 *Tensile Properties*

Tensile strength is probably the most important property of a paper. Tensile strength is especially important for printing papers, bag and wrapping papers and adhesive tape papers. For paper, tensile strength is expressed in breaking load/unit width. The unit is N/m in SI system, kg/m in metric system and lb/inch in English system.

Paper is not a homogeneous material. Therefore, there is tensile strength variation in paper which depends on the location. Stress in a local region can be calculated as follows [33]:

$$\sigma_1 = b\sigma_a \qquad (6.7)$$

where σ_1 is the local stress, b is the stress

concentration factor and σ_a is the average stress acting on the cross-section of the sheet.

Sheet strength depends on two components: intrinsic fiber strength and bond strength. For well-bonded paper, sheet strength is directly related to fiber strength. Improving fiber bonding increases the strength contribution from the fibers. The specific strength of paper increases with density [6,34,35,36]. The relationship between fiber and sheet strength can be written as:

$$T_s = \Phi T_f \tag{6.8}$$

where

T_s = sheet strength
Φ = intensity function which shows the efficiency of fiber strength utilization in the structure
T_f = fiber strength

Sheet failure occurs when a local stress exceeds the fiber failure limit in that location. Paper physicists developed relatively simple models to predict the sheet stress based on individual fiber strength. Treating the sheet as a two phase structure made of fibers and void volume, sheet strength can be approximated as follows:

$$T_s = (1/3)V_f T_f \tag{6.9}$$

where V_f is the fiber volume fraction which is equal to the ratio of sheet density to fiber density. Theoretically it was found that maximum tensile strength of a random 2-D sheet is equal to one third of strength of individual component fibers [34].

In practical end use, sheet strength is judged by its strength per unit width for a given grammage which is called tensile index. The tensile index is equivalent to expressing the sheet strength as stress divided by density. Therefore, the tensile index or the specific sheet tensile strength is given by:

$$T_s^* = T_s/\rho_s \tag{6.10}$$

where ρ_s is the sheet density.

Similarly, the specific fiber strength can be calculated as:

$$T_f^* = T_f/\rho_f \tag{6.11}$$

where ρ_f is the fiber density.

As a result, the relation between sheet specific strength and fiber specific strength can be written as:

$$T_s^* = (1/3)T_f^* \tag{6.12}$$

This equation refers to the case where the full fiber strength is utilized in the sheet. Considering the efficiency function, Φ^*,

$$T_s^* = (1/3)\Phi^* T_f^* \tag{6.13}$$

It was experimentally shown that for normal paper grades, Φ^* is between 0.2 and 0.7. The value increases with fiber beating until a certain point (Figure 1.8) and with the intensity of wet pressing of the wet web [37,38].

Figure 6.17 shows typical tension and compression curves of a paper made of bleached kraft board. The area under the tensile stress-strain curve is defined as the tensile energy absorption (TEA) of the sheet. It is a measure of toughness (J/m²) which is important for grocery bags. Increasing stress or strain or both increases toughness.

Tensile strength of a paper is affected by the cellulose content (i.e. pulp yield) in the fibers. Increasing sheet density by beating increases the modulus and tensile strength of the sheet to a certain extent. This is because the fiber strength is better utilized towards the sheet strength. Increasing density also increases the strain to failure. However, it is important that fiber length loss is controlled during beating.

Paper stretch is an indication of toughness.

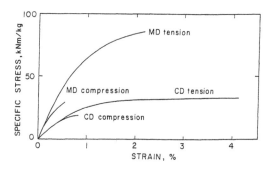

FIGURE 6.17. Tension and compression stress-strain curves of a bleached kraft paperboard [13].

Stretch becomes especially important for papers that are subject to stress during the end-use such as twisting papers, toweling, cable wrapping paper and corrugating papers. Stretch is increased by beating and reduced by short fibers, filling, heavy pressing and high tension during drying. Breaking strength of paper is approximately proportional to grammage.

Fiber orientation has a big influence on paper tensile properties such as elastic modulus and tensile strength (Section 6.3.1); however, it does not affect the strain to failure. In general, tensile strength is greater in MD than CD because of fiber alignment and strain developed during drying. Typical MD/CD tensile ratio is 2 for fine paper and 0.5–1 for tissue. Drying restraints basically influence the strain to failure and elastic modulus. For a freely dried sheet, the strain to failure is linearly related to the sheet shrinkage during drying and independent of the fiber orientation [16,39].

Wet Tensile Strength

Moisture affects tensile strength. Maximum tensile is obtained between 20–30% relative humidity (RH). Below this range stress intensification and above this range plastification make the paper weaker. Figure 6.18 shows

the effect of moisture content on tensile strength of kraft linerboard. At complete saturation, the tensile strength of paper becomes practically zero. Sizing provides temporary wet strength to the paper. Some papers must have permanent wet strength due to their applications. Permanent wet strength is provided by wet-strength agents such as melamine-formaldehyde (MF), urea-formaldehyde (UF), Kymene (polyamide-polyamine-epichlohydrin) and others. TAPPI T-456 is designed to measure the wet tensile breaking strength of paper and paperboard.

Z-Direction Tensile Strength (Internal Bond Strength)

Z-direction tensile test measures the internal bonding strength of paper. Internal bonding strength is especially important for paperboard for glue bonding at carton side seams, delamination on scoring and use of high tack coatings.

TAPPI T-541 is designed to measure the internal bond strength of paperboard. A one-inch square paper is attached to metal blocks of the jaws on each side by double sided tape. A tensile load is applied to the paper using a standard material testing system. This method can also be used to evaluate coated papers. Figure 6.19 shows typical stress-strain behavior of paper in Z-direction.

Zero-Span Tensile Strength

This test is basically done to determine a strength index of the longitudinal structure of individual fibers in a pulp test handsheet (TAPPI T-231). A zero-span jaw attachment is used in a regular tensile testing system (Figure 6.20). The ratio of normal to zero-span breaking lengths provides an index of fiber cohesiveness. The method is also useful to determine the relative degree of orientation of fibers in MD and CD.

6.5.2 *Compression Behavior*

Compression stresses may cause the paper to fail because of possible structural instabilities. Compression properties of paper are

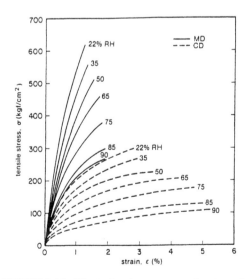

FIGURE 6.18. Effect of moisture on tensile strength of kraft linerboard [40].

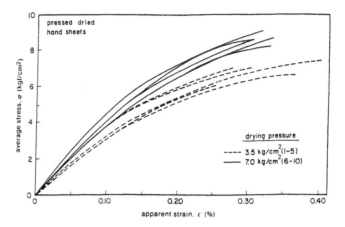

FIGURE 6.19. Typical stress-strain behavior of paper in Z-direction [40].

especially important for cartons and corrugated boxes since they are supposed to protect their contents against compression forces during packing, storage and distribution. Bending stiffness and edgewise compression strength are the two most important material properties for packaging boards [13,41–44].

Figure 6.17 shows the compression stress-strain curves of a typical bleached kraft paperboard. Since the elastic modulus is the same in tension and compression, the initial load bearing capacity is the same in both modes. However, the yield point, strain to failure and tenacity (stress at failure) are lower in the compression mode. Flaws in the paper structure are detrimental to compression strength. Compression failure usually does not cause any bond or disruptive fiber failure in the structure.

Increasing sheet density by beating increases the modulus and compression strength of the sheet due to better utilization of fiber strength towards the sheet strength. As the density increases, the bonding also increases which causes the failure to be fiber dominated. Compression strength is independent of the pulp yield which means that both cellulose and lignin bear the compression load. Compression strength increases with increasing wet pressing. Adsorption of moisture reduces the compression strength [13,42].

Different principles of compression testing have been developed including ring crush test, short span test, blade support test and plate support test. Edgewise compressive properties are critical for the performance of corrugated containers. Figure 6.21 shows a specimen, which is supported between two fixed parallel

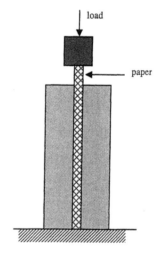

FIGURE 6.21. Schematic of edgewise compression test.

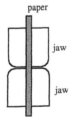

FIGURE 6.20. Arrangement for zero span test.

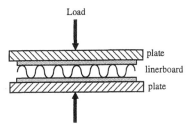

FIGURE 6.23. Schematic of flat crush test of corrugated linerboard.

side walls of the corrugations. Figure 6.24 shows the load-deformation curve for a corrugated linerboard.

Short Column Crush (SCC) Test

The SCC test method is developed to test the in-plane crush resistance of side panels of a corrugated container under a stacking load (Figure 6.25). This test gives an indication for the stacking strength performance of boxes.

6.5.3 *Visco-Elastic Behavior*

Like textile materials, paper is a visco-elastic material which means that its elongation-recovery properties are time dependent. For example, if a paper breaks in 4 seconds under a 10 kg load, it takes 11 min to break under 9 kg, 14 hours under 8 kg and 220 days under 4 kg [2].

Paper creeps with time under constant load. There are two types of creep (Figure 6.26): primary creep which is recoverable and secondary creep which is non-recoverable

A spring-dashpot model is representative

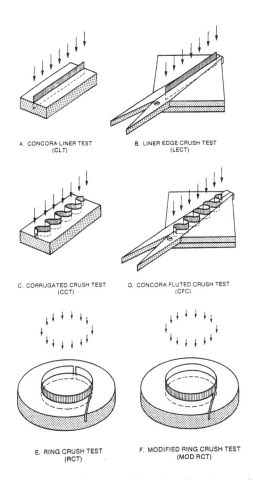

FIGURE 6.22. Various configurations for edge crush test [40].

plates, under load in edgewise compression testing. Other various test configurations are shown in Figure 6.22. Increasing basis weight increases edge crush resistance.

Flat Crush Tests

Flat crush test measures the flute rigidity of corrugated board. TAPPI has developed two test methods to measure the crush resistance of corrugated materials: T-808 (Flat Crush Test of Corrugated Board, Flexible Beam Method) and T-809 (Flat Crush of Corrugated Medium, CMT Test). A compressive load is applied normal to the plane of corrugating medium as shown in Figure 6.23. Maximum compressive force is measured for the failure which is defined as the collapse of the

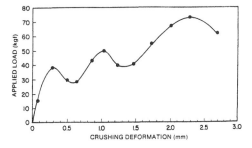

FIGURE 6.24. Typical flat crush load-deformation curve of a corrugated linerboard [40].

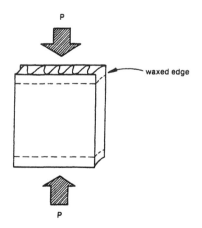

FIGURE 6.25. Short column crush test [40].

of the visco-elastic behavior of paper (Figure 6.27). Upon application of load, springs elongate and dashpot does not move. As the load is increased, the dashpot also starts moving. Due to the dashpot in the system, creep is completely recovered.

6.5.4 Bending Stiffness (Flexural Rigidity)

Bending involves both compression and tension (Figure 6.28). The bending moment for a given curvature is related to the stress-strain properties. Flexural rigidity for a material is the product of elastic modulus and moment of inertia. Paper stiffness is expressed in two ways [2]:

(1) Paper stiffness: flexural rigidity/ grammage

$$(2) \qquad Paper\ Stiffness = \frac{ET^3c}{12d^2} \qquad (6.14)$$

where

E = elastic modulus
T = thickness of sample
c = width of sample
d = length of sample

Most stiffness measurements are based on either measurement of bending angle of a paper strip under a force or measurement of force required to bend the strip at a certain angle. TAPPI methods to measure stiffness are T-451 (Flexural Properties of Paper, Clark Stiffness), T-489 (Stiffness of Paper and Paperboard, Taber-type Stiffness Tester) and T-543 (Stiffness of Paper, Gurley Type Stiffness Tester).

Figure 6.29 shows the bending moment-curvature curves for bleached and unbleached sulfate board beams. The curves are analogous to tension and compression curves in Figure 6.17. Bending stiffness is represented by the initial slope. The maximum moment is the bending strength.

Bending stiffness is especially important in most packaging applications where paperboard is used. For these applications high bending stiffness both in MD and CD is required. Papers that must stand up during use, such as index Bristol, card, typing papers, playing cards, etc., require good stiffness. Corrugating medium and linerboard also need to have good flexural rigidity since they are used in structural constructions. Stiffness is

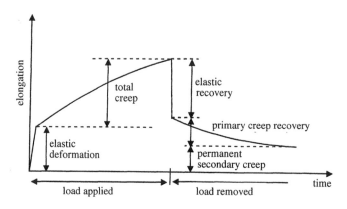

FIGURE 6.26. Primary and secondary creep.

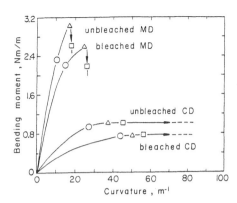

FIGURE 6.29. Bending moment versus curvature for bleached and unbleached paperboard [13].

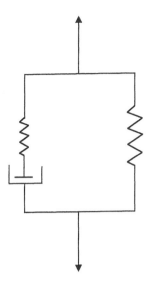

FIGURE 6.27. Spring-dashpot model for visco-elastic behavior of paper.

not wanted in tissue, toweling, printing papers and labels.

Bending stiffness depends on thickness, sheet construction, mass fraction of mixed pulps and degree of beating. Beating (refining) and additives, such as starch, increase stiffness. In general, multilayer sheet structures have greater bending stiffness than single layer structures [23,45].

6.5.5 *Bursting Strength*

Bursting strength is a measure of resistance against rupture of paper. It is defined as the hydrostatic pressure to rupture the paper as the pressure is increased at a constant rate. Bursting strength is expressed in kiloPascal (kPa) or psi (pounds per square inch).

Fiber length and bonding affect the bursting

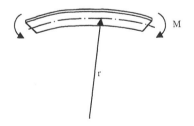

FIGURE 6.28. Bending of a paper strip; *r* = radius of curvature, *M* = bending moment.

resistance. Beating and pressing density increase burst strength. Loading conditions in bursting is similar to tensile loading. Therefore, if tensile strength increases, bursting resistance also generally increases. The failure occurs along the MD direction in bursting which is the direction with the least amount of stretch.

TAPPI has developed several test methods for bursting resistance of different papers: T-403 (Bursting Strength of Paper) which is also known as Mullen Burst test, T-807 (Bursting Strength of Paperboard and Linerboard), T-810 (Bursting Strength of Corrugated and Solid Fiberboard).

6.5.6 *Tear Resistance*

Long and coarse fibers increase tear resistance. During tearing a paper, fibers are broken and pulled out of the structure. Increasing grammage and flexibility increase tear resistance. Creped sheets also have higher tear strength.

TAPPI test method T-414 (Internal Tearing Resistance of Paper, Elmendorf Type Method) is designed to measure tear resistance of paper.

6.5.7 *Folding Endurance*

Folding endurance is defined as the number of folds that a paper will withstand before failure. Folding endurance is important for the papers that are subject to bending, folding and creasing. TAPPI method T-423 is designed to measure folding endurance of papers with a thickness of 0.25 mm or less.

6.5.8 Formation

The term "formation" is used to describe the uniformity of fiber distribution in a paper sheet. The more uniform is the fiber distribution, the better is the formation. Formation is usually evaluated by inspecting the paper against a light source and estimating the variations in the transmitted light intensity. Formation testers, based on light transmission, are generally used for light weigth grades. Formation generally correlates to other end use properties such as printability, smoothness or mechanical strength. Sheet formation affects the wet pressing and drying processing efficiencies.

A dimensionless Formation Number, N_f, is defined as an indication for formation. Formation number is based on local grammage variations and is defined as the coefficient of variation of local grammage W_A:

$$N_f = \frac{\sigma(W_A)}{\overline{W}} \qquad (6.15)$$

where A indicates the size of the measuring area. As the uniformity in the sheet increases, the formation number decreases. A more accurate way of characterizing the formation is the formation spectrum (floc-size spectrum) which shows the contribution of different floc sizes to the variations in local grammage [46].

6.5.9 Dimensional Stability, Curl and Twist of Paper

Dimensional stability of paper is critical for most applications including papers for printing, maps, templates, wallboard tape, abrasion, etc. Dimensional instability can be reduced by several methods [2]:

- using fibers that do not expand
- running a bulky sheet
- reducing internal bonding
- using bonding agents
- coating the paper to make it

impervious to water
- restraining CD shrinkage during drying

Ambient moisture changes can cause the paper and paperboard to curl or twist by deviating from the flat structure (Figure 6.30). This may cause problems in converting or printing operations where a flat material is required for good runnability.

Curling may be due to a moisture gradient in thickness direction or an asymmetrical structure through the thickness. If a swelling agent like water or water vapor is absorbed, the sheet volume dilates. Restraining of swelling causes stress development in the material [23,47]. Curl free paper is necessary for photosensitive papers, printing papers, coated papers and adhesive papers such as labels. TAPPI methods to test for curl are T-466 (Degree of Curl and Sizing of Paper) and T-520 (Curl of Gummed Flat Papers).

Twisting of paperboard may take place due to nonsymmetrical orientation of the layers above and below the midplane as shown in Figure 6.30.

6.5.10 Grammage (Basis Weight)

"Grammage" is a relatively new term introduced in the metric system to replace the conventional "basis weight". Grammage is the

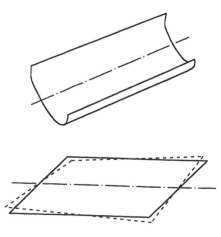

FIGURE 6.30. Paper curl (top) and twist (bottom).

weight of the paper (in grams) per unit area (one square meter). In the Standard system, the conventional basis weight (ream weight or substance) is based on the mass in pounds of a ream for a given sheet size (a ream can have 480 or 500 sheets). Different paper grades have different ream sizes. Different end use of each type of paper generally requires different grammage. Table 6.3 lists typical grammage values for different types of paper.

Grammage of paper is very important because all physical properties and many optical and electrical properties are affected by the change in grammage. Paper is sold by weight; therefore, the lowest grammage paper that will do the job is usually selected by the customer.

There are always grammage variations in a typical paper. If the fiber distribution in the paper were random, then the local grammage would be represented by a Poisson Distribution. However, the process of sheet formation causes deviations from such a distribution [46].

The grammage variation is determined by three major methods: weighing, absorption of light and absorption of beta radiation. Grammage measurement is done on-line with beta radiation. Grammage measurement is affected by the relative humidity. The standard conditions in the US are 50% relative humidity (RH) at 23°C. It is reported that a 10% change in RH around the standard RH can cause 1% change in grammage [2,46].

Grammage variations can cause less effective water removal and density fluctuations due to uneven wet pressing. Another effect of grammage variation is uneven drying conditions and variations in drying stresses.

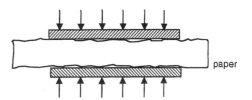

FIGURE 6.31. Measurement of sheet thickness.

6.5.11 *Thickness (Caliper)*

Uniformity of paper thickness is important especially for printing and writing. Sheet thickness is the perpendicular distance between the two surfaces of the sheet. The thickness is measured by placing the sheet between two parallel plates under a certain pressure (Figure 6.31). The pressure is 50 kPa in the standard TAPPI test method, T-411 Thickness (Caliper) of Paper, Paperboard and Combined Board. In Europe, 100 kPa pressure is applied. The unit for thickness is micrometer (μm) in the metric system and points (0.001 inch) in the Standard system. Table 6.4 shows typical caliper values for different papers. In general, as the grammage of paper increases, the thickness also increases as shown in Figure 6.32. Sheet thickness affects physical, optical and electrical properties.

6.5.12 *Sheet Density and Specific Volume*

Sheet density is a structural parameter that affects mechanical properties of paper. Sheet density is affected by both the total retention and the uniformity of retention of the fiber and fillers in the furnish. Total retention affects the finished optical properties of the sheet, and the

TABLE 6.3. Values of Grammage for Different Paper Types [2,48].

Paper	Grammage (g/m^2)
Tissues and paper towels	16 to 57
Glassine	30 to 75
Newsprint	45 to 70
Grocery bags	49 to 98
Fine paper	60 to 150
Cement bag	120 to 195

TABLE 6.4. Examples of Thickness Values for Different Papers [2,48].

Paper Sample	Thickness (μm)
Glassine	25–70
Facial tissue	55–75
Newsprint	75–95
Typing or ditto paper	80–120
Cement bag	230–635
Hardbound book cover	700–770

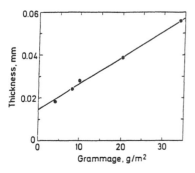

FIGURE 6.32. Thickness versus grammage for a typical paper [22].

uniformity of the fiber-fines-filler distribution can have an effect on top to bottom uniformity of density.

Sheet density is determined from the thickness and grammage measurements of the paper [49]:

$$\rho_s = W/t \qquad (6.16)$$

where ρ_s is the density (g/cm^3), W is the grammage and t is the thickness. Increased beating of pulp increases the density of the sheet. The typical relation between sheet density and grammage is shown in Figure 6.33. In low grammage region, density is proportional to the grammage. In the higher grammage region, grammage does not have much effect on density. Short, flexible and highly fibrillated fibers give a dense sheet. Density is also affected by the amount of fiber bonding, amount of void volume and calendering.

Every optical and physical property are affected by sheet density. Sheet density is related to porosity, rigidity, hardness and

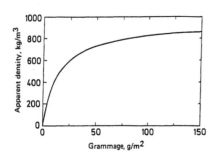

FIGURE 6.33. The relation between sheet density and grammage [22].

TABLE 6.5. Specific Volume (bulk) of Some Papers [2].

Paper Type	Specific Volume (cm^3/gr)
Bulky groundwood	3.0
Unbeaten sulfite	1.8
Glassine	1.0
Cellulose	0.65

strength of paper. Tensile and burst strengths are directly related to density; tear and porosity are inversely related. Specific volume or bulk is the reciprocal of density. Table 6.5 shows typical specific volumes of some papers.

6.5.13 Porosity and Permeability

Paper is a porous structure formed by a network of fibers. Therefore, paper is a two-phase structure in which pores and voids between the fibers are an important part of the structure. The total apparent pore volume can be calculated from the solid phase density and apparent density of the sheet as follows:

$$Porosity = \frac{pore\ volume}{total\ volume} \qquad (6.17)$$
$$= 1 - \frac{bulk\ density}{solid\ density}$$

Although pure cellulose has a density (specific gravity) of 1.5, paper has a low density of 0.5–0.8 due to its porous structure. Commercial papers contain about 70% air.

Pore volume is related to equivalent pore radius (EPR) which is used to compare paper porosities. Equivalent pore radius is defined as the radius of a single pore of length equal to the sheet thickness that would give the same flow as the average value for all the pores in a unit area of the paper [2]. The pore structure of a typical paper can be represented by a logarithmic normal distribution. This distribution is affected by beating, calendering and mixture of fibers in the pulp [12].

Pores affect the optical properties of the paper since they separate surfaces that controls light scattering. Pores also influence liquid absorption which is critical in many applications. Porosity is important for bag papers, cigarette papers, filter papers and insulating papers.

Air Permeability of Paper

Flow of air through paper is directly proportional to the pressure difference, time, effective area of the specimen and the fourth power of the pore radius and indirectly proportional to sheet thickness. Air resistance is defined as the time necessary for a specific air volume to pass through the sheet, or, the air volume that passes in a specific amount of time. Increasing solid fraction increases air resistance. Air resistance depends on fiber type. TAPPI T-460 is developed to measure the air resistance of paper.

6.5.14 *Surface Texture (Smoothness and Roughness)*

Absence or presence of surface irregularities makes the paper smooth or rough, respectively. Gloss is related to smoothness both of which can be produced by supercalendering. Machine glaze (MG) in the paper can be obtained via Yankee dryer. Other factors affecting sheet smoothness are fiber source, pulping process, beating, shake of fourdrinier tables, forming and press fabric surface, wet pressing, filling, surface sizing and pigment coating. In general, use of short slim fibers in the furnish, promotes smoothness in the paper sheet.

Smoothness is important for quality printing, transfer of adhesives and non-slip packages, among other things. Smoothness can be measured in various ways including direct methods, optical contact methods and air leak methods. Most common methods are Sheffield (macro) and Parker Print (micro) smoothness. TAPPI T-479 is designed to measure the smoothness of paper with Bekk Method.

6.5.15 *Absorbency and Swelling Properties (Water Resistance Tests)*

Paper is a hygroscopic material which means that it reaches an equilibrium with the surrounding humidity by taking up or giving off water with time. The sorptivity of paper comes from the fact that the paper structure is porous and the fibers can take up liquids by swelling. Thus, both phases in a paper structure are absorbent. The quantity of liquid absorbed is given by:

$$q = q_p + q_f \qquad (6.18)$$

where q_p is the sorption into the pores and q_f is the sorption into the fibers. The liquid enters the pore network by capillary flow. The sorption of water into the fibers takes place by diffusion.

It has been shown that when the paper is immersed in a liquid, all the pore volume is accessible to the liquid through gaps between the fibers in the surface of the paper. The sorption of liquids by paper depends on the paper structure. In the case of aqueous liquids, chemical factors also become important. Liquid sorption increases the thickness of the paper due to swelling. The sorption mechanisms of sized and unsized papers are different. Paper sizing is done to make the paper hydrophobic. In unsized paper, there is a simultaneous pore and fiber take-up until the liquid penetrates to the reverse side of the paper with all the pores filled. After the liquids penetrate the paper, further sorption takes place due to continued swelling. In sized paper, water can not easily enter the pore system. Therefore, the only sorption mechanism is diffusion into and swelling of the fiber system [12].

There is a wetting delay before the water begins to be absorbed by the paper. It was reported that the wetting delay decreased with an increase in moisture content of the paper. Aging or heat treatment increases the wetting time. Pure cellulose has a wetting time of about 8 ms [50]. Results have also shown that the short time sorption of oils follows a square-root dependence on time.

Wettability

Measurements of contact angle of liquid drops on paper surface has been used to determine wettability of paper (Figure 6.34). TAPPI T-458 is designed to measure the surface wettability of paper (angle of contact

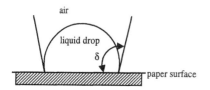

FIGURE 6.34. Contact angle between liquid drop and paper surface.

method). A small drop of water or ink is placed on the paper surface and contact angle after a certain amount of time is measured using a microscope. As the hydrophobicity of the surface increases, contact angle, δ, also increases. Wettable surfaces have smaller values of contact angle.

6.5.16 *Optical Properties*

Paper can reflect, transmit and absorb light (Figure 6.35). Optical properties of paper, such as opacity, reflectivity and transmittance, depend on type of pulp, amount of bleaching, fillers, coating, dyes and colored pigments, finishing operations and grammage. These in turn affect specific scattering coefficient and specific absorption coefficient of the paper. Increasing light scattering coefficient increases reflectivity and opacity and reduces transmittance.

Paper materials can be classified as optically thin or thick. A material is optically thin if most of the light is transmitted. If reflectivity is high (more than 50%) and the transmittance

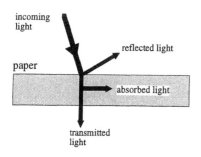

FIGURE 6.35. Interaction of light with sheet.

is low (less than 20%), then the material is said to be optically thick. Most papers belong to optically thick category [51,52].

For a single-layer structure, the reflectivity is increased with a high scattering coefficient or with a low absorption coefficient. Reflectivity is, by definition, independent of grammage. Opacity is increased by increasing grammage, light scattering coefficient and light absorption coefficient. However increasing absorption coefficient reduces reflectivity. Coating increases reflectivity and opacity. Type of filler also affects opacity.

A fine paper made of bleached chemical pulp has a high reflectivity and a low opacity due to its low scattering and absorption coefficients. To increase opacity of fine papers, usually a filler is included in the structure to increase light scattering coefficient. A paper with mechanical pulp has a higher opacity than a paper with bleached chemical pulp because of its high value of light scattering coefficient. However, its reflectivity is generally lower due to a higher value of absorption coefficient. Filler in a mechanical pulp would have a slight effect on opacity.

Coating is an effective way for improving both reflectivity and opacity. This is because a coating pigment has a high scattering coefficient and a fairly low absorption coefficient. Choice of pigment is important for the optical properties of the coating layer. It should be noted that not all coatings contain pigments.

Optical properties can be more readily altered by producing paper in layers. For example, a dark secondary or mechanical pulp may be placed in the center and covered with bleached chemical pulp for brightness. Playing cards contain a very dark center to ensure high opacity whereas the top layers have a high scattering coefficient for good reflectivity. In paperboard, a layered structure is generally used with a dark center and brighter top layer to increase reflectivity. In fact, paper is generally considered as a layered structure for the analysis of the optical properties. Kubelka-Munk Theory is a fairly accepted model to analyze the optical properties of multilayer paper structures. In

TABLE 6.6. Test Methods for Opacity and Brightness [53].

	Opacity	Brightness
International Organization for Standardization	ISO 2471	ISO 2470
USA	TAPPI T 425	TAPPI T 452
	T 519	
	ASTM D 589	ASTM D 985
Japan	JIS P-8138	JIS P-8123
Canada	CPPA E.2	CPPA E.1
Scandinavia	SCAN P8	SCAN P3

a multi layer structure, the optical properties are the result of contributions from each individual layer. Reflectivity of a multi layer structure depends on grammage [51].

Brightness

Brightness refers to the lightness of overall reflectivity of the paper. Brightness is a function of spectral reflectance, energy distribution of the illuminant and viewing conditions. TAPPI T-452 is used to measure the brightness of pulp, paper and paperboard. Table 6.6 shows standard test methods for opacity and brightness in different countries.

Opacity

Opacity is the property of the paper that prevents objects on the other side to be seen through the paper. The higher the opacity the less the objects can be seen through the paper. For example, if the print can not be seen through the page, the paper has high opacity. Opacity level is important especially for printing papers which require high opacity to prevent show-through. Bond, newsprint and writing papers also require high opacity. Some papers such as black photographic wrapping paper and paperboard are considered to be truly opaque. Low opacity is required in tracing and glassine papers. TAPPI T-425 (Opacity of Paper) and T-519 (Diffuse Opacity of Paper) are used to measure paper opacity. Opacity increases with grammage. The opposite of opacity is transparency which is defined as the ability of paper not to scatter light.

Gloss

Gloss is the lustrous or shiny appearance of the paper surface. Glossiness depends on the type of illumination, angle of incidence and reflection, and relative position of paper and observer. Rough surface may not seem as glossy. High gloss may make reading difficult because of glare. The opposite of gloss is matte. TAPPI T-480 (Specular Gloss of Paper and Paperboard at 75°) and T-653 (Specular Gloss of Paper and Paperboard at 20°) are used to measure gloss.

Color

Color for paper and paperboard is important for aesthetic purposes and marketing. Color can be characterized as a numerical value. TAPPI -527 is used to measure color of paper and paperboard.

6.5.17 Hand Properties (Softness and Hardness)

Softness is especially important for papers that come in contact with the skin such as sanitary and facial tissues and toweling.

There are several manufacturing parameters that affect softness. Minimum amount of bonding reduces stiffness and improves softness. High fiber strength is required to have adequate strength in soft sheets. Sulfite is considered a superior pulp for softness. Dry or wet creping methods are used in manufacturing to increase softness. Humectants (e.g. glycerin, diethylene glycol) and debonders (e.g. alkyl amine salts, amides) are the additives that are used for softening. Sheets with nearly equal MD/CD tensile ratios also produce softer sheets.

TAPPI method T-451 (Flexural Properties of Paper) is used to measure softness. Softness is related to sheet density, rigidity, compressibility and surface smoothness.

Hardness can be defined as the resistance to indentation. Hardness is usually related to density. As a result, highly beaten pulps produce hard paper. Compressibility is the opposite of hardness. Compressibility is determined by the change in the thickness of the sheet.

6.5.18 Electrical Properties

Paper is used in a variety of electrical devices and related applications including capacitors, transformers, electronic circuit boards, wires, coil windings and cables. Paper has good dielectric properties. The electrical properties of paper are also important in direct electrophotographic and dielectric printing methods. Examples of other less known applications of paper are electrostatically assisted rotogravure, radio frequency or microwave dryers, and radio frequency or microwave moisture gages that involve the interaction of electromagnetic fields with paper and water systems [54].

Electrical properties of paper are anisotropic. The important electrical properties of paper and paperboard are volume and surface conductivities or resistivities, dielectric constants and loss, permittivities and dielectric breakdown strength. Dielectric constant is affected by the apparent density, fiber configuration and crystalline cellulose [55].

6.5.19 Thermal Properties

As the development and application areas of papers increase, the thermal properties of papers are also getting more attention. Important thermal properties of paper are specific heat, thermal conduction, thermal expansion, combustion and thermal decomposition [56]. The specific heat of paper is similar to natural cellulosic fiber. Moisture content of paper affects these properties significantly. Thermal conductivity of paper depends on apparent density and absorbed moisture. Temperature and moisture changes affect paper expansion and shrinkage significantly. Of the two, moisture has more effect than temperature.

Paper and paperboard has a high flammability. This is an advantage for disposal by burning. However, flammability is a disadvantage for paper for applications where fire hazard is critical such as interior decoration. Methods to measure flame resistance of paper are TAPPI T-461 and ASTM D-777.

Heating changes the physical properties of paper leading to charring and pyrolysis. Increasing temperature increases the rate of deterioration and degradation. The thermal stability of paper can be improved by inclusion of inorganic materials.

6.5.20 Finish

Finish is a broad and relatively subjective term used to describe surface characteristics of paper that affect paper appearance and feel. It involves smoothness, softness, gloss and other properties that are explained above.

6.6 TAPPI Test Methods

The following is the breakdown of the TAPPI test methods [32]:

- Fibrous Materials and Pulp Testing
  T 1-200 Series
- Paper and Paperboard Testing
  T 400-500 Series
- Nonfibrous Materials Testing
  T 600-700 Series
- Container Testing
  T 800 Series
- Structural Materials Testing
  T 1000 Series
- Testing Practices
  T 1200 Series

Appendix E lists the major paper testing and research laboratories [32].

6.7 References

1 Paper Structure and Properties, Eds. Bristow, J. A., and Kolseth, P., *International Fiber Science and Technology Series/8,* Marcel Dekker, Inc., 1986.

2 Ring, J. F. G., Introduction to Pulp and Paper Properties and Technology, Course Notes, The Paper Science Department, University of Wisconsin-Stevens Point, 1991.

3 *The Dictionary of Paper,* 4th Edition, American Paper Institute, Inc., 1980.

4 Mark, R. E., "Fiber Structure" in *Handbook of Physical and Mechanical Testing of Paper and Paperboard, Vol. 2,* Ed. R. E. Mark, Marcel Dekker, Inc., 1984.

5 Kolseth, P., and Ehrnrooth, E. M. L., "Mechanical Softening of Single Wood Pulp Fibers", in *Paper Structure and Properties,* Eds. J. A. Bristow and P. Kolseth, Marcel Dekker, Inc., 1986.

6 Page, D. H., et al., "Elastic Modulus of Single Wood Pulp Fibers", *TAPPI,* 60(4), 1977.

7 Ehrnrooth, E. M. L., and Kolseth, P., "The Tensile Testing of Single Wood Pulp Fibers in Air and in Water", *Wood Fiber Sci.,* 16(4), 1984.

8 Page, D. H., and Seth, R. S., "The Elastic Modulus of Paper II. The Importance of Fiber Modulus, Bonding and Fiber Length", *TAPPI,* 63 (6), 1980.

9 Hollmark, H., et al, "Mechanical Properties of Low Density Sheets", *TAPPI,* 61(9) , 1978.

10 TAPPI T-233, Fiber Length of Pulp by Classification, *TAPPI Test Methods, 1996–1997,* TAPPI Press, Atlanta, GA.

11 TAPPI T-234, Coarseness of Pulp Fibers, *TAPPI Test Methods, 1996–1997,* TAPPI Press, Atlanta, GA.

12 Bristow, J. A., "The Pore Structure and the Sorption of Liquids," in *Paper Structure and Properties,* Eds. J. A. Bristow and P. Kolseth, Marcel Dekker, Inc., 1986.

13 Fellers, C., "The Significance of Structure for the Compression Behavior of Paperboard," in *Paper Structure and Properties,* Eds. J. A . Bristow and P. Kolseth, Marcel Dekker, Inc., 1986.

14 Bristow, J. A., "The Paper Surface in Relation to the Network" in *Paper Structure and Properties,* Eds. J. A. Bristow and P. Kolseth, Marcel Dekker, Inc., 1986.

15 Mardia, K. V., *Statistics of Directional Data,* Academic Press, New York, 1972.

16 Setterholm, V., and Kuenzi, E., "Fiber Orientation and Degree of Restraint During Drying-Effect on Tensile Anisotropy of Paper Handsheets", *TAPPI,* 53(10), 1970.

17 Ruch, H., and Krassig, H., "The Determination of Fiber Orientation in Paper", *Pulp Pap. Mag. Can.,* 59(6), 1958.

18 Prusas, Z. C., "Laboratory Study of the Effects of Fiber Orientation on Sheet Anisotropy," *TAPPI,* 46(5), 1963.

19 Htun, M., and Fellers, C., "The In-Plane Anisotropy of Paper in Relation to Fiber Orientation and Drying Restraints," in *Paper Structure and Properties,* Eds. J. A. Bristow and P. Kolseth, Marcel Dekker, Inc., 1986.

20 Lim, Y. W., et al, "Light Scattering by Cellulose II. Oriented Condenser Paper", *TAPPI,* 53(12), 1970.

21 Kallmes, O. J., "Technique for Determining the Fiber Orientation Distribution Throughout the Thickness of a Sheet," *TAPPI,* 52(3), 1969.

22 Rigdahl, M., and Hollmark, H., "Network Mechanics" in *Paper Structure and Properties,* Eds. J. A . Bristow and P. Kolseth, Marcel Dekker, Inc., 1986.

23 Carlsson, L., "The Layered Structure of Paper" in *Paper Structure and Properties,* Eds. J. A. Bristow and P. Kolseth, Marcel Dekker, Inc., 1986.

24 Jones, A. R., "An Experimental Investigation of the In-Plane Elastic Moduli of Paper," *TAPPI,* 51(5), 1968.

25 Mann, R. W., "Elastic Wave Propagation in Paper," *TAPPI,* 62(8), 1979.

26 Kallmes, O., and Bernier, G., "The Structure of Paper IV. The Free Fiber Length of a Multiplanar Sheet," *TAPPI,* 46(2), 1963.

27 Perez, M., and Kallmes, O. J., "The Role of Fiber Curl in Paper Properties," *TAPPI,* 48(10), 1965.

28 Perez, M., "A Model for the Stress-Strain Properties of Paper," *TAPPI,* 53(12), 1970.

29 Page, D. H., and Seth, R. S., "The Elastic Modulus of Paper III. Effects of Dislocations, Microcompressors, Curl, Crimps and Kinks," *TAPPI,* 63(10), 1980.

30 Htun, M., "Internal Stress in Paper", in *Paper Structure and Properties,* Eds. J. A. Bristow and P. Kolseth, Marcel Dekker, Inc., 1986.

31 Li, J. C. M., "Dislocation Dynamics in Deformation and Recovery," *Can. J. Physics,* 45, 1967.

32 *TAPPI Test Methods, 1996–1997,* TAPPI Press, Atlanta, GA.

33 De Ruvo, A., Fellers, C., and Kolseth, P., "Descriptive Theories for the Tensile Strength of Paper," in *Paper Structure and Properties,* Eds. J. A . Bristow and P. Kolseth, Marcel Dekker, Inc., 1986.

34 Page, D. H., et al, "Behavior of Single Wood Fibers Under Axial Tensile Strain," *Nature,* 229, 1971.

35 Page, D. H., et al, "The Mechanical Properties of Single Wood-Pulp Fibers I. A New Approach," *Pulp Pap. Mag. Canada,* 73(8), 1972.

36 El-Hosseiny, F., and Page, D. H., "The Mechanical Properties of Single Wood Pulp Fibers: Theories of Strength," *Fibre Sci. Technol.,* 8(1), 1975.

37 Cowan, W. F., and Cowdrey, E. J. K., "Evaluation of Paper Strength Components by Short-Span Tensile Analysis," *TAPPI,* 57(2), 1974.

38 Boucai, E., "Zero-span Tensile Test and Fibre Strength," *Pulp Pap. Mag. Can.,* 72(10), 1971.

39 Gates, E. R., and Kenworthy , I. C., "Effects of Drying Shrinkage and Fiber Orientation on Some Physical Properties of Paper," *Pap. Technol*, 4(5), 1963.

40 Johnson, J. A., et al, "Failure Phenomena" in *Handbook of Physical and Mechanical Testing of Paper and Paperboard,* Ed., R. E. Mark, Vol. 1, Marcel Dekker, Inc., 1983.

41 Fellers, C., "Edgewise Compression Strength of Paper" in *Handbook of Physical and Mechanical Testing of Paper and Paperboard,"* Eds. R. E. Mark and K. Murakami, Vol. 1, Marcel Dekker, Inc., 1983.

42 Fellers, C., et al, "Edgewise Compression Properties: A Comparison of Handsheets Made from Pulps of Various Yields," *TAPPI,* 63(6), 1980.

43 Fellers, C., and Carlsson, L., "Measuring the Pure Bending Properties of Paper. A New Method," *TAPPI,* 62(8), 1979.

44 Carlsson, L., et al, "The Mechanism of Failure in Bending of Paperboard," *J. Mater. Sci.,* 15, 1980.

45 Carlsson, L., and Feller, C., "Flexural Stiffness of Multiply Paperboard," *Fibre Sci. Technol.,* 13(3), 1980.

46 Norman, B., "The Formation of Paper Sheets" in *Paper Structure and Properties,* Eds. J. A. Bristow and P. Kolseth, Marcel Dekker, Inc., 1986.

47 Carlsson, L., "Out-of-Plane Hygroinstability of Multiply Paperboard," *Fibre Sci. Technol.,* 14(3), 1981.

48 Burdette, J., Conway, L., Ernst, W., Lanier, E., and Sharpe, J., *The Manufacture of Pulp and Paper: Science and Engineering Concepts,* TAPPI Press, 1988.

49 Fellers, C., Andersson, H., and Hollmark, H., "The Definition and Measurement of Thickness and Density," in *Paper Structure and Properties,* Eds. J. A. Bristow and P. Kolseth, Marcel Dekker, Inc., 1986.

50 Lyne, M. B., and Aspler, J. S., "Wetting and the Sorption of Water by Paper Under Dynamic Conditions," *TAPPI Journal,* 65(12), 1982.

51 Pauler, N., "Opacity and Reflectivity of Multilayer Structures," in *Paper Structure and Properties,* Eds. J. A. Bristow and P. Kolseth, Marcel Dekker, Inc., 1986.

52 Billmayer, F. W., and Richards, L. W., "Scattering and Absorption of Radiation by Materials," *J. Color Appearance,* 11(2), 1973.

53 Borch, J. , "Optical and Appearance Properties" in *Handbook of Physical and Mechanical Testing of Paper and Paperboard,* Ed. R. E. Mark, Vol. 2, Marcel Dekker, Inc., 1984.

54 Baum, G. A., "Electrical Properties: I Theory" in *Handbook of Physical and Mechanical Testing of Paper and Paperboard,* Ed., R. E. Mark, Vol. 2, Marcel Dekker, Inc., 1984.

55 Matsuda, S., "Electrical Properties: II Practical Considerations and Methods of Measurement of Electrical Properties" in *Handbook of Physical and Mechanical Testing of Paper and Paperboard,* Ed., R. E. Mark, Vol. 2, Marcel Dekker, Inc., 1984.

56 Nakagawa, S., and Shafizadeh, F., "Thermal Properties" in *Handbook of Physical and Mechanical Testing of Paper and Paperboard,* Ed., R. E. Mark, Vol. 2, Marcel Dekker, Inc., 1984.

General References

Aspler, J. S., et al, "The Dynamic Wettability of Paper. Part I. The Effect of Surfactants, Alum, and pH on Self-sizing," *TAPPI J.,* 67(9), 1984.

Aspler, J. S., and Lyne, M. B., "The Dynamic Wettability of Paper. Part II. Influence of Surfactant Type on Improved Wettability of Newsprint," *TAPPI J.,* 67(10), 1984.

Baum, G. A., et al, "Orthotropic Elastic Constants of Paper," *TAPPI,* 64(8), 1981.

Bernie, J., and Douglas, W. J. M., "Local Grammage Distribution and Formation of Paper by Light Transmission Image Analysis," *TAPPI Journal,* Vol. 79, No. 1, January 1996.

Blokhius, G., and Kalff, P. J., "Dynamic Smoothness Measurements of Papers and Print Unevenness," *TAPPI,* 59(8), 1976.

Borch, J., and Scallan, A. M., "A Comparison Between the Optical Properties of Coated and Loaded Sheets," *TAPPI,* 58(2), 1975.

Borch, J., and Scallan, A. M., "An Interpretation of Paper Reflectance Based Upon Morphology-The Effect of Mass Distribution," *TAPPI,* 59(10), 1976.

Bristow, J. A., and Kolseth, P., Eds., *Paper Structure and Properties,* Marcel Dekker, Inc., New York, 1986.

Bristow, J. A., "Optical Properties of Pulp and Paper—New Standardization Proposals," *TAPPI Journal,* Vol. 78, No. 12, December 1995.

Byrd, V. L., "Effect of Relative Humidity Changes on Compressive Creep Response of Paper," *TAPPI,* 55(11), 1972.

Byrd, V. L., "Web Shrinkage Energy: An Index of Network Fiber Bonding," *TAPPI,* 57(6), 1974.

Byrd, V. L., et al, "Method for Measuring the Interlaminar Shear Properties of Paper," *TAPPI,* 58(10), 1975.

de Ruvo, A., et al, "The Biaxial Strength of Paper," *TAPPI,* 63(5), 1980.

Diaz-Kotti, M., "Test Methods to Evaluate the Contamination Resistance of Forming Fabrics Subjected to Secondary Fibers," *TAPPI Journal,* Vol. 78, No. 7, July 1995.

Dinus, R.J. and Welt, T., "Tailoring Fiber Properties to Paper Manufacture: Recent Developments," *TAPPI Journal,* April 1997.

Dumbleton, D. F., "Longitudinal Compression of Individual Pulp Fibers," *TAPPI,* 55(1), 1972.

El-Hosseiny, F., and Aberson, D., "Light Scattering and Sheet Density," *TAPPI,* 62(10), 1979.

Fahey, M. D., "Mechanical Treatment of Chemical Pulps," *TAPPI,* 53(11), 1970.

Grundstrom, K. J., et al, "High-Consistency Forming of Paper," *TAPPI,* 56(7), 1973.

Hagerty, G. A., et al, "Design of a Modern Air Permeability Instrument Comparable to Gurley-Type Instruments," *TAPPI Journal,* Vol. 76, No. 2, February 1993.

Hollmark, H., "Evaluation of Tissue Paper Softness," *TAPPI Journal,* 66(2), 1983.

Htun, M., and de Ruvo, A., "Correlation Between the Drying Stress and the Internal Stress in Paper," *TAPPI,* 61(6), 1978.

Htun, M., and Fellers, C., "The Invariant Mechanical Properties of Oriented Handsheets," *TAPPI,* 65(4), 1982.

Kapoor, S. G., et al, "A New Method for Evaluating the Printing Smoothness of Coated Papers," *TAPPI,* 61(6), 1978.

Karenlampi, P., et al, "The Effect of Pulp Fiber Propertis on the In-Plane Tearing Work of Paper," *TAPPI Journal,* Vol. 79, No. 5, May 1996.

Karenlampi, P., et al, "Opacity, Smoothness, and Toughness of Mechanical Printing Papers: The Effect of Softwood Kraft Pulp Propreties," *TAPPI Journal,* April 1997.

Komori, T., and Makishma, K., "Numbers of Fiber-to-fiber Contacts in General Fiber Assemblies," *Textile Research Journal,* 47(1), 1977.

Komori, T., and Makishima, K., "Geometrical Expressions of Spaces in Anisotropic Fiber Assemblies," *Textile Research Journal,* 49(9), 1979.

Koning, J. W., and Stern, R. K., "Long-Term Creep in Corrugated Finerboard Containers," *TAPPI,* 60(12), 1977.

Koubaa, A., and Koran, Z., "Measure of the Internal Bond Strength of Paper/Board," *TAPPI Journal,* Vol. 78, No. 3, March 1995.

Lyne, B., "Measurement of the Distribution of Surface Void Sizes in Paper," *TAPPI,* 59(7), 1976.

Mark, R. E., Ed., *Handbook of Physical and Mechanical Testing of Paper and Paperboard, Vol. 2,* Marcel Dekker, Inc., New York, 1984.

Naito, T., "Torsional Properties of Single Wood Pulp Fibers," *TAPPI,* 63(7), 1980.

Nguyen, H. V., and Durso, D. F., "Absorption of Water by Fiber Webs: An Illustration of Diffusion Transport," *TAPPI J.,* 66(12), 1983.

Nissan, A.H., and Batten, G.L. Jr., "The Link Between the Molecular and Structural Theories of Paper Elasticity," *TAPPI Journal,* April 1997.

Oliver, J. F., "Dry-creping of Tissue Paper—A Review of Basic Factors," *TAPPI,* 63(12), 1980.

Page, D. H., et al, "The Elastic Modulus of Paper I. The Controlling Mechanisms," *TAPPI,* 62(9), 1979.

Page, D. H., and Seth, R. S., "The Elastic Modulus of Paper II. The Importance of Fiber Modulus, Bonding and Fiber Length," *TAPPI,* 63(6), 1980.

Panshin, A. J., and de Zeeuw, C., *Textbook of Wood Technology,* 4th ed., McGraw-Hill, New York, 1980.

Perez, M., "A Model for the Stress-Strain Properties of Paper," *TAPPI,* 53(12), 1970.

Perkins, R. W., and Mark, R. E., "On the Structural Theory of the Elastic Behavior of Paper," *TAPPI,* 59(12), 1976.

Perkins, R. W., and McEvoy, R. P., "The Mechanics of the Edgewise Compressive Strength of Paper," *TAPPI,* 64(2), 1981.

Piggott, M. R., "A Theoretical Framework for the Compressive Properties of Aligned Fibre Composites," *J. Mater. Sci.,* 16, 1981.

Prud'homme, R. E., and Robertson, A. A., "Composite Theories Applied to Oriented Paper Sheets", *TAPPI,* 59(1), 1976.

Rigdahl, M., et al, "Analysis of Cellulose Networks by the Finite Element Method," *J. Mater. Sci.,* 19(12), 1984.

Roberts, J. C., Ed., Paper Chemistry, Chapman and Hall, New York, 1991.

Robinson, J. V., "Optical Properties of Paper as Affected by Wet-End Chemistry," *TAPPI,* 59(2), 1976.

Rosen, A., "Effect of Calendering Variables on the Smoothness and Density of Blade Coating Raw Stock," *TAPPI,* 54(11), 1971.

Salmen, N. L., and Back, E. L., "Moisture-Dependent Thermal Softening of Paper, Evaluated by Its Elastic Modulus," *TAPPI,* 63(6), 1980.

Sastry, C. B. R., et al, "Measurement and Prediction of Fiber Coarseness in Western Hemock," *TAPPI,* 56(4), 1973.

Scallan, A. M, and Borch, J., "An Interpretation of Paper Reflectance Based Upon Morphology I. Initial Considerations," *TAPPI,* 55(4), 1972.

Scallan, A. M., and Borch, J., "An Interpretation of Paper Reflectance Based Upon Morphology: General Applicability," *TAPPI,* 57(5), 1974.

Seth, R. S., "Measurement of Fracture Resistance of Paper," *TAPPI,* 62(7), 1979.

Sharma, A., et al, "Physical and Optical Properties of Steam-Exploded Laser-Printed Paper," *TAPPI Journal,* Vol. 79, No. 5, May 1996.

Thorpe, J. L., et al, "Mechanical Properties of Fiber Bonds," *TAPPI,* 59(5), 1976.

Uesaka, T., et al, "Biaxial Tensile Behavior of Paper," *TAPPI,* 62(8), 1979.

Van Den Akker, J. A., "Structure and Tensile Characteristics of Paper," *TAPPI,* 53(3), 1970.

Van Liew, G. P., "The Z-direction Deformation of Paper," *TAPPI,* 57(11), 1974.

Wentao, L., and Carlsson, L. A., "Micro-Model of Paper, Part 2: Statistical Analysis of the Paper Structure," *TAPPI Journal,* Vol. 79, No. 1, January 1996.

Yamauchi, T., and Murakami, K., "Acoustic and Optical Measurements of Paper Under Strain," *TAPPI Journal,* Vol. 76, No. 2, February 1993.

6.8 Review Questions

1 What is the most important property of paper? Why?

2 What properties of wood fiber make it the most suitable fiber for papermaking? Explain.

3 What are the differences between hardwood and softwood tree fibers?

4 Calculate the maximum compacting force

between two fibers of 1 inch length (refer to Figure 6.16). Surface tension of water is 72 dynes per cm.

5 Are the fibers oriented randomly in the sheet? What are the factors that affect fiber orientation? Explain.

6 What is hydrogen bonding? How does it provide strength in the paper?

7 What are the possible solutions to the two sidedness of the sheet?

8 Which one is stronger: fiber or paper? Explain the strength relationship between them.

9 Explain visco-elasticity. What are the advantages and disadvantages of visco-elastic behavior of paper?

10 How does the grammage affect the other properties of paper?

11 What can be done to alter the optical properties of paper?

Total Quality Management (TQM)

Total Quality Management (TQM) is a management strategy that is designed to continuously improve performance in all areas of responsibility at every level in an organization. The very definition of TQM indicates three major components [1,2,3,4,5]:

(1) The *totality* implies that the system should encompass all functions and groups in the organization such as manufacturing, engineering, research and development, sales and marketing and management at every level.

(2) The *quality* is the end result that can be evaluated by two primary factors: performance and producer-user mutual satisfaction

(3) The *management* implies the degree of communication among different units in the organization to provide the necessary advise and feedback which are essential to high quality

In a highly competitive environment, commitment to quality is not an option but an absolute necessity. In today's business environment, quality is a relentless pursuit without an end. To stay competitive, quality should come second only to health, safety and environmental factors. As in every industry, product quality has become one of the most critical factors in paper machine clothing and paper industries. Today, the same equipment and machinery are generally available to every manufacturer. Yet, the quality of products from different manufacturers may differ. The reasons for this difference are various but include expertise and know how of the technical people as well as the management philosophy and style of the company.

7.1 Historical Development of TQM

The purpose of TQM is to ensure high quality products and services. Various methods and systems have been developed and implemented in the past to achieve this goal. The concept of quality dates back to 3000 BC Babylon where it was practiced to impose uniformity of units for weights and measures [6]. Throughout time, fitness for use of a product has been measured in regards to its uniformity and quality. Preliminary standards were used to ensure uniformity in goods produced which were measured by means of product physical attributes which were judged by the inspector's individual apprehension of these attributes. Prior to industrial revolution, quality was mostly managed by individuals that were involved in various levels of the product life cycle. In 1875, the principles of scientific management was introduced by F. W. Taylor. Around this time, assembly line concept was developed. Products were manufactured in several steps and usually different person(s) were involved at each step. Therefore, the workers no longer had the opportunity to fabricate from beginning to end, personally monitoring and controlling the quality and productivity of their efforts. They became responsible only for a portion of the complete manufacturing cycle without seeing direct effects of their efforts on the final product and its required function. This caused the development of barriers or walls between the departments and workers.

During the early 20th century, written standards showing the product's critical statistics were used. In 1911, Taylor suggested "time" as the simplest measurement of performance. His efforts focused on development of time study analysis, work standards and wage incentives. Prior to 1920s, the implementation of quality control was still somewhat subjective. Product standards, as they are known, became more common after the World War I, and the only interaction between the standards and the product was the physical inspection. This era can be classified as small size business environment with low production and labor intensive manufacturing. There was an intimate relationship between the manufacturer and customer. International competition was relatively small.

In 1920s, Quality Control (QC) started to appear as a specific discipline and became one of the central activities in manufacturing. In-process and final inspection became two essential functions of quality control. Statistical quality control was developed for this purpose in this period. Inspection and sampling were evaluated using statistical control charts and sampling techniques such as random sampling.

Between 1930s and 1970s, quality control was mostly done by sampling inspection prior to shipping to customers. Off-line testing methods were developed for this purpose and statistics were used in sampling inspection. Government regulations and military standards during the World War II demanded sampling inspection. The American Society for Quality Control (ASQC) was established in 1945. Since the late 1970s, the emphasis has been on making quality products and delivering quality services. Quality control expanded to include evaluation of product design, manufacturing and end-use [6]. Toward the end of this period, the international competition started to surface.

The 1980s witnessed the revival of quality control which was driven by global competition. Statistical Process Control (SPC) was introduced. The computer revolution has improved quality and quality control techniques.

In the mid 1980s, the concept of Total Quality Management (TQM) began to develop. Quality control and quality engineering were integrated with quality management. Efforts are being made to remove the barriers and walls among departments within the organization. Giant steps in information technology resulted in development of expert systems, integrated information systems, neural networking and fuzzy logic methods. The concept of concurrent engineering was developed. Continuous improvement became the way of life and different ways of wage incentives were developed.

7.2 Quality Philosophies

Quality today is measured in many different ways by many different experts. Webster's dictionary defines quality as "a. An inherent feature, b. Degree of excellence" [7]. Quality masters such as Dr. W. E. Deming, Dr. J. M. Juran and Dr. D. A. Garvin all have their definitions and explanations of causes associated with quality. Quality can be viewed as a matter of whether or not a product is useable and the extent to which that product is fit for use. This "fitness for use" [8] is gauged by the customer in different areas from price to ethics. Quality standards such as ISO give manufacturers assurance of minimized variance and adherence to specifications.

Quality is the main ingredient in a product that delights the customer by either meeting or exceeding expectations [9]. Thus, it should be paramount to any business, to meet or even exceed expectations of its customers. Another necessity for quality is the cost. Production costs can be very high when quality is having to be reworked into a satisfiable product. It is reported that over 30% of production costs can be due to reworking or replacing defective parts or products [10]. If quality could be made into the product the first time, a company could save this 30%, which could improve quality without raising the price of the item out of the competitive range.

Customers today are more willing to pay for quality. It is reported that 87% of consumers said quality is their number one criterion when making a purchase and 84% of consumers said that they would be willing to pay more for top quality [11].

The benefits of quality to the companies are as follows [12,13]:

(1) *Greater Market Share* is caused by the customer's perceived better quality of an item and their repeated buying and reference of the product to others.

(2) *Higher Growth Rate* is directly proportional to the growth in the market share due to the role that a company's market share plays on the growth rate of that company.

(3) *Premium Price* is a great benefit of perceived quality by the customer. A premium price can be charged for a product due to the users' willingness to pay more for what is perceived as better quality

(4) *Loyal Customers* are produced when they see that a product has not let their expectations down and has fulfilled the promise of quality that the premium price sometimes dictates. This benefits the company greatly on the basis of advertising and reputation due to the loyalty of these repeat customers

(5) *Highly Motivated Employees* come from having pride in the fact that the product they produce is perceived to have the highest quality.

These benefits translate into long lasting prosperity for these companies, as well as a healthy bottom line at the end of the financial quarter.

There are many dimensions of quality such as performance, features, reliability, conformance, durability, service ability and perceived quality [14]. Aesthetic, which is often considered as a measure for quality, is secondary importance for industrial textiles such as paper machine clothing. These dimensions of quality are independent of one another while at the same time one may be sacrificed for the benefit of the other. A product or service may be high in one dimension while being low in another without the two being directly related. This is what gives these dimensions such an encompassing view of quality and its application.

Perceived quality is where the definition of quality gets somewhat vague. It seems to hold true that quality can be defined only in terms of the person assessing the item or service. This raises the question: who is the final judge of quality? Is it the customer or the manager? The manager would assume quality in his product if it meets manufacturing standards and production quotas. The customer has a different view of quality and its meaning. The factors that affect customer satisfaction for quality are price, technology, psychology, time orientation, contractual and ethical [9]. Customers tend to associate quality with higher price. There is some evidence that price is used by consumers in quality estimates and that for some products consumers' estimates of quality are affected by price [15].

In the manufacture of paper machine clothing, technology indicates factors such as fabric and seam strength and other properties that are affected by the state of technology in the industry. Psychological effect generally means aesthetic requirements of customers which have secondary significance in paper machine clothing. Time orientation includes durability. Contractual refers to a product guarantee, the refund policy, etc. Most major purchases made creates some type of anxiety about whether or not the choice was right. Refund policies also tell the customer that the producer stands behind his/her promise that the product will meet the customer's needs or the customer will receive a refund. Ethical refers to honesty of advertising, courtesy and responsiveness of sales/service personnel, etc.

7.2.1 *W. Edwards Deming's View of Quality*

According to Dr. Deming, quality begins with the intent, which is fixed by the management. The intent must be converted by engineers and others into plans, specifications, tests and production [16]. Deming supports this view with his 14 obligations of management which are listed in Table 7.1.

Deming's description of the steps to improve quality show mainly managerial techniques that might be ignored by companies who hope for quality to appear through such programs as management by objective or management by sheer numbers. This does not ensure that everyone in the organization is committed to quality or their intent is to build quality into the item itself. Without this intent for inherent quality, consistent perceived quality by both the customer and management will fall short.

Dr. Deming sees quality as a triangle of interaction among:

(1) The product itself
(2) The user and how he uses the product, how he installs it, how he takes care of it, what he was led to expect

(3) Instructions for use, training of customer and training of repairmen, service provided for repairs and availability of parts

Dr. Deming supports standardization as a way of improving production. He clarifies regulation versus standards by explaining how standards made voluntarily keep federal regulations down to a minimum. With an industry overly regulated by the government there can be little room for that business to compete with unregulated businesses overseas.

7.2.2 *Joseph M. Juran's View of Quality*

Dr. Juran sees quality as "fitness for use" [8] which seems to have gained widespread acceptance as the short definition of quality. Dr. Juran describes quality as having a twofold meaning; that of product features that meet customer needs and a product's freedom from deficiencies. In the eyes of the customer, the fewer the deficiencies the higher the quality. Juran shows that product features mostly impact sales and increasing product features increases the cost. When product deficiencies are reduced, the cost of quality is decreased.

TABLE 7.1. Deming's 14 Points of Quality Management.

1. Create constancy of purpose for the improvement of product or service
2. Adopt the new philosophy
3. Cease dependence on mass inspection for quality control
4. End the practice of awarding business on the basis of price tag
5. Improve constantly and forever the system of production and service, to improve quality and productivity, and thus constantly decrease costs
6. Institute more thorough, better job-related training
7. Institute leadership
8. Drive out fear, so that everyone may work effectively for the organization
9. Break down barriers between departments
10. Eliminate slogans, exhortations and targets for the work force that ask for zero defects and new levels of productivity without a means of accomplishment
11. Eliminate work standards on the factory floor
12. Remove the barriers that rob employees at all levels in the company of their right to pride of workmanship
13. Institute a vigorous program of education and self improvement
14. Put everybody in the organization to work to accomplish the transformation

7.2.3 *David A. Garvin's View of Quality*

Dr. Garvin, a Harvard expert on quality, states that quality has five definitions or approaches to definitions which are as follows [17]:

- *Transcendent* is a view of quality that believes in the "innate excellence" of a product. This view sometimes contends that quality is found in craftsmanship and not in mass production. This view is that of unanalyzable philosophical properties.
- *Product Based* is the view that quality is a precise and measurable variable, wherein differences in quality must be based on the components of that product.
- *User Based* definitions of quality take the cliche "Quality lies in the eyes of the beholder" approach.
- *Manufacturer Based* definitions are mainly based on the engineering and manufacturing side of the supply for the product.
- *Value Based* definitions are those that define quality in terms of costs and prices.

According to Garvin, these five definitions show all sides of quality. The definitions of Dr. Juran and Dr. Deming would fall into the user based and manufacturing based definitions, respectively.

7.2.4 *Other Views of Quality*

There have been many other academicians and scientists who have developed quality philosophies. Philip B. Crosby defined quality as the "conformance to requirements". His method to achieve quality is based on prevention of defects. Performance standards are established for zero defects. According to him, the quality performance measurement is the cost of quality.

According to Genichi Taguchi, quality must be designed into every product and corres-

ponding process. He defines the quality as "the (minimum) loss imparted by the product to the society from the time the product is shipped."

7.3 Core Concepts of TQM

TQM is an integrated quality improvement system. There are certain components of TQM that should be well defined, understood and implemented by the organization that is implementing it. First of all, the organization must define the quality either by adopting one of the existing quality definitions or by developing a new one to suit the company's needs. Before defining quality, a careful study of customer expectations, requirements and priorities would be helpful.

For the TQM to be successful, top management should be committed to the development and implementation of the quality program at every level of the company. Management is responsible for providing help and proper facilities, equipment, etc., for the company associates. Motivation and appreciation of the quality control activities are also crucial.

Training is an important part of TQM. Each person in the organization should get proper training in his/her area. The necessary tools of quality control including Statistical Process Control (SPC) should be taught in theory and practice. Training of employees is a long term investment for the company. Lack of training reduces the probability of success in quality manufacturing.

Job satisfaction encourages employees to make voluntary contributions to the company in the form of suggestions, initiatives, etc. Education, not criticism, should be the form of retribution in case of any problems.

Communication is one of the keys for the success of any company. Communication is especially critical between production and marketing. Optimization of product mix should be considered very carefully when planning production.

Modern textile manufacturing machinery minimize the labor involvement. They also require less frequent maintenance than in the past. State of the art equipment can be a valuable asset for quality manufacturing. However, well run older equipment may be a better investment in the short term than poorly run high tech equipment. Computerized machinery and equipment increases accuracy and quality of production. Integrated automation involves automation of material flow and transportation, machine settings, machine operation, maintenance, data recording and analysis. Preventive maintenance reduces the risk of machine downtime. An effective plant housekeeping schedule should also be implemented.

Information systems provide on-line and real-time information about various production parameters including production rate, efficiency, speed and idle time. Mechanical, electronic and optical sensors can be installed for measuring several product properties such as thickness, uniformity, tension and surface characteristics. In addition to on-line information systems, off-line testing systems can provide detailed data analysis using computers and software packages [4].

The concepts and procedures of TQM are not industry specific, i.e. they can be applied in any industry, small or large. Quality improvements must continue even after a quality assurance program is implemented.

7.4 Basic Management Elements of TQM

Commitment, awareness, planning, accountability and recognition are the main managing elements in a TQM environment. Commitment level indicates where the management stands on quality. Awareness involves an ongoing education and training of people on the progress of quality improvement. Employees should be made aware of the relationship between the company and customer (partnership).

Good planning is one of the most critical managing elements in TQM. Quality improvement must be a result instead of a reaction. All quality improvement efforts must be distributed through different levels of the organization with careful planning. Levels of planning ranges from simple statements of goals and specific actions to more detailed approaches that include graphical and analytical tools such as flow diagrams, cause-and-effect diagrams, etc.

Accountability involves the measure, tracking and review of the performance of individuals and teams in improving quality. Recognition naturally follows the accountability. Proper recognition of the efforts of individuals and teams is an effective way of motivation.

7.5 Statistical Process Control

Statistics is a branch of mathematics dealing with the collection, analysis, interpretation and presentation of masses of numerical data. Statistical Process Control (SPC) is an analytical tool for TQM with a manufacturing focus on process variation. SPC is a method by which one identifies the capabilities of a given process and by using statistical methods, identifies certain aspects of this process that can be improved. The end result of a statistical process control program is reduced variability, increased productivity and cost reduction [18].

Applied statistics is a tool that can be used to answer a variety of questions and the use of statistical procedures is, therefore, always a middle step. To answer a question, first an investigation must be formed. The typical steps in an investigation are:

(1) A substantive question is formulated and refined and a plan is developed to obtain relevant evidence.

(2) When appropriate, a statistical model is chosen to assist in organizing and analyzing the data to be collected. At this point, a statistical question may be developed, the answer to which may be expected to throw light on the substantive question. A statistical question differs from a substantive question in that it always concerns a statistical property of that data such as the average of the set of measures.

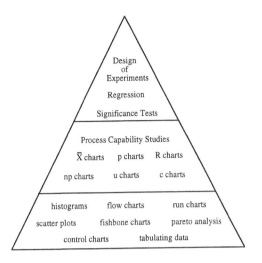

FIGURE 7.1. Statistical techniques.

(3) Upon applying the statistical procedure, one arrives at a statistical conclusion.

(4) Finally, a substantive conclusion is drawn.

There are numerous statistical techniques that have been available and well developed both in theory and in practice. Figure 7.1 shows some of the most common statistical techniques in use today. The degree of complexity of the technique increases towards the top. A detailed description of these methods is not among the purposes of this book. Therefore, only some of the more basic techniques will be described here. The reader is referred to the literature for the other techniques.

7.5.1 *Data*

Data are facts or figures from which conclusions can be drawn. Data are the raw materials of statistics. Data can be gathered from any number of sources. The term *population* refers to the complete set of observations or measurements about which we would like to draw conclusions. Single observations or measurements are called *elements*. An element is also referred to as a *sample*. Suppose we were interested in the average permeability of Type A dryer fabrics manufactured during a three month period. Our research reveals that 50

fabrics were manufactured during this time period and that a number of final permeabilities were produced. The results of our research are listed in Table 7.2.

The entire collection is referred to as the population and any individual final permeability as an element of that population. Assume that the target permeability was 75 cfm. Analysis of data shows that there are some number of permeabilities close to the target, some higher and some lower. Also note that there are some permeabilities in the 60's and some in the 90's, and that they are few in number, and that no permeabilities are lower or higher.

There are different purposes of collecting the data. When a particular method of doing a job is introduced, it is natural to consider whether the method is appropriate or not. The decision is usually based on past results and experience, or perhaps some conventional method. However, in the case of factory work where data are collected through the actual manufacturing process, the procedural methods are introduced on the basis of the information obtained. The manufacturing procedure will be most effective if a proper evaluation is made and on-the-job data are essential for making a proper evaluation. Data and subsequent evaluation will form the basis for actions and decisions. As factory operations will vary with the manufacturing procedure involved, data should be classified in terms of the various purposes. These purposes are:

(1) Data to assist in understanding the actual situation: For example, these data are collected to check the extent of the dispersion

TABLE 7.2. The Final Permeability of Type A Dryer Fabrics Manufactured in One Quarter.

84	82	72	70	72
80	62	96	86	68
68	87	89	85	68
87	85	84	88	89
86	86	78	70	81
70	86	88	79	69
79	61	68	75	77
90	86	78	89	81
67	91	82	73	77
80	78	76	86	83

in part sizes coming from a machining process or to examine the percentage of defective parts contained in lots received. As the number of data increased, they can be arranged statistically for easier understanding. Estimates and comparisons can then be made concerning the condition of lots received as well as the manufacturing process, utilizing specified figures, standard figures, target figures, etc.

(2) Data for analysis: Analytical data may be used, for example, in examining the relationship between a defect and its cause. Data are collected by examining past results and making new tests. In this case, various statistical methods are used to obtain correct information.

(3) Data for process control: After investigating product quality, this kind of data can be used to determine whether or not the manufacturing process is normal and in control. Control charts are used in this evaluation and action is taken on the basis of these data.

(4) Regulating data: This is the type of data used, for example, as the basis for raising or lowering the temperature of an electronic furnace so that a standardized temperature level may be obtained. Actions can be prescribed for each datum and measures taken accordingly.

(5) Acceptance or rejection data: This form of data is used for approving or rejecting parts and products after inspection. On a basis of the information obtained it can be decided what to do with the parts or products.

Data serve as the basis for action. After evaluating actual conditions, as we have other data, proper action can be taken. After data are collected, they are analyzed, and the information is extracted through the use of statistical methods. Finally, data are collected because the population (the work situation) from where data are drawn is constantly changing.

To determine the magnitude of that change (in determining whether or not action must be taken) we statistically analyze that data to determine if the variation is part of the natural variation for that process, or if the variation is due to some cause outside the normal variation of that process. After collecting data, it is possible to see that items seemingly produced in exactly the same way turn out differently. In many of the cases, this dispersion occurs because of differences in:

- the raw materials
- the tools, machinery or equipment
- the work method or process
- the measurement

Raw materials differ slightly in composition according to the source of supply and size differences will occur within accepted limits. Machines may seem to be functioning uniformly, but dispersion can arise from differences within parts of the machine itself. Likewise, a piece of equipment may be operating optimally only part of the time. Work methods, although programmed according to prescribed processes, can lead to perhaps even greater variations. Finally, a random task such as measurement is not always exactly achieved. Even slight differences can add up to a great deal of product quality dispersion.

7.5.2 Frequency Distributions

Statistical analysis can not proceed until the data of interest in an investigation are similarly organized in a useful manner. The simplest way to facilitate this kind of analysis is to put the permeabilities in order. To do so, locate the highest and lowest permeability values. Then, record all possible permeability values (including the two extremes) in descending order. Among the data in Table 7.2 the highest permeability is 96, and the lowest permeability is 61. The recorded sequence is 96, 95, 94, . . . , 61 as shown in Table 7.3 along with the number of occurrence (frequency) for each sample. Such an array showing the permeabilities and their frequency of occurrence is called a *Frequency Distribution*.

Thus organized, many features of interest are easily perceived. It is seen that the target permeability of 75 cfm is a little below the

TABLE 7.3. Permeabilities from Table 7.2, Organized in Order of Magnitude.

Perme- ability	Fre- quency	Perme- ability	Fre- quency
96	1	78	3
95	0	77	2
94	0	76	1
93	0	75	1
92	0	74	0
91	1	73	1
90	1	72	2
89	3	71	0
88	2	70	3
87	2	69	1
86	6	68	3
85	2	67	1
84	2	66	0
83	1	65	0
82	3	64	0
81	2	63	0
80	2	62	1
79	2	61	1

middle of the distribution. Although permeabilities range from 61 to 96, the bulk of the distribution lies between 67 and 91. There is one permeability near the top of the tolerance range and two permeabilities near the bottom of the tolerance range.

When data range widely, reducing individual data to a smaller number of groups makes it easier to display the data and to grasp their meaning. Table 7.4 shows two ways in which the permeabilities in Table 7.2 may be grouped into *class intervals*.

There are, however, two matters that should be kept in mind when considering grouped scores.

(1) When scores are grouped some information is lost.

(2) A set of raw scores does not result in a unique set of grouped scores.

Table 7.4 shows, for example, two different sets of grouped scores that may be formed from the raw data in Table 7.2. In converting raw data to grouped data there are several principles that should be kept in mind:

- A set of class intervals should be mutually exclusive. That is, intervals should be chosen so that one data point can not belong to more than one interval.
- All intervals should be the same width.
- Intervals should be continuous throughout the distribution. In part "A" of Table 7.4, there are no terms in an interval 64–66. To omit this interval and "close ranks" would create a misleading impression.
- The interval containing the highest value is placed at the top.
- The number of classes is dependent upon the amount of data.

Fewer class intervals mean greater interval width, with consequent loss of accuracy. The larger the number of classes in a frequency distribution, the more detail is shown. If the number of classes is too large, though, the classification loses its effectiveness for summarizing the data. Too few classes, on the other hand, condenses information so much as to leave little insight into the pattern of the distribution. The best number of classes in a frequency distribution often needs to be determined by experimentation. Usually an effective number of classes is somewhere between 4 and 20.

TABLE 7.4. Permeabilities from Table 7.2, Converted to Grouped Data Distributions.

A: Class Interval Width = 3		B: Class Interval Width = 5	
Perme- ability	Fre- quency	Perme- ability	Fre- quency
94–96	1	95–99	1
91–93	1	90–94	2
88–90	6	85–89	15
85–87	10	80–84	10
82–84	6	75–79	9
79–81	6	70–74	6
76–78	6	65–69	5
73–75	2	60–64	2
70–72	5		
67–69	5		
64–66	0		
61–63	2		

The choice of the class width (or class interval) is related to the determination of the number of classes. Many class intervals result in lesser convenience. It is generally best if all the classes have the same width. If the classes are not equally wide, one often can not tell readily whether the differences in class frequencies are due mainly to differences in the concentration of items, or to differences in the class width. For choice of class limits, it is usually suggested as good statistical practice that the class limits be chosen so that the midpoint of each class is approximately equal to the arithmetic average of the values falling in that class. The class limit is often referred to as the class boundary.

The next step is to translate a set of raw scores to a grouped data frequency distribution. We can illustrate the procedure with the data in Table 7.4. The first step is to find the lowest score, 61, and the highest score, 96. The difference between these two figures is the range of scores. Therefore, the range is $96 - 61 = 35$. Since there are to be from 4 to 20 class intervals we must find an interval width satisfying this condition. Often, this can be done by inspection. Next, we must determine the starting point of the bottom class interval. Since the lowest permeability is 61, the lowest interval should be 59–61, 60–62 or 61–63. Often, it is convenient to make the lower score limit a multiple of the class interval width. Next, refer to the collection of raw data in whatever order they may be, and tally their occurrence one by one against the list of class intervals. Note that when it comes to the selection of width of the class interval a choice is usually available. The number of intervals desired is one factor in choosing, but convenience is sometimes another. Choosing a width of five or ten or some multiple thereof makes both the construction and interpretation of a distribution easier.

The steps in constructing a frequency distribution are summarized below:

(1) Find the value of the lowest score and the highest score.

(2) Find the range by subtracting the lowest score from the highest.

(3) Determine the width of the class interval needed to yield four to twenty class intervals.

(4) Determine the point at which the lowest interval should begin.

(5) Record the limits of all class intervals placing the interval containing the highest value at the top. Intervals should be continuous and of the same width.

(6) Using the tally system enter the raw data in the appropriate class intervals.

A frequency distribution can be presented as a table or a graph. A graph is based entirely on the tabled data and therefore can tell no story that can not be learned by inspecting the table. However, graphic representation often makes it easier to see pertinent features of a set of data. Graphic representation is one way of presenting all kinds of quantitative information. There are many different kinds of graphs. For the representation of frequency distributions, there are three main types of graphs: the *histogram*, its variant, the *bar diagram*, and the *frequency polygon*.

7.5.3 Histogram

The histogram consists of a series of rectangles, each of which represents the frequency of data in one of the class intervals in a tabled distribution. A sample histogram is illustrated in Figure 7.2. The rectangle is erected so that

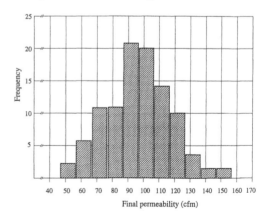

FIGURE 7.2. Histogram: Final permeability of Type A dryer fabric manufactured in a year.

its two vertical boundaries coincide with the exact limits of the particular interval, and its height is specified by the frequency of scores for that interval. Either frequencies or proportionate frequencies may be represented by the histogram. There are some details in construction that apply to the histogram.

- The graph has two axes: horizontal and vertical. The intersection of the two axes represents the 0 (zero) point on both scales.
- It is customary to represent scores (the measured data) along the horizontal axis and frequency along the vertical axes.
- Convenient units should be chosen to identify position along the axes. Limits of class intervals or interval midpoints usually do not form the most convenient frame of reference for the horizontal axis.
- Whether the graph of a frequency distribution appears squat or slender depends on the choice of scale used to represent position along the two axes.

Since it is desirable that similar distributions should appear similar when graphed, it is customary to choose a relative scale such that the height will be no less than about half or no more than about 3/4's of the width. Some trial and error may be necessary to create a graph suitable in size and convenient in scale. Construction of a histogram is very similar to that of a frequency distribution. The details are given below:

(1) Calculate the range of all the data: $X_1 - X_s$ where X_1 is the largest value and X_s is the smallest value in the group.
(2) Determine the number of classes (the number of histogram bars) as follows:

Number of Data	Approximate Number of Classes (k)
under 50	7–10
50–100	10–15
100–250	15–20
over 250	20–25

(3) Calculate the class interval (width of class)

$$h = (X_1 - X_s)/k \qquad (7.1)$$

(round h to a convenient number)
(4) Determine class boundaries (class limits)
(5) Tally points falling into each class
(6) Construct histogram

7.5.4 *Bar Diagram*

The bar diagram is very similar to the histogram, except that space is inserted between the rectangles, thus properly suggesting the essential discontinuity of the several categories. An example of a bar diagram is given in Figure 7.3. Since the categories have no necessary order they may or may not be arranged in order of magnitude of frequency. Bar diagrams can be vertical or horizontal.

7.5.5 *Frequency Polygon*

The same data plotted as a histogram in Figure 7.2 has been represented by frequency polygon in Figure 7.4. In this type of graph, a point is plotted above the midpoint of each class interval at a height commensurate with the frequency of the scores in that interval. These points are then connected with straight lines.

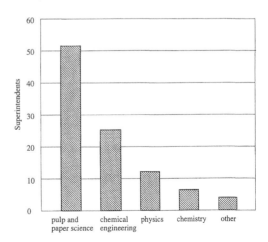

FIGURE 7.3. Bar diagram: comparative frequency of academic majors of machine superintendents in fine paper mills in Alabama.

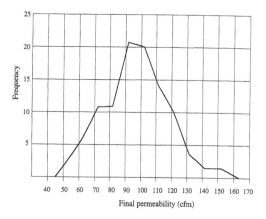

FIGURE 7.4. Frequency polygon: final permeability of Type A dryer fabrics manufactured in one year.

7.5.6 Pie Chart

In pie charts, the data are presented in a circle (Figure 7.5). The entire circle represents 100% of the data. The circle (pie) is divided into percentage slices that show the shares of data. This is useful in the same way as a Pareto Chart.

7.5.7 Flow Chart

A flow chart is used to identify the steps of a process. Flow charts are useful in showing the sequence and relationship of the process steps. Flow charts allow to uncover loopholes which are potential sources of problems in the process. Flow charts can be applied to any type of process. Figure 2.26 shows the flow chart for forming fabric manufacturing.

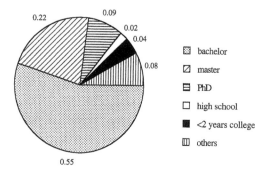

FIGURE 7.5. Pie chart: highest level of education in paper industry.

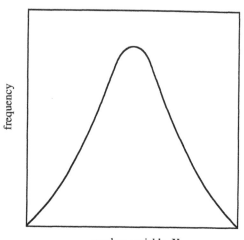

FIGURE 7.6. Normal curve.

7.5.8 Normal Curve

The "normal curve" is shown in Figure 7.6. It represents a distribution of any random variable that is symmetrical.

7.5.9 Measures of Central Tendency

There are several measurements that summarize important characteristics about the data sets that are being considered. These measurements fall into two broad categories: measures of central tendency and measures of dispersion (variability).

Measures of central tendency (also known as measures of position) are intended to provide a single summary figure which describes the level (high or low) of a set of observations. The three common measures of central tendency are arithmetic mean, mode and median.

In statistical terminology, the capital letter X is used as a collective term to specify the particular set of data being considered. An individual score in a set may be identified by a subscript such as X_1, X_2, etc. For example, consider the set of three scores 31, 21 and 41. We may identify them as follows:

$$X_1 = 31, X_2 = 21, X_3 = 41$$

A set of scores in a sample may be represented as follows:

$$X : X_1, X_2, X_3, \ldots, X_n$$

When the scores in a set are to be summed, the capital Sigma (Σ) indicates that this operation is to be performed. It should be read "the sum of" (whatever follows).

Mean

The arithmetic mean (or average) $\overline{X}$ of a set of values $X_1, X_2, \ldots, X_n$, is the sum of the values divided by the number of items, n:

$$\overline{X} = \frac{\sum_{i=1}^{n} X_i}{n} \qquad (7.2)$$

Example: Find the mean of the following group of numbers.

17, 9, 23, 18, 6, 42, 15 $n = 7$

$$\overline{X} = \frac{\sum_{i=1}^{7} X_i}{n}$$
$$= \frac{(17 + 9 + 23 + 18 + 6 + 42 + 15)}{7}$$
$$= \frac{130}{7} = 18.57$$

The mean has a number of unique properties that should be considered when deciding on an appropriate measure of position.

- The sum of the values of a set of items is equal to the mean multiplied by the number of items.
- The sum of the deviations of the X_n values from their mean $\overline{X}$ is 0.

Mode (Mo) and Median (Mdn)

The *mode* is a score that occurs with the greatest frequency. In grouped data, it is taken as the midpoint of the class interval that contains the greatest number of scores. The symbol is Mo. The median (Mdn) of a set of items is the value of the middle item when all items are arranged by magnitude.

Mdn: — Arrange the items (numbers) in numerical order (ascending or descending)
— For an odd-numbered group the median is the middle value
— For an even-numbered group the median is the average of the two middle numbers

Example: Find the median of the following groups of numbers.

A. 17, 9, 23, 18, 6, 42, 15 becomes 6, 9, 15, 17, 18, 23, 42.
   17 is the median.

B. 17, 9, 23, 18, 6, 42, 15, 25 becomes 6, 9, 15, 17, 18, 23, 25, 42.
   The median is $(17 + 18)/2 = 17.5$.

Properties of Mode, Median and Mean

The mode is easy to obtain, but it is not very stable. Further, when the data are grouped, the mode may be strongly affected by the width and location of class intervals. Another problem is that there may be more than one mode for a particular set of scores.

In distributions that are perfectly symmetrical, that is, the left half is a mirror image of the right half, the mean, median and mode will yield the same value. It is important to note that the widely occurring "normal distribution" falls in this category. We have earlier referred to this phenomenon as the normal curve. Figure 7.7 shows what happens to

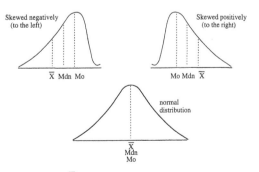

FIGURE 7.7. $\overline{X}$, Mo and Mdn in the normal distribution and in skewed distributions.

mean, median and mode in skewed distributions, as compared with the normal distribution.

Equality of mean and median does not guarantee that the distribution is symmetrical, although it is not likely to depart very far from that condition. On the other hand, if the mean and median have different values, the distribution can not be symmetrical. Furthermore, the more skewed, or lopsided the distribution is, the greater the discrepancy between these two measures. In a smooth negatively skewed distribution the mode has the highest score value and the median falls at a point about 2/3's of the distance between that and the mean. The mean, as might be expected, has been specially affected by the fewer, but relatively extreme values in the tail. In a positively skewed distribution, exactly the opposite situation is obtained. As a consequence, the relative position of the median and the mean may be used to determine a direction of skewness in the absence of a look at the entire distribution. At a rough descriptive level, the magnitude of the discrepancy between the two measures will afford a clue to the degree of departure from a state of symmetry.

The median is the point that divides the upper half of the scores from the lower half. In doing so, the median responds to how many scores lie below or above it, but not to how far away the scores may be. A little below the median, or a lot, both count the same in determining its value. Since the median is sensitive to the number of scores below it, but not to how far they are below, the median is less sensitive than the mean to the presence of a few extreme scores. Consider this distribution:

$$X : 5, 6, 7, 8, 24$$

The median is 7, the mean is 10. The value of the top score, 24, is quite different from that of the remainder of scores. It strongly affects the total, and hence the mean, but it is just another score above the median. If the score had been 9 rather than 24, the median would be the same.

Sometimes one encounters a distribution that is reasonably regular except for several quite deviant scores at one end. If it is feared that these scores will carry undue weight in determining the mean, a possible solution is to calculate the median. The median will respond to their presence, but no more than to others that lie on that side of its position. Similarly, in distributions that are strongly asymmetrical, the median may be the better choice if it is desired to represent the bulk of the scores and not give undue weight to the relatively few deviant ones. Of the measures of central tendency under consideration, the median stands second to the mean in ability to resist the influence of sampling fluctuation in ordinary circumstances.

The mean, unlike any of the other measures of central tendency, is responsive to the exact position of each score in the distribution. The mean may be thought of as a balance point of the distribution. The mean is more sensitive to the presence (or absence) of scores at the extremes of a distribution than are the median and the mode. When a measure of central tendency should reflect the total of the scores, the mean is the choice since it is the only measure based on this quantity. When further statistical computation is to be done, the mean is likely to be the most useful of the measures of central tendency.

A characteristic of great significance concerns sampling stability. Suppose that from a large population of scores samples were completely drawn at random. Means of such samples would have similar but not identical values. This would also be true for the medians of the same samples and for their modes. However, for distributions encountered most frequently in statistical work, the means vary least among them. Thus, under ordinary circumstances, the mean best resists the influence of sampling fluctuation.

7.5.10 Measures of Dispersion (Variability)

Measures of variability express quantitatively the extent to which the data in a set scatter about or cluster together. Whereas a measure of central tendency is a summary description of the level of performance of a group, a measure of variability is a summary description of the spread of performance. Measures of variation include the range, deviation, variance and the standard deviation.

Range

The range is the difference between the largest and smallest values in a set of items. The range indicates the maximum extent of variation in a set. A limitation of the range as the measure of variation of a set of items is that it depends only on the extreme items, and does not consider the others. Thus, a data set might contain items quite close to each other with the exception of one extreme value.

Despite the concentration of almost all the items, the range should be large since it is based only on the extreme values. Another limitation of the range as a measure of variability is that it depends on the number of items in the set. The larger the number of items, the larger the range tends to be. Like other measures of variability, the range is a distance and not, like measures of central tendency, a location. In a frequency distribution, the range may be figured as the difference between the value of the lowest raw scores that could be included in the bottom class interval and that of the highest raw scores that could be included in the upper-most class interval.

$$\text{Range: } R = X_l - X_s \qquad (7.3)$$

where l denotes the largest number in the group, and s denotes the smallest number in the group.

Example: Determine the range, R, for the following group of numbers

$$17, 9, 23, 18, 6, 42, 15$$

$$X_l = 42$$
$$X_s = 6$$
$$R = 42 - 6 = 36$$

The range is a rough and ready measure of variability. In most situations, including those involving samples drawn from a normal distribution, the range varies more with

sampling fluctuation than other measurements do. Only the two outermost scores of a distribution affect this value. The remainder could lie anywhere between them. It is thus not very sensitive to the total conditions of the distribution. In addition, a single errant score is likely to have a more substantial effect on the range than on the standard deviation, which is particularly sensitive to conditions at the extremes of a distribution. Like other measures, with the exception of the standard deviation, it is of little use beyond the descriptive level. In many types of distributions, including the important normal distribution, the range is dependent on sample size, being greater when sample size is larger.

Deviation

A deviation score expresses the location of a score by indicating how many score points it lies above or below the mean of the distribution. The deviation score is symbolized by the lowercase letter x to distinguish it from a raw score which is symbolized by the upper case letter X. In symbols, the deviation score is defined as:

$$x = (X - \overline{X}) \qquad (7.4)$$

The deviation score states the position of scores relative to the mean.

Variance

A commonly used measure of dispersion is called the *variance*. It is a measure that takes into account all the values in a set of items. It can be defined as the mean of the squares of the deviation scores. The defining formula for the variation of a sample (variance) is:

$$Variance = S^2 = \frac{\displaystyle\sum_{i=1}^{n} (X_i - \overline{X})^2}{n - 1} \qquad (7.5)$$

The variance is a measure which provides quantified information about the variability in a data set. The first step in calculating the variance is to square the deviations around the mean. Then, to obtain an overall measure of variability per item, the average of the squared deviation is taken. This mean squared deviation is called the variance. The units of the variance is the square of the variable units.

The wider the dispersion of the values around the mean, the greater the variance. If there is no dispersion of the values (if all are equal and hence all are equal to the mean), then the variance is equal to zero.

For calculation purposes, equation (7.5) can be written as:

$$S^2 = \frac{\sum_{i=1}^{n} X_i^2 - \frac{1}{n}\left(\sum_{i=1}^{n} X_i\right)^2}{n-1} \qquad (7.6)$$

Standard Deviation

The most important and widely used measure of variability is the standard deviation, s. It is the square root of the variance.

$$s = \sqrt{\frac{\sum_{i=1}^{n}(X_i - \overline{X})^2}{n-1}} \qquad (7.7)$$

Example: Determine the standard deviation for the following group of numbers.

$$17, 9, 23, 18, 6, 42, 15$$

Calculation:

$$\overline{X} = \frac{\sum X_i}{n}$$

$$= \frac{(17 + 9 + 23 + 18 + 6 + 42 + 15)}{7}$$

$$= \frac{130}{7} = 18.57$$

$$\sum X_i^2 = 17^2 + 9^2 + 23^2 + 18^2 + 6^2 + 42^2 + 15^2$$

$$= 3248$$

$$\frac{\left(\sum X_i\right)^2}{n} = \frac{130^2}{7} = 2414.28$$

$$S^2 = \frac{3248 - 2414.28}{7 - 1} = 138.95$$

$$s = \sqrt{138.95} = 11.78$$

The standard deviation, like the mean, is responsive to the exact position of every score in a distribution. If a score is shifted to a position more deviant from the mean, the standard deviation will be larger than before. If the shift is to a position closer to the mean, the magnitude of the standard deviation is reduced. Consequently, we may expect that the standard deviation is more sensitive to the exact condition of the distribution than measures that do not share this property. Because of this characteristic sensitivity, the standard deviation may not be the best choice among measures of variability when the distribution contains a very few extreme scores, or when the distribution is barely skewed. One of the most important points favoring the standard deviation is its resistance to sampling fluctuation. In repeated sampling of the same population, the numerical value of the standard deviation tends to jump about less than would other measures computed from the same samples.

Coefficient of Variation (CV)

$$CV\% = \frac{Standard\ Deviation}{Mean} \times 100 \quad (7.8)$$

$$= s \times \frac{100}{\overline{X}}$$

Measures of position are concerned with the location around which items are concentrated, whereas measures of dispersion consider the extent to which the items vary. Measures of skewness are still another kind of measure summarizing the extent to which the items are symmetrically distributed. One way to determine the nature of skewness of a frequency distribution is to compare the values of the mode, median and mean. The three frequency polygons in Figure 7.8 show the comparative positions of the mean, median and mode in unimodal (one mode) frequency distributions with differing degrees and directions of asymmetry.

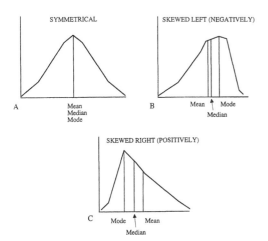

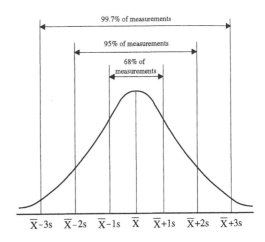

FIGURE 7.8. Examples of symmetrical and skewed unimodal frequency distributions.

FIGURE 7.9. Relative frequency of cases contained within certain limits in the normal distribution.

Figure 7.8(A) is an example of a symmetrical unimodal frequency distribution. The values of the mean, median and mode in any section distribution are identical. Figures (B) and (C) show, respectively, an example of a unimodal frequency distribution skewed to the left, or skewed negatively, and one skewed to the right, or skewed positively. These figures indicate the relationship typically existing between the mean, median and mode in unimodal frequency distributions that are moderately skewed: the mean is furthest out toward the tail of the distribution and the median is between the mean and the mode.

Let us now consider measures of variability and the normal distribution. Since the measures of variability are defined in different ways we would expect them to yield different values when calculated for the same distribution. Their relative magnitudes can not be exactly specified in terms of a generalization true for all distributions, because special characteristics of some distributions have a different effect on the different measures. The normal distribution, however, is so frequently useful that properties of these measures in that distribution ought to be given attention.

In the ideal normal distribution, the interval $\pm 1s$ contains about 68% of the scores. The interval $\pm 2s$ contains about 95% of the scores and the interval $\pm 3s$ contains about 99.7% of the scores. These relationships are pictured in Figure 7.9.

Of course, in any sample drawn from a normal distribution, there will be a certain degree of irregularity, and these statements will be only approximately true. Figure 7.9 also shows that the more extreme the score the greater the rarity of its occurrence in the normal distribution. Consequently, when a sample of limited size is drawn at random from a large number of scores that are normally distributed, very extreme scores may not be encountered.

It clearly would be advantageous if substantially all the information contained in a frequency distribution could be packed into two figures, such as a measure of central tendency and a measure of dispersion. Under favorable circumstances it is possible to come close enough to this ideal for many purposes. Mathematical statisticians have discovered that the best measure of central tendency for this purpose is the arithmetic mean. A useful measure of dispersion for this purpose is the standard deviation. The ideal would be to be able to say, given any average and standard deviation, just what proportion of the measurements fell within any specified limits. These two numbers sufficiently define the normal curve that has been mentioned earlier. Many observed frequency distributions of measured qualities of manufactured products and many other frequency distributions found in nature, do correspond roughly to this normal curve. Although

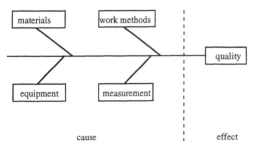

cause | effect

FIGURE 7.10. Cause-and-effect diagram.

most of the area under this bell shaped symmetrical form is included within the limits $\bar{X}$ ±3s, the curve extends from minus infinity (−∞) to plus infinity (+∞). The curve is fully defined by $\bar{X}$ and s.

Many of the useful actions that may be taken in manufacturing are related to future production rather than completed lots of items. The knowledge that must be gained from the sampling procedure is knowledge of the pattern of variation of the production process from which the sample was drawn. We have already defined population as the complete set of observations or measurements about which we would like to draw conclusions. In order to draw conclusions about this population it is necessary to rely on numerical values derived

from samples drawn from that population. Such numerical values, which include the mean, median, standard deviation and range, summarize the information contained in the sample data.

7.5.11 *Cause-and-Effect Diagram*

The causal factors of dispersion diagrammed in Figure 7.10 point out the cause-and-effect relationship. The objective of improving the quality of the output first must be approached by an analysis of the causal factors. But it is necessary to know both cause and effects in greater detail and in more concrete terms in order to illustrate their relationship on a diagram and make them more useful. The variables which can cause the dispersion can be called factors. Quality characteristics describe the effect. A cause-and-effect diagram is useful in sorting out the causes of dispersion and organizing mutual relationships. These same cause-and-effect diagrams can assist in determining from which areas the data should be collected.

Figure 7.11 shows a cause-and-effect diagram for print quality variables. Because of their shapes, cause-and-effect diagrams are

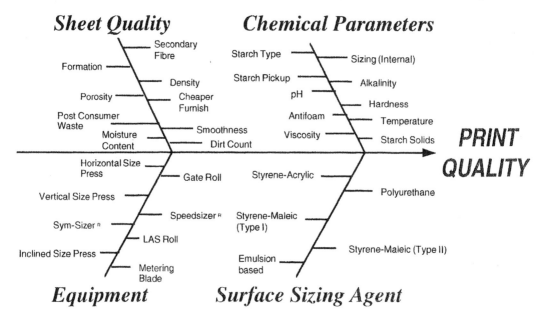

FIGURE 7.11. A cause-and-effect diagram for print quality variables of paper [19].

also called fishbone diagrams. To construct a cause-and-effect diagram, first determine the quality characteristic which is what to be improved and controlled. Second, write the main factors which may be contributing to the quality characteristic. It is recommended to group the major possible causal factors of dispersion into such items as raw materials, equipment (machines or tools), method of work (workers), measuring method (inspection), etc. Each individual group will form a branch. Third, onto each of these branch items, write in the detailed factors which may be regarded as the causes. Continue to write—in even more detailed factors. Defining and linking the relationships of the possible cause of factors should lead to the source of the quality characteristic.

Cause-and-effect diagrams are drawn to clearly illustrate the various causes affecting product quality by sorting out and relating the causes. There are several ways to use a cause-and-effect diagram, but primarily, the cause-and-effect diagram is a guide to concrete action. The cause-and-effect diagram reveals those areas or aspects which influence quality.

7.5.12 *Pareto Diagram*

Once the causes or factors producing a particular result are determined, the next step is to determine which factor to work on first. A Pareto diagram is a useful tool for just this purpose. A pareto diagram is a special form of vertical bar diagram. As a hypothetical example, Table 7.5 lists specific complaints encountered in joining pin seams.

TABLE 7.5. *Pin Seam Joining Problems (Hypothetical).*

Type Problem	Number of Complaints	Percent Distribution
Wrong wire	4	10.5
Zip tape failure	5	13.2
Missing loops	3	7.9
Loops too small	15	39.5
Irregular loop size	11	28.9

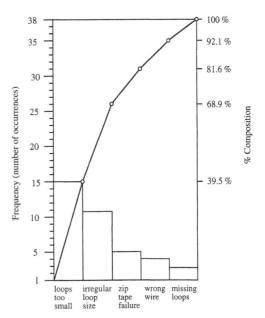

FIGURE 7.12. Pareto diagram.

The number of specific complaints is shown as is the percent distribution of that specific complaint. Data from this table has been made into a bar graph as shown in Figure 7.12.

In the Figure, the left vertical axis shows the number of defectives for each defective category and the right vertical axis shows the percent distribution for each category over the total defectives. The horizontal axis lists the defective items starting with the most frequent one on the left progressing to the least frequent on the extreme right and the rest arranged by order of magnitude. The cumulative total of the number of defectives for each category is shown by the line graph.

A *Pareto* diagram, such as this, indicates which problem should be solved first in eliminating defects and improving the operation. According to the graph "loops too small" should be tackled first because it forms the tallest bar. The next most significant defective item is the second tallest bar. This may appear very simple, yet bar graphs are extremely useful in factory quality control. It is much easier to see which defects are most important with the bar graph than by using only a table of numbers.

Establishment of a Pareto Diagram

(1) Step 1. Ascertain the classifications that will be used in the graph. For example, graphs could list items according to kinds of defectives, defects, work groups, products, size, damage, etc.

(2) Step 2. Decide on the time period to be covered on the graph. There is no prescribed period of time, so naturally the period will vary according to the situation.

(3) Step 3. Total the frequency of occurrence for each category for the period. Each total will be shown by the length of the bar.

(4) Step 4. Draw horizontal and vertical axes on graph paper and demarcate the vertical axis in the proper units.

(5) Step 5. Draw in the bars, beginning on the far left with the most frequent defective items. The height of the bar will correspond to the value on the left vertical axis. Keep the width of the bars the same and in contact with its neighbor. If there are several categories with limited frequencies, they can be grouped together as "others" and placed at the right most bar on the graph.

(6) Step 6. Under the horizontal axis label each of the bars.

(7) Step 7. Plot a line showing the cumulative total reached from the addition of each category.

(8) Step 8. Title the graph and briefly write the source of the data on which the graph is based.

The Pareto diagram clearly and distinctly exposes the relative magnitude of defects and provides a base of common knowledge from which to operate. At a glance one can see that the two or three taller bars account for the majority of problems, the smaller bars being lesser causes.

Experience has shown that it is easier to reduce a tall bar by half than to reduce a short bar to zero. If we can reduce the tallest bars, the ones which account for the most defectives, it will have been a considerable accomplishment. If it requires the same effort to reduce the tall bar by half as to reduce a short bar by half, there is no doubt which should be selected as the target.

The value of Pareto diagrams is that they indicate which factors are most prevalent and therefore deserve concentrated efforts for improvement.

7.5.13 Run Charts and Control Charts

Histograms consolidate the data to show the overall picture. Pareto diagrams are used to indicate problem areas. These methods group the data for a specified period and express them in static form. However, in a manufacturing environment, one also wants to know more about the nature of the changes that take place over a specified period of time, that is, the dynamic form. This means that we not only have to see what changes in data occur over time; but we must also study the impact of the various factors in the process that change over time.

Thus, if the materials, the workers, or the working methods or equipment were to change during this time, we would have to note the effect of such changes on production. One way of following these changes is by using graphs called run charts.

As an example, Table 7.6 lists coil diameter measurements taken over a 20 hour period. Four measurements were taken each hour. The mean ($\bar{X}$) and the range (R) for each subgroup have been calculated. Additionally, an "average of the average" ($\bar{\bar{X}}$) and an "average of the range" ($\bar{R}$) have been determined.

Using this data, a run chart (Figure 7.13) was drawn indicating the average hourly value and the hourly range.

This run chart shows that the values were low at the outset but showed a tendency to rise over time. Table 7.6 does not reveal this trend. In other words, new information is discovered by looking at the movement of the data.

Now the problem is to find out whether the points on the run chart are abnormal or not. For example, the first 4 points of $\bar{X}$ could be normal or they may be below normal. Such determination can not be made

TABLE 7.6. Spiralmesh Coil Diameter Measurements (in mm) Taken at Coiling Machine over a 20-hour Period.

Sampling (Subgroup)	X_1	X_2	X_3	X_4	$\bar{X}$	R
1	19.8	21.4	19.4	20.1	20.2	2.0
2	19.5	18.7	20.3	18.8	19.3	1.6
3	19.8	19.3	20.5	21.2	20.2	1.9
4	18.0	19.0	21.0	20.0	19.5	3.0
5	20.4	21.0	20.8	21.0	20.8	0.6
6	20.0	20.9	20.6	19.7	20.3	1.2
7	21.8	21.3	20.3	19.3	20.7	2.5
8	20.2	19.1	21.4	19.4	20.0	2.3
9	20.6	19.1	20.4	20.6	20.2	1.5
10	18.2	19.1	20.1	20.0	19.4	1.9
11	20.4	20.0	19.0	21.0	20.1	2.0
12	20.6	20.0	19.5	21.4	20.4	1.9
13	18.9	21.4	20.4	20.2	20.2	2.5
14	19.1	18.7	20.6	21.1	19.9	2.4
15	20.4	18.9	19.1	20.2	19.7	1.5
16	21.3	19.7	20.7	19.6	20.3	1.7
17	20.5	18.7	20.3	20.1	19.9	1.8
18	22.2	20.7	20.9	20.0	21.0	2.2
19	19.9	20.2	20.9	20.2	20.3	1.7
20	21.5	21.0	19.0	19.9	20.4	2.5

$$\bar{\bar{X}} = 402.8/20 = 20.14$$
$$\bar{R} = 38.7/20 = 1.9$$

unless standards of evaluation are set. Without such standards, one is liable to make arbitrary judgment of the ones favorable to oneself and the graph can not be meaningful. When irrational evaluations are made, necessary action may be missed or unsuitable action may be taken in haste, thus causing confusion. This will result in inappropriate conclusions being drawn, thus lowering quality and efficiency.

Limit lines can be drawn on run charts to indicate the standards for evaluation. These lines indicate the dispersion of data on a statistical basis and indicate if an abnormal situation occurs in production. If we add limit lines to Figure 7.13, the graph in Figure 7.14 is obtained which is called a control chart.

The upper control limit corresponds to $+3s$. The lower control limit is equivalent to $-3s$.

The purpose of drawing a control chart is to detect any changes in the process, signaled by any abnormal points on the graph from which the data has been collected. Therefore each point on the graph must correctly indicate from which process the data were drawn.

For example, in making control charts the daily data are averaged out in order to obtain an average value for that day. Each of these values then becomes a point on the control chart which represents the characteristics of that given day. Or, data may be taken on a lot by lot basis. In this case, data must be collected in such a way that the point represents the given lot.

The points on a control chart represent arbitrary divisions in the manufacturing process. The data broken down into these divisions are referred to as subgroups. In Figure 7.13, the four measurements made in one hour constitute one subgroup. In other words, the production process is divided into units of one hour, and daily production (the 20 hour period) has been presented by points on a control chart. We can now determine whether the process is in a controlled state or not.

Control charts for $\bar{X}$ and R supply basis

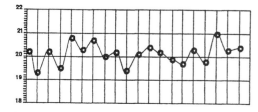

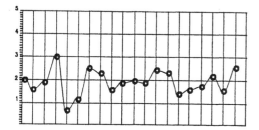

FIGURE 7.13. Run chart.

for judgment on a major question of practical importance. This question might be phrased in different ways, such as, "Does this graph indicate a stable pattern of variation?" or "Is this variation the result of a specific cause that must be corrected?"

Any rule that might be established for providing a definite " yes" or "no" answer to these questions is bound to give the wrong answer part of the time. The decision where to draw the line between a "yes" and a "no" answer must be based on the expected action to be taken if each answer is given.

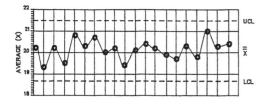

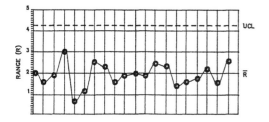

FIGURE 7.14. Control chart with limit lines.

In quality control in manufacturing, the answer, "No, this is not normal variation", leads to a hunt for an assignable cause of variation in an attempt to remove it, if possible. The answer, "Yes, this is normal variation", leads to leaving the process alone, making no effort to hunt for causes of variation. The rule for establishing the control limits that will determine the "yes" or "no" answer in any case should strike an economic balance between the cost due to two kinds of errors—the error of hunting for trouble when it is absent, and the error of leaving the process alone when an error really is present.

Any rule for establishing control limits for use in manufacturing should be a practical one based on this point of economic balance. It is common practice to use the 3s limits. Experience indicates that in most cases 3s limits do actually strike a satisfactory economic balance between the two questions.

Measured quality of manufactured product is always subject to a certain amount of variation as a result of chance. Some stable system of chance causes is inherit in any particular scheme of production and inspection. Variation within the stable pattern is inevitable. The results of a variation outside this stable pattern may be discovered and corrected. The power of the control chart lies in its ability to separate out these assignable causes of quality variation. This makes the diagnosis and correction of many production troubles possible, and often brings substantial improvements in product quality and reduction in rejects and rework. Moreover, by identifying certain of the quality variations as inevitable chance variations, the control chart tells when to leave a process alone, and thus prevents unnecessarily frequent adjustments that tend to increase the variability of the process rather than to decrease it.

Control charts provide more information than mere data plotted in a chronological sequence to indicate how the influences of various factors change over a period of time. Assume that the process analysis has been made and that a control state has been achieved. Standardization of working methods is necessary to obtain this state. Control charts with control limit lines enable us to see if the standardization applied was correct, and if it is being maintained. If so, then all points in the chart thereafter should be within the control limit lines. If points appear on the control chart outside these limits, then some change must have occurred on the assembly or manufacturing line. The cause must be investigated and the proper action taken. This use of charts is called process control. By setting standards of evaluation to define a range of acceptability, we can observe abnormal or unacceptable values, investigate the probable cause and initiate action for process correction.

Making an $\overline{X}$-R Control Chart

An $\overline{X}$-R control chart is one that shows both the mean value, $\overline{X}$, and the range, R. This is the most common type of control chart using continuous measurements. The $\overline{X}$ portion of the chart mainly shows any changes in the mean values of the process, while the R portion shows any changes in the dispersion process.

This chart is particularly useful because it shows changes in mean value and dispersion of the process at the same time, making it a very effective method for checking abnormalities in the process. The steps to make an $\overline{X}$-R control chart are as follows:

(1) Collect the data
(2) Divide the data into subgroups. The data should be divided into subgroups in keeping with the following conditions.
   a. The data obtained under the same technical conditions should form a subgroup
   b. A subgroup should not include data for a different lot or of a different nature.
(3) Record the data in such a manner that it allows for easy calculation of the mean and the range for each subgroup.
(4) Find the mean value, $\overline{X}$, for each subgroup.
(5) Find the range, R.

(6) Find the overall mean, $\overline{\overline{X}}$.
(7) Compute the average value of the range, $\overline{R}$.
(8) Compute the control limit lines ($\pm 3s$). The formulas for $3s$ control limits on charts for $\overline{X}$ are given below.

Central Line CL $= \overline{\overline{X}}$
Upper control limit UCL $= \overline{\overline{X}} + A_2\overline{R}$
Lower control limit LCL $= \overline{\overline{X}} - A_2\overline{R}$

The equations for the upper and lower control limits for the Range are given below.

Central Line CL $= \overline{R}$
Upper control limit UCL $= D_4\overline{R}$
Lower control limit LCL $= D_3\overline{R}$

The constants A_2, D_3 and D_4 are derived from statistics, the mechanics of which are beyond the scope of this book. Table 7.7 gives these constants for various subgroups. The use of the above equations

TABLE 7.7. Constants of A_2, D_3 and D_4.

Subgroup Size (n)	A_2	D_3	D_4
2	1.880	0.0	3.267
3	1.023	0.0	2.574
4	0.729	0.0	2.282
5	0.577	0.0	2.114
6	0.483	0.0	2.004
7	0.419	0.076	1.924
8	0.373	0.136	1.864
9	0.337	0.184	1.816
10	0.308	0.223	1.777
11	0.285	0.256	1.744
12	0.266	0.283	1.717
13	0.249	0.307	1.693
14	0.235	0.328	1.672
15	0.223	0.347	1.653
16	0.212	0.363	1.637
17	0.203	0.378	1.622
18	0.194	0.391	1.608
19	0.187	0.403	1.597
20	0.180	0.415	1.585
21	0.173	0.425	1.575
22	0.167	0.434	1.566
23	0.162	0.443	1.557
24	0.157	0.451	1.548
25	0.153	0.459	1.541

provides a quick and easy method of calculation of the 3s control limits.

(9) Construct the control chart. Draw in the control lines and label them with their appropriate numerical values. The central line is a solid line and the limit lines are dotted lines.

(10) Plot the $\overline{X}$ and R values as computed for each subgroup.

Reading and Using Control Charts

The purpose of making a control chart is to determine, on the basis of the movements of the points, what kind of changes have taken place in the production process. Therefore, to use the control chart effectively, we have to set the criteria for evaluating what we consider an abnormality. When a production process is in a controlled state, it means that:

(1) All points lie within the control limits
(2) The point grouping does not assume a particular form

We would therefore know that an abnormality has developed if:

(1) Some points are outside the control limits (including points on the limits) which means that the process is not under control.
(2) The points assume some sort of particular form even though they are all within the control limits. Four terms describe the patterns that might appear on a control chart.

- When several points line up consecutively on one side of the center line, this is called a *run*. The number of points in that run is called the *length of run*. In evaluating runs, if the run has a length of seven points, we conclude that there is an abnormality in the process. Even with a run of less than 6, if 10 out of 11 points or 12 out of 14 points lie on one side, we consider that there is an abnormality in the production process.

- If there is a continued rise or fall in a series of points, it is considered a *trend*. In evaluating trends, we consider that if 7 consecutive points continue to rise or fall, there is an abnormality. Often, however, the points will extend beyond the control limits before reaching 7.

- If the points show the same pattern of change (for example, rise or fall) over equal intervals, then there is *periodicity*. Unfortunately, when it comes to evaluating periodicity there is no simple method as with runs and trends.

- When the points on the control chart stick close to the central line or to the control limit line, it is called *hugging of the control line*. Often, in this situation, a different type of data or data from different factors have been mixed into the subgroup. It is therefore necessary to change the subgrouping, reassemble the data, and redraw the control chart.

Some of the patterns described above are illustrated in Figure 7.15. A sample control chart showing points grouped near the control limits is also included.

When points fall outside the control limits we say that the process is "out of control." This is equivalent to saying that factors outside normal variation are present. With 3s limits, we can make the statement with considerable confidence. In contrast to this, when all points fall inside the control limits we can not say with the same assurance, "No assignable causes of variation are present." No statistical test can give us this positive assurance. When we say, "This process is in control," the statement really means, "For practical purposes, it pays to act as if no assignable causes of variation are present."

7.5.14 *Process Capability*

Any process in total quality management must be repeatable and consistent in addition to being under control. This is characterized

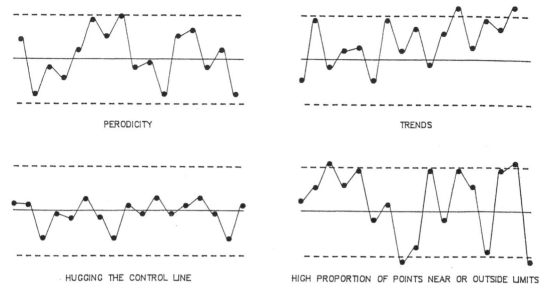

PERODICITY

TRENDS

HUGGING THE CONTROL LINE

HIGH PROPORTION OF POINTS NEAR OR OUTSIDE LIMITS

FIGURE 7.15. Patterns indicative of abnormal situations that might appear on a control chart.

by process capability which is indicated by the process capability index, C_p:

$$C_p = (USL - LSL)/6s \qquad (7.9)$$

where

USL = upper specification limit
LSL = lower specification limit
s = standard deviation of the process

If C_p is equal to or greater than 1.0, then it is concluded that the process is capable of operating within the process tolerance ($6s$) or within specifications (Figure 7.16). If C_p is less than 1.0, then the process variation exceeds process tolerance.

The process capability index C_p, does not show how well the process average, $\overline{\overline{X}}$, is

centered to the target value. Therefore, in practice, C_{pk} is used to predict process capability:

$$C_{pk} = \min \{C_{pl}, C_{pu}\} \qquad (7.10)$$
$$C_{pu} = (USL - \overline{\overline{X}})/3s \qquad (7.11)$$
$$C_{pl} = (\overline{\overline{X}} - LSL)/3s \qquad (7.12)$$

where

C_{pu} = upper half capability index
C_{pl} = lower half capability index
$\overline{\overline{X}}$ = process average (target value).

Although it is arbitrary, in general, if C_{pk} is equal to or greater than 1.33, then the process is considered as "capable" [20, 21, 22].

7.6 Regulations and Standards

7.6.1 *ISO 9000 Quality Standards*

ISO 9000 is a series of standards prepared by the International Organization for Standardization (ISO) headquartered in Geneva, Switzerland. ISO formed a technical committee on quality (TC 176) in 1980 which

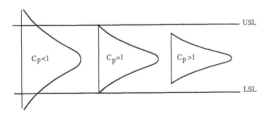

FIGURE 7.16. Process capability index in relation to USL and LSL.

authored five standards that were released to the world in 1987: ISO 9000-9004. ISO 9000 developed from the British Standards, BS 5750. The European equivalent to ISO is the EN 29000 series and the US equivalent is ANSI/ASQC Q90 series.

The ISO 9000 series provides a set of good practice procedures in order to gain consistency in production, design or provision of services. It is a set of international standards designed to be used for establishing and maintaining quality management and systems for company use or to satisfy outside contracts. It provides standards for registering quality systems and is accepted worldwide. Numerous standardization organizations have adopted the ISO 9000 series bringing most of the world under a universal standardization system.

The ISO 9000 series is not product specific, but instead is a generic quality management system model that supplements specifications. ISO may be used by both larger and smaller companies as a tool to help insure quality throughout the organization.

ISO 9000 series is reviewed and revised on a periodic basis to insure that it is updated. The ISO Technical Committee TC 176 works on a five year cycle extending the scope and detail for the series.

Documentation is at the center of the ISO 9000 series. Management must document all changes and decisions in order to be consistent. The ISO 9000 series operates under the assumption that if all personnel were suddenly replaced, the new people, properly trained, could use the documentation to continue making the products or providing the services as before [23].

Contents of ISO 9000 Series

The ISO series of standards consists of five standards (US equivalents are shown in parenthesis):

- ISO 9000 (ANSI/ASQC Q90)
- ISO 9001 (ANSI/ASQC Q91)
- ISO 9002 (ANSI/ASQC Q92)

- ISO 9003 (ANSI/ASQC Q93)
- ISO 9004 (ANSI/ASQC Q94)

ISO 9000, the first standard in the series, is the roadmap for choosing which contractual standard (9001, 9002 or 9003) is best suited for which type of business organization. It also provides key definitions for the use of the standards. Table 7.8 lists the specifications for 9001-9003.

ISO 9001 specifies a quality system model for a contract between two parties that requires the demonstration of a supplier's capability to design, produce, install and service the product. This certification is the highest an organization can acquire.

ISO 9002 is specifically designed for process and manufacturing business. ISO 9002 is distinguished from ISO 9001 due to the fact that it does not cover service and development and/or design of a product. ISO 9002 covers the remaining of the 9001 elements.

ISO 9003 deals mainly with documentation, detection and controlling non-conforming materials.

ISO 9004 addresses the elements to be used by an organization in order to develop and implement a quality management system. Selection of the appropriate elements contained in ISO and the extent to which these elements are adopted and applied by an organization depends upon the factors such as markets being served, nature of products, production processes and customer needs.

An organization using total quality management uses all resources to control processes and services that affect the quality of the end product. It is important that management systems be documented when variation occurs in order to analyze the cause and implement a possible solution to correct the variation.

Procedure for ISO Certification

A company that wants to register for ISO 9000 must be certified by an accredited third party registrar agency. A national accreditation body gives accreditation to each registrar. In the US, the "American National

TABLE 7.8. Specifications of ISO 9001–9003 Standards.

	ISO 9001	ISO 9002	ISO 9003
1. Management responsibility	x	x	x
2. Quality system	x	x	x
3. Document control	x	x	x
4. Product identification and traceability	x	x	x
5. Inspection and testing	x	x	x
6. Inspection, measuring and test equipment	x	x	x
7. Inspection and test status	x	x	x
8. Control of nonconforming product	x	x	x
9. Handling, storage, packaging and delivery	x	x	x
10. Quality records	x	x	x
11. Training	x	x	x
12. Statistical techniques	x	x	x
13. Contract review	x	x	
14. Process control	x	x	
15. Purchasing	x	x	
16. Purchaser supplied product	x	x	
17. Corrective action	x	x	
18. Internal quality audits	x	x	
19. Design control	x		
20. Servicing	x		

Accreditation Program for Registrars of Quality Systems" is established by the American National Standards Institute (ANSI) and the Registrar Accreditation Board (RAB) for accrediting registrars.

The commitment from the top management of an organization is crucial for the successful implementation of any ISO quality system. The main steps to ISO certification are as follows:

- comparison of existing quality procedures against the requirements of ISO 9001-9003 standards
- determination of the corrective action needed to conform with ISO 9000
- preparation of a quality assurance program
- preparation of definitions and documentation of new procedures and implement them
- preparation of a quality manual
- initial audit by the registrar. Take corrective measures based on findings by the registrar's audit team

- final visit by the registrar and official registration

Benefits of ISO Certification

The ISO certification process in the US is relatively new and the long term benefits of ISO certification is yet to be judged. However, from the early results, it is reported that the firms employing ISO quality systems experienced a decrease in seconds, as well as in returns. These same firms reported a marked decrease in rejects of raw materials from suppliers due to the strict regulations that suppliers must meet before doing business with an ISO registered firm. Companies that implemented ISO system reported that their total costs went down with regard to maintaining quality [24].

Although ISO does not guarantee quality itself, it can help a company keep up with its quality. It gives guidelines for keeping quality constantly high and improving the processes that are essential for the desired quality standards. When management becomes responsible for quality, which is required by ISO, then everyone in the organization is usually made aware of the significance of the quality for their job. ISO quality system can also become a good marketing tool for a company. Many customers look at ISO certified companies as having already taken care of quality.

Reduction of the number of suppliers to a business reduces the chance for variation in raw materials which causes problems in mass-production industries. The ability to use just one supplier is feasible due to the ISO certification since the buyer will know the rigorous standards and auditing that are encompassed in the ISO series. This is not the case however with most companies since the reliance on one supplier is rather risky. More than one certified supplier may be used by a company, yet one may be used more than the other. ISO certification, for suppliers and production organizations alike, may lead to the competitive advantages in markets and the improved productivity which is vital for international survival in today's economic market.

7.6.2 *ISO 14000 Environmental Standards*

ISO 14000 is an international standard that is developed for Environmental Management System (EMS). What ISO 9000 is to quality management is what ISO 14000 to environmental management. Therefore, there are some similarities between ISO 9000 and 14000 in concept. In fact, some of the elements of both systems, such as document control and statistical methods, are the same. ISO 9000 and 14000 can exist simultaneously to form one uniform management system for quality and the environment. ISO 14000 has the potential to apply to every manufacturing company. It is supposed to help any company in any country to meet the goal of "sustainable development" and environmental friendliness [25].

As in the case of ISO 9000, ISO 14000 is a series of standards. The family of ISO 14000 standards are listed in Table 7.9.

The members of the ISO approved ISO 14001 and 14004 as official international standards in the summer of 1996. ISO 14001 is the registration standard for an Environmental Management System (EMS) and ISO 14004 is the guideline standard for more guidance on implementation. These two new standards have also been approved by the European Union as a European standard. The official approval of the complete ISO 14000 family of standards is still a couple of years away. However, many countries already adopted the complete series. ISO 14000 was adopted by the US as a national standard in early 1996. ISO Technical Committee TC 207 is responsible for the development and revision of the ISO 14000 standards.

Implementation of ISO 14000

The following steps are useful for the implementation process of ISO 14000.

- Senior management must support the implementation of the system.
- Identify the external environmental regulations to which your company must comply
- Internal audit of the environmental system
- Identify the environmental regulations and issues
- Develop and write a strategic environmental plan
- Start implementing the strategic plan
- Continually monitor the progress

ISO 14000 can be used as a powerful marketing tool since environmental friendliness is popular with individuals, companies, state and federal agencies. However, it should be noted that ISO 14000 does not guarantee an environmentally friendly company as ISO 9000 does not guarantee quality. Moreover, ISO 14000 does not automatically provide compliance with the local, state or federal regulations.

TABLE 7.9. ISO 14000 Series of Standards.

ISO 14001	Environmental Management Systems—Specification with guidance for use
ISO 14004	Environmental Management Systems—General guidelines on principles, systems and supporting techniques
ISO 14010	Guidelines for Environmental Auditing—General principles on environmental auditing
ISO 14011/1	Guidelines for Environmental Auditing—Audit Procedures, Auditing of Environmental Management Systems
ISO 14011/2	Guidelines for Environmental Auditing—Compliance audits
ISO 14011/3	Guidelines for Environmental Auditing—Statement audits
ISO 14012	Guidelines for Environmental Auditing—Qualification criteria for environmental auditors
ISO 14013	Guidelines for Environmental Auditing—Management of environmental audit programs
ISO 14014	Guidelines for Initial Reviews
ISO 14015	Guidelines for Environmental Site Assessments
ISO 14020	Principles of Environmental Labeling
ISO 14021	Environmental Labeling—Terms and definitions of self-declaration environmental claims
ISO 14022	Symbols of Environmental Labeling
ISO 14023	Testing and Verification of Environmental Labeling
ISO 14024	Environmental Labeling—Guiding principles, practices and certification procedures for multiple criteria
ISO 14030	Environmental Performance Evaluation
ISO 14040	Principles and Guidelines of Life Cycle Assessment
ISO 14041	Life Cycle Assessment—Life cycle inventory analysis
ISO 14042	Life Cycle Assessment—Impact Assessment
ISO 14043	Life Cycle Assessment—Evaluation and interpretation
ISO 14050	Terms and Definitions

7.6.3 *Associations for Standards, Regulations and Specifications*

—American National Standards Institute (ANSI)
11 West 42nd Str., 13th Floor
New York, NY 10036
USA
Phone: (212) 642-4900
Fax: (212) 302-1286
http://www.ansi.org

—American Society for Quality Control (ASQC)
P.O. Box 3005
Milwaukee, WI 53201
USA
Phone: 800-248-1946
http://www.asqc.org

—American Society of Testing and Materials (ASTM)

100 Barr Harbor Drive
West Conshohocken, PA 19428-2959
USA
Phone: (610) 832-9585
Fax: (610) 832-9555
http://www.astm.org

—American Association of Textile Chemists and Colorists (AATCC)
P.O. Box 12215
Research Triangle Park, NC 27709
USA

—Environmental Protection Agency (EPA)
National Headquarters
401 M-Street, SW
Washington, DC 20460
USA
Phone: (202) 260-5917
Fax: (202) 260-3923
http://www.epa.gov

—Industrial Fabrics Association
International (IFAI)
345 Cedar St., Suite 800
St. Paul, MN 55101-1088
USA
Phone: (612) 222-2508
Fax: (612) 222-8215

—International Standards Organization
(ISO)
Rue de Varembe 1
CH-1211 Geneva 20
Switzerland
Phone: 41 22 749 0111
Fax: 41 22 733 3430
http://www.iso.ch

—National Institute of Standards and
Technology (NIST)
US Department of Commerce
Administration Building, Room A629
Gaithersburg, MD 20899
USA
Phone: (301) 975-5923
http://www.nist.gov

—National Standards Association
1200 Quince Orchard Boulevard
Gaithersburg, MD 20878
USA

—Occupational Safety and Health
Administration (OSHA)
401 M Street SW
Washington, DC 20460
USA
Phone: (202) 219-4667
http://www.osha.gov

• Region IV (Alabama, Florida, Georgia,
Kentucky, Mississippi, North Carolina,
South Carolina, Tennessee)
1375 Peachtree Street, NE, Suite 587
Atlanta, GA 30367
USA
Phone: (404) 347-3573

—Registrar Accreditation Board (RAB)
611 East Wisconsin Avenue
P.O. Box 3005
Milwaukee, WI 53201-3005

USA
Phone: (414) 272-8575
Fax: (414) 765-8661

7.7 References

1 *Total Quality Management,* Prentice Hall, 1995.
2 Saylor, J. H., *TQM Simplified: A Practical Guide,* McGraw Hill, New York, 1996.
3 Duncan, W. L., *Total Quality: Key Terms and Concepts,* AMACOM, New York, 1995.
4 ElMogahzy, Y., *Total Quality Management in the Textile Industry,* Quality Tech, 1996.
5 Hradesky, J. L., Total Quality Management Handbook, McGraw Hill, New York, 1995.
6 Hashim, M., and Khan, M., "Quality Standards—Past, Present and Future", *Quality Progress,* Vol. 23, June 1993.
7 *Webster's Ninth Collegiate Dictionary,* Merriam-Webster, Inc., 1988.
8 Juran, J. M., *Juran on Leadership for Quality,* MacMillan, New York, 1989.
9 Mehta, P. V., *An Introduction to Quality Control for the Apparel Industry,* ASQC, Quality Press, Milwaukee, 1992.
10 "Better Quality", *Business Week* Special Edition, June 8, 1987.
11 Baldwin, P., "Ad Firm Charts Trends in What Shoppers Want in 1990s", *The Dallas Morning News,* Jan. 20, 1990.
12 "Quality: Top Management Takes Up the Challenge", *Business Week* Special Report, Nov. 1, 1982.
13 Scheoffler, S., et al, "Impact of Strategic Planning on Profit Performance", *Harvard Business Review,* March–April 1974.
14 Garvin, D. A., *Managing Quality,* MacMillan, New York, 1988.
15 Gardner, D. M., "An Experimental Investigation of the Price/Quality Relationship", *Journal of Retailing,* Vol. 46, No. 3, Fall 1970.
16 Deming, W. E., *Out of Crisis,* MIT Press, Sept. 1991.
17 Garvin, D. A., *Managing Quality,* MacMillan, New York, 1988.
18 Crosby, T. P., *Statistical Process Control (SPC) Handbook,* Asten-Hill Company, 1987.
19 Guerro, G. J., and Bazaj, R. K., "Chemical Technologies for Optimizing Machine Variables", *PaperAge,* May 1996.
20 Rutherford, J. H., "How to Perform Process Capability Studies", *TAPPI Journal,* Vol. 78, No. 1, January 1995.
21 Carlsson, O., and Rydin, S., "Some Notes on the Computation and Interpretation of Capability Indices", *TAPPI Journal,* Vol. 78, No. 1, January 1995.
22 The Memory Jogger™, GOAL/QPC, 1988.

23 Manquest, D. W., "ISO 9000: A Universal Standard for Quality", *Management Review,* 1992.

24 Adanur, S., and Allen, B., "First Results on the Effects of ISO 9000 in the US Textile Industry", *Benchmarking for Quality Management & Technology,* Vol. 2, No. 3, 1995.

25 Clements, R. B., *Complete Guide to ISO 14000,* Prentice Hall, 1996.

General References

Aubrey II, C. A., and Felkins, P. K., *Teamwork: Involving People in Quality and Productivity Improvement,* Quality Press, American Society for Quality Control, 1988.

Broh, R. A., *Managing Quality for Higher Profits,* McGraw Hill, New York, 1982.

Cartin, T. J., *Principles and Practices of TQM,* ASQC Quality Press, White Plains, NY, 1993.

Cortada, J. W., *TQM for Sales and Marketing Management,* McGraw Hill, New York, 1993.

Cortada, J. W., *The McGraw-Hill Encyclopedia of Quality Terms and Concepts,* McGraw-Hill, 1995.

Crosby, P. B., *Quality Is Free,* New York, New American Library, 1979.

Deming, W. E., *Elementary Principles of the Statistical Control of Quality,* 2nd Ed., Tokyo, Japan: Nippon Kagaku Gijutsu Remmei, 1952.

Deming, W. E., *The New Economics for Industry, Government, Education,* MIT Center for Advanced Engineering Study, 1994.

Edwards, C. D., "The Meaning of Quality", *Quality Progress,* Oct. 1968.

Ferguson, K. H., "International Quality Standards May Affect Industry's Effort in Europe", *Pulp and Paper,* March 1991.

Gevirtz, C. D., *Developing New Products with TQM,* McGraw Hill, New York, 1994.

Hutchins, G., *ISO 9000: A Comprehensive Guide to Registration, Audit Guidelines and Successful Certification,* Oliver Wight Publications, 1993.

Ishikawa, K., *Guide to Quality Control,* Asian Productivity Organization, Tokyo, Kraus International Publications, White Plains, NY, 1982.

ISO 9000 is Here, E.I. DuPont de Nemours & Co., June 1992.

Juran, J. M., *Quality Control Handbook,* 3rd Ed., McGraw Hill, New York, 1974.

Leffler, K. B., "Ambiguous Changes in Product Quality", *American Economic Review,* Dec. 1982.

Lofgren, S., et al, "ISO 9000 Requirements for Pulp and Paper Testing Equipment Maintenance", *TAPPI Journal,* Vol. 76, No. 2, February 1993.

Mahoney, F. X., *The TQM Trilogy: Using ISO 9000, the Deming Prize, and the Baldridge Award to Establish a System for Total Quality Management,* American Management Association, New York, 1994.

McLaughlin, G. C., *Total Quality in Research and Development,* St. Lucie Press, 1995.

Menon, H. G., *TQM in New Product Manufacturing,* McGraw Hill, 1992.

Questions and Answers about the UL ISO Registration Program, Underwrites Laboratoriers, Inc., 1989.

Rabbitt, J. T., *The ISO 9000 Book: A Global Competitor's Guide to Compliance and Certification,* Quality Resources, 1994.

Rhodes, S., "International Environmental Guidelines to Emerge as the ISO 14000 Series", *TAPPI Journal,* Vol. 78, No. 9, September 1995.

Rothery, B., *ISO 14000 and ISO 9000,* Gower, 1996.

Sayre, D., *Inside ISO 14000,* St. Lucie Press, 1996.

Suzaki, K., *The New Manufacturing Challenge, Techniques for Continuous Improvement,* The Free Press, 1987.

Swift, J. A., *Introduction to Modern Statistical Quality Control and Management,* St. Lucie Press, 1995.

Tibor, T., and Feldman, I., *ISO 14000,* Irwin Professional Publishing, 1996.

Tuchman, B. W., "The Decline of Quality", *New York Times Magazine,* Nov. 2, 1980.

Wolkins, D. O., *Total Quality: A Framework for Leadership,* Productivity Press, 1996.

7.8 Review Questions

1 Explain what you understand from the term *TQM.*

2 Can you make your own definition of quality?

3 Does ISO guarantee quality products and services? Explain why.

4 Is there a relation between "consistency" and quality in products and services. Explain.

5 Explain the core concepts and basic elements of TQM.

6 The following data indicate the air permeabilities of a single layer forming fabric type and its variations.

605, 680, 675, 550, 590, 580, 740, 665, 650, 630, 625, 675, 660, 560, 500, 545, 580, 660, 700, 725, 605, 695, 665, 650, 680, 550, 600, 610, 590, 595.

Calculate:

a. mean
b. mode
c. median
d. range
e. standard deviation
f. coefficient of variation (CV%).

7 Establish a cause-and-effect diagram for seam quality in dryer fabrics.

8 What are the characteristics of a normal curve?

9 Among the product quality, safety and environment, which one should have the first priority in a manufacturing environment? Which is second? Explain your reasons.

10 Summarize the major steps in ISO qualification process.

Computers in Paper Machine Clothing

There has been a computer revolution within the last two decades that has changed every aspect of life drastically. One of the profound effects of this change has been in the area of product and process design, development and manufacturing. Like in every other industry, today computers are widely used in every stage of papermaking and paper machine clothing from product concept design to shipping and receiving.

Parallel to the expansion of computer application areas in industry, the computer technology itself went through impressive developments within the last two decades. The development of computer technology, both software and hardware, is ever increasing without any pause in the foreseeable future. Computers are getting faster and more powerful in terms of memory, versatility, adaptability, performance and supporting software programs. The change is so fast that any specific description of the current computer hardware and software may be outdated in a few years or even months. Therefore, the purpose of this chapter is not to discuss the computer technology but to demonstrate the current and potential application areas of computers.

Probably the most exciting development that is taking place last few years is in the information area. Using World Wide Web (WWW) on the Internet, users can connect to any other computer in the world (provided that it is also connected to the net), browse a web page, send e-mail, participate in discussions and video conferences, etc. People use this technology to exchange ideas, solve problems and discuss issues. Federal and state institutions, private and public companies and individuals are developing their own homepages full of information about themselves, their products, companies and services. Homepages can be designed in any format by the programmer. Figure 8.1 shows a homepage developed for the pulp and paper industry.

By pointing the cursor and clicking any item on the homepage, one can go into a specific topic for more information.

8.1 Definitions

In order to fully understand this chapter, a glossary of terms follows.

- account (Computer Account)—The name used to uniquely identify users of a computer resource. Referred to variously as GID (global ID), computing account, userid or simply "ID."
- assembly language—This level of language is between high level language and machine language. The commands are in English letters but they may not have any literal meaning since they are abbreviated. Execution of assembly language programs are

**Welcome to
Pulp and Paper.Net**

**An Easier Way to Get
More Information
in Less Time.**

About | Become a Supplier | Become a Member | To
Advertise | Feedback

Suppliers and Products
- Browse Segments
- Search for Company or Product

Classified Ads
- Job Opportunities
  Used Equipment -
  Coming soon

Trade Shows
- Tradeshow Search
  Tradeshow Scrapbook

From the 'Net
  Pulp and Paper
Jumplist
  Newsgroup Topics

Business News
  Walden Paper Report
  Walden Fiber and
Board Report   The
Green Paper/Fortrans

You must register to
access these areas.

New Pulp and Paper.Net Welcomes Voith Sulzer

Suppliers and Products

This section is where you can easily find information on companies, products, chemicals or services that serve the pulp and paper industry. Whether you know exactly what you are looking for or just want to browse, our navigation system brings you the information you are seeking.

Browse Market Segments
Search for Company or
Product

TradeShows

Pulp and paper industry tradeshows are another valuable resource for industry information. Pulp and Paper.Net wants to keep you informed about tradeshows and seminars that may be of interest to you. This section offers you information on upcoming tradeshows and seminars and includes images from past tradeshows we have attended.

Tradeshow Search
Tradeshow Scrapbook

New Business News

In the information age, staying up to date on industry information is not an easy task. This section provides pulp and paper industry information. Whether it is unique information we bring to you or other pulp and paper on-line resources, we want to help you find the information quicker and easier.

Walden Paper Report
Walden Fiber and Board
Report
The Green Paper/Fortrans

Classified Ads

Whether you are in the market for a new job or just curious about available openings, this section of Pulp and Paper.Net allows you to search the Job Opportunities database. If you locate a posting that interests you, Pulp and Paper.Net will provide the information you need to confidentially contact the company posting the job opportunity.

Job Opportunities
Coming soon - Used
Equipment

From The 'NET

The Internet is a valuable resource for information, but finding that information is not always easy. Pulp and Paper.Net's goal is to make pulp and paper producers' jobs easier by putting them closer to the information they need. This section of Pulp and Paper.Net provides information and links to other sites on the Internet containing industry information. Make a bookmark for Pulp and Paper.Net and begin your search for pulp and paper industry information with us.

"The Pulp and Paper
Jumplist"
Newsgroup Topics

FIGURE 8.1. A homepage for the pulp and paper industry (http://www.pulpandpaper.net/).

faster than high level language programs because there is no need for compilation. That is why video games, for which speed is critical, are written in assembly language. Assembly language is more difficult to write but gives more control of computer to the user.

- bitnet—a worldwide computer network
- browser (Web Browser)—It is a software program that provides a graphical user interface to the Internet's World-Wide Web (WWW). There are several browser programs currently in use including Netscape and Mosaic.
- CD (compact disk)—Digital disk for data storage.
- CPU (central processor unit)—A microchip inside the computer that runs all the software and operations of the computer.
- cyberspace—A generic name for the computer environment that includes e-mail, WWW and networks, etc.
- desktop—A personal computer that can fit on top of a desk.
- digitizer—A device that converts analog data, figures, etc., to digital form.
- download—Transferring information or files from one computer to another. Downloading generally implies that the data are moving from a larger, more powerful machine to a smaller one.
- DPI—dots per inch on the computer monitor.
- EDI (electronic data interchange)—Transfer of data via computers from place to place.
- e-mail (electronic mail)—Mail that is sent from computer to computer via telecommunications media.
- expert system—computer software packages that are based on scientific data with empirical and theoretical knowledge of experts in the field. Expert systems are also called "intelligent systems."
- floppy disk—Portable computer disk for data storage.

- FTP (file transfer protocol)—Used to transfer files from one computer to another over the Internet.
- gopher—An on-line information retrieval system available on the Internet.
- hard disk—High storage capacity permanent disk inside the computer
- high level language—Programming languages such as Basic, FORTRAN, Pascal and 'C' that consist of special commands to write computer programs. The commands are in English language which is why it is called 'high level', i.e. level that a person can understand. A computer code written in English is converted to assembly language first (compilation) and then to machine language before it is executed by the computer. 'C' language has several versions and it has gained wide acceptance in recent years. Figure 8.2 shows comparison of high level language programs to calculate the area of a rectangle.
- Homepage (or Page)—In the context of WWW, it refers to the entry point of a Web site. It is similar to a main menu which may consist of any number of screens full of information and hyper-links to related items (Figures 8.1 and 8.3).
- HTML (hyper text markup language)—HTML documents are text files that contain HTML tags to develop a home page. HTML tags tell a web browser (such as Netscape or Mosaic) how to format or display the text of the document. HTML tags are set off from the rest of the text by left and right angle brackets, i.e. <p> for paragraph tag. Although most HTML tags are paired, some are not. Paired tags have a beginning tag, such as <title>, and an ending tag such as </title>. Text affected by the tag is included between the paired tags (<title>My home page </title>). HTML is not case sensitive, i.e., <title> is the same as <Title> or <TITLE>. Tags that are not paired

PROGRAM IN FORTRAN

```
        PROGRAM    CALCULATING    AREA
        PRINT *,'INPUT A, B'
        READ(*,*) A,B
        AREA = A*B
        WRITE (*,20) AREA
20      FORMAT (F10.2)
        STOP
        END
```

PROGRAM IN BASIC

```
10 print "enter values for a and b"
20 input a,b
30 let area = (a*b)
40 print "a=";a,"b=";b,"area=";area
60 end
```

PROGRAM IN C

```
#include <stdio.h>
#include <math.h>

main()
{

double a,b,area;
scanf("%lf %lf\n",&a,&b);
printf("a=%g,b=%g\n",a,b);
area=a*b;
printf("area=%g\n",area);

}
```

FIGURE 8.2. High level language program codes to calculate the area of a rectangle.

include paragraph tag <p>, the horizontal rule tag <hr>, and the line break tag <br>. There are several HTML editors and add-ons that can be used to develop a homepage. Figure 8.3 shows the homepage for the White House.

- Internet—A collection of networks and gateways around the world. Internet is the World's largest computer network comprised of thousands of interconnected computer networks worldwide.

- IP addresses—A unique number that identifies each host computer or computer on a network. IP stands for Internet Protocol.

- laptop— A small personal computer that can fit on one's lap or briefcase.

- machine language—The commands in this language are made of 0's and 1's that computer can understand. Normally, programs are not written in machine language because of its difficulty. However machine language is very powerful compared to assembly and high level languages.

- main frame—A host computer that can run a large number of personal computers or workstations. Main frames are used in large corporations, universities and R&D centers.

- midi frame—A host computer smaller than main frames that is used in small to medium size companies to run personal computers and workstations

- mini frame—A host computer that can control and run several personal computers or workstations usually in an office.

- Mosaic—Mosaic is a web browser that provides a graphical user interface to the Internet's WWW.

- MS-DOS—Micro Soft Disk Operating System.

- Netscape—Netscape is a web browser that provides a graphical user interface to the Internet's WWW.

- network systems—These are communication systems that connect different information systems together all over the world. Examples of the worldwide network systems are Internet and Bitnet.

- Newsgroup—An on-line forum for discussion of related topics, accessible by a newsreader. Some newsgroups allow postings or messages from

Good Morning

Welcome to the
White House

 The President & Vice President: Their accomplishments, their families, and how to send them electronic mail

 What's New: Remarks by the President: White House Commission on Aviation Safety and Security

 Interactive Citizens' Handbook: Your guide to information and services from the Federal government

 White House History and Tours: Past Presidents and First Families, Art in the President's House and Tours

 The Virtual Library: Search White House documents, listen to speeches, and view photos

 The Briefing Room: Today's releases, hot topics, and the latest Federal statistics

 White House Help Desk: Frequently asked questions and answers about our service

 White House for Kids: Helping young people become more active and informed citizens

Please sign the White House electronic guest book.
Or comment on this service at feedback@www.whitehouse.gov.

FIGURE 8.3. Homepage of the White House (http://www.whitehouse.gov/WH/welcome.html)

anyone, while others may be moderated, i.e. postings are screened.

- Newsreader—Software that allows access to Usenet News and the newsgroups available therein.
- PC (personal computer)—A computer that has all the necessary operating system programs to run by itself, i.e. without connecting to a large computer. IBM (International Business Machines) and Apple Computer, Inc., are examples of the largest personal computer manufacturers in the world.
- PLC (programmable logical controller)—Device that is used to control and operate machines and equipment electronically
- Programming language—set of commands to instruct the CPU to perform a specific job. There are three levels of programming languages: high level language, assembly language and machine language.
- RAM (random access memory)—Memory that is used by CPU to run the software programs. RAM size should be larger than the total size of the programs that will be run. Turning the computer off will erase everything in the RAM.
- ROM (read only memory)—A permanent memory location in the computer that has the necessary commands to run the computer. ROM can not be accessed by the user and therefore can not be erased.
- scanning—digitizing a picture or text and loading it to the computer screen (i.e., opposite to printing).
- source (source code)—Source is the HTML codes for a Web document.
- Sun—A computer or workstation manufactured by Sun Microsystems.
- super computers—Extremely fast and powerful mainframes that are used mainly for research purposes in large universities and institutions such as NASA.
- Telnet—Telnet software allows a user

to login to a remote computer and use its resources.

- Unix—An operating system used by Sun and other companies. Unix is the most widely used multi-user general purpose operating system in the world.
- upload—Transferring information or files from one computer to another. Upload implies that the data is moving from a smaller machine to a larger, more powerful one.
- URL (universal resource location)—The electronic address for an information source on the WWW or Gopher, such as an FTP site, gopher service or web page. For example, the URL for the Auburn University home page is http://www.auburn.edu and the American Forest and Paper Association is http://www.afandpa.org. Each address has a prefix which tells the Browser what kind of resource it is viewing. FTP site will have the prefix ftp://, a Gopher site will have the prefix gopher:// and an HTML file that is read from the hard drive or diskette will have the prefix file://.
- virus (computer virus)—A computer virus is a computer program that is developed with malicious intent. These viruses can have very bad effects on the programs in standalone computers or networks. For example, viruses can turn a file that is created with a wordprocessor into a garbage. Viruses can replicate themselves and spread from computer to computer via floppy disks or through file downloading in a network. There are computer programs developed to scan computers and disinfect them if they are hit by a virus. There are many types of viruses and different disinfectant programs. Probably the best protection from viruses is to frequently make clean (i.e. free from virus) backup copies of working files.
- work station—A computer that is connected to a large central computer. A work station usually does not have a

CPU of its own. However, computers with CPUs can be used as workstations as well.

- WWW (World Wide Web)—A client-server hypertext distributed information retrieval system available over the internet. WWW was developed by CERN High-Energy Physics laboratories in Geneva, Switzerland.

- zip drive—A removable hard disk capable of holding anywhere from 100 MB (megabyte) to 1 GB (gigabyte) of data on a removable cartridge depending on the model of the drive. They are typically slower than hard disks and magneto-optical drives, but faster than tapes and CD ROMs.

8.2   Application Areas of Computers

8.2.1   *Computer Aided Design (CAD) and Modeling*

Computers have been widely used in design, development, modification and analysis of various products and processes. A typical CAD system consists of computer, monitor, scanner, plotter, printer and digitizer.

Paper machine clothing is relatively expensive. Design and development of a new fabric by trial and error can increase product development costs and delay the introduction of a product into the market. Computers can be used to develop a product with correct geometrical and mechanical properties such that potential performance of the product can be estimated and a decision about the product can be made before actual manufacturing trials. More realistic computer programs make the decision making process simpler and faster thus reducing the cycle time from conception of an idea to actually making the product whether it is a fabric or paper.

Product design is an important application of CAD. CAD programs are used to design and characterize polymers, fibers, yarns, fabrics, nonwovens and paper. Use of CAD shorten the development time of new, innovative products. Figure 8.4 shows a fabric structure developed with a CAD system.

Modeling of a product in computer involves geometric modeling and mechanistic model-

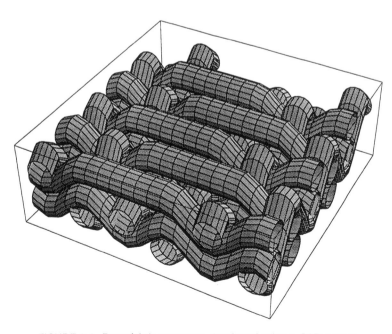

FIGURE 8.4. Dryer fabric structures developed using a CAD system.

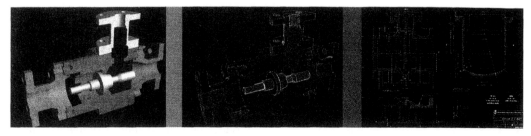

FIGURE 8.5. Steps in computer aided design of a product: drawing, 3D meshing and 3D modeling (photo courtesy of Research Engineers, Inc.).

ing. Realistic modeling of a product structure is critical for success. Therefore, modeling and computer aided design must be based on theoretical and empirical data available in the literature or through research. Generally, finite element analysis (FEA) method is used to model and analyze the structures. Paper machine clothing materials have nonlinear, visco-elastic properties which make modeling relatively more complicated. The major steps in computer aided design and modeling of a product are drawing, 3D meshing and 3D model generation as shown in Figure 8.5.

Various CAD software programs have been developed including AutoCAD®, VersaCAD®, Algor®, MSC, etc. A good CAD system might be quite expensive and should be evaluated very carefully. The issues to consider when purchasing a CAD system are simplicity, customer support, availability of user groups, availability of manuals, software update options, flexibility and service.

8.2.2 Computer Integrated Manufacturing (CIM)

Computers are used in almost every manufacturing stage of paper machine clothing and

FIGURE 8.6. Computer control room for preneedling of press fabrics.

papermaking. Their usage includes testing of incoming raw materials, process control, property testing, etc. Most of the mechanisms of modern textile machines are electronically controlled and operated using programmable logical controllers (PLCs). With the PLCs, the performance of machines can be monitored by measuring picks per minute, stop and start times, etc. Figure 8.6 shows the computer control room for a preneedling machine for press fabric manufacturing.

CIM is an integrated computerized production system that is used to improve operations and product quality in a manufacturing environment. Benefits include savings in energy, labor, cost, maintenance and product quality improvements. Factory automation has been a major driving force behind CIM. CIM allows us to collect data in real time and eliminate waste and bottlenecks between processes by controlling each process while monitoring the data.

In general, CIM consists of three building blocks: process technology, control technology and knowledge technology. Process technology includes production machinery and automation to minimize manufacturing labor. Control technology uses computers, microprocessors and programmable controllers for self-regulatory manufacturing process control. Knowledge technology requires computers capable of interpreting information and decision making in real time. Decision science, expert systems, artificial intelligence and knowledge engineering are the tools to construct a CIM environment in a plant.

Production planning and control systems are used as part of Management Information Systems (MIS) to integrate planning, control and execution of manufacturing processes. These systems allow planning of quantities, checking delivery dates, monitoring order progress, materials management, capacity planning and scheduling. These are important elements for the success of Just in Time (JIT) and Quick Response (QR) systems. The purpose of JIT is to have the right product quantity at the right time and in the right place. QR shortens cycle (delivery) time of a product

and allows fast flow of information from supplier to the customer.

8.2.3 *Process Simulation and Testing*

Computers are widely used to simulate the process conditions either during manufacturing or during end use. For example, the behavior of a forming fabric during heatsetting can be predicted using computer simulation programs. Similarly, drainage characteristics of a forming fabric on the paper machine can be obtained using a fluid flow simulation program.

Computer testing of products such as fabrics and paper products can be done by first modeling the product and then subjecting it to the test conditions. For example, test simulation for tension, compression, bending, etc., of fabrics can be done using software packages that are specially developed for this purpose.

Today, most of the testing equipment are computerized for accuracy, reliability and speed. Figure 8.7 shows computerized tensile testing of yarns. For example, with this system, the modulus of yarn is calculated directly by the computer from the stress-strain curve which eliminates the human error and subjectivity.

8.2.4 *Image Technology*

High resolution computers are ideal tools for microscopic view of materials. A high resolution camera is connected to a computer for this purpose. Magnified views of products can be viewed on the computer monitor and printed for a hard copy (Figure 8.8). Imaging technology allows documents such as forms, drawings, pictures and charts to be scanned and digitized such that they can be manipulated electronically.

Computer imaging technology can be used to quantify various fabric properties such as void volume, surface smoothness and contact, etc.

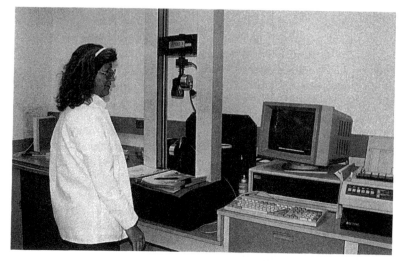

FIGURE 8.7. Computerized tensile testing.

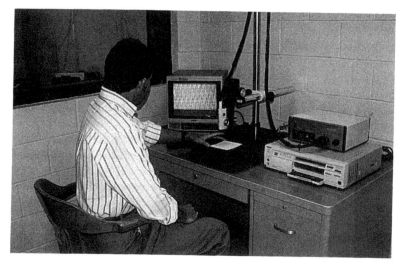

FIGURE 8.8. Image analysis of a dryer fabric.

8.3 References

Adanur, S., *Wellington Sears Handbook of Industrial Textiles,* Technomic Publishing Co, 1995.

CAD in Clothing and Textiles: A Collection of Expert Views, BSP Professional Books, 1992.

Dugnani, G., *CAD in the European Textiles,* BTTG, Manchester, England, 1989.

O'Brien, J., "Process Control Critical to the Industry's Future", *PaperAge,* August 1995.

Ruzicka, Y., "Pulp and Paper Resources on the Internet", *PaperAge,* September 1995.

Sundaresan, J., *Knowledge Management and Problem Solving in Textiles,* University of Microfilms International, Ann Arbor, Michigan, 1990.

8.4 Review Questions

1 Discuss the effects of computer revolution on paper machine clothing and paper manufacturing.

2 What are the technical parameters to consider when designing a forming fabric using a CAD system in order to make it a realistic representation of the real fabric.

3 Give 10 examples for the application of computers in paper machine clothing.

9

Future of Papermaking and Paper Machine Clothing

With the current speed of technology development, it is hardly possible to predict the future of any industry. Nevertheless, some attempt is made in this chapter to predict some of the future trends in papermaking and paper machine clothing.

In the 1800s, a paper machine was producing a sheet of 5 feet wide at a speed of 100 fpm. Today, paper machines are close to 500 inches in width. The speed is at 7,000 fpm. Some pilot paper machines are already running at 8,000 fpm. It is expected that in a decade or so, the machine speeds will reach 10,000 fpm. The cost of a new machine can be over $400 million. Parallel to the improvements in paper machine technology, paper machine clothing manufacturing also made giant steps within the last few decades.

With the introduction of the first synthetic forming fabric in 1957, the research and development efforts in paper machine clothing made a dramatic turn. Metal wires did not allow development of design varieties. Synthetic fabrics allowed more complicated designs and structures in forming, press and dryer fabrics. However, pressures and high temperatures still present a challenge for paper machine clothing manufacturers which will require new products to meet the demands. Demand for quality and uniformity will continue for all grades of paper and paperboard. Ever increasing recycled fiber content will require development of more sophisticated fabrics to solve the problems encountered in forming, pressing and drying.

In 1995, the forest, wood and paper industry executives released the Agenda 2020 to set the future goals for these industries. The Agenda identifies six priority areas for research: sustainable forest management, environmental performance, energy performance, improved capital effectiveness, recycling and sensors and control [1].

The search will continue for new fabrics and processes that will improve paper quality and increase paper machine productivity. Paper machine manufacturers, papermakers, paper machine builders and universities need to continue working together to take the paper machine clothing and papermaking to new heights. Development of paper machine clothing with the input from paper manufacturers will be more successful. The key in paper machine clothing business will be to combine custom designed products with a good technical service.

9.1 Paper Machines and Papermaking

New concepts for forming, pressing and drying are expected to be developed in the future. Environmental issues will continue to be a concern that will affect the future directions in papermaking and paper machines. More rational use of energy and water will become necessary. More automation including fuzzy logic controls will be implemented.

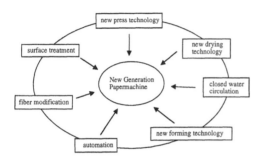

FIGURE 9.1. Factors that will affect the development of future paper machines [2].

Figure 9.1 shows the factors that will affect the future paper machine development. Layered headbox designs that will allow furnish layering will increase. The machine speeds and widths will continue to increase, the former being more likely. As the speed and tension of fabrics increase, more abrasion resistant materials will be developed for paper machine clothing. More hydraulic headboxes will be used in the future. Dilution control of basis weight rather than slice control allows for finer adjustments in basis weight profile without affecting fiber orientation [3]. Use of single fourdriniers will decline. Gap forming and twin wire machines will dominate the market for tissue and many paper grades. Adjustable blades are used in the latest generation of gap and blade formers. More multiple formers will be used for board grades. Former types will continue to evolve to meet the requirements of wider paper grades. Development work such as transfer belt technology will continue to have a fully supported web throughout the paper machine. High consistency forming will get more attention. Papermaking will move even more towards being science and engineering rather than an art and science.

Ever continuing progress of technology and lifestyles will necessitate new products with more demanding properties. More engineered papers including composite paper structures will be developed for high performance applications. Use of paper and paperboard products in industrial applications such as architecture and construction, medical technology, safety and protection, etc., will increase. This will close the gap between nonwoven textiles and paper even further. As a result, it is expected that the percent of nonwood/nonfibrous materials will increase in paper and board products [4].

Recycled fiber consumption in North America was doubled between 1980-1993. The current recovery rate is 40%. Use of recycled fibers will continue to grow. The US Paper Industry set 50% paper recovery goal by 2000. The current recycling rate for Europe is 43% with Sweden leading at 50%. However, from an economical and technological point of view, there may be a limit to recycling increase in the future [5–8].

More two-sided coated grades will be made in bleached board. Liquid packaging (aseptic) grades are growing rapidly. Demand for recycled fiber in non-food packaging grades will increase. Use of chlorine as a bleaching agent will decrease. However, non-chlorine methods may reduce brightness levels. Use of oxygen bleaching agents such as ozone (O_3), hydrogen peroxide (H_2O_2) and others will increase for fine paper. Recycle fiber content will increase for brown paper. Departure from standard basis weight requirements will take place. Better printability of the sheet will be required. For tissue, increased wet strength (polyamide resins) will be required.

Automation in manufacturing of paper and paper machine clothing will continue to increase. This will require new sensors and new control strategies. Computer aided design and computer aided manufacturing will be utilized further.

9.2 Forming

Most forming fabric development efforts during the 1970's and 1980's focused on increasing drainage capacity and structural integrity including stability and life. These areas are of relatively lesser concern today because of improvements made in these areas. However, increasing machine speeds and widths will require even more stable and uniform fabrics. Further stability in both MD and CD may be achieved by incorporating larger strands and new fabric designs. The use of polyamide yarns in the wear surfaces will become the norm. Increased use of abrasive calcium carbonate in alkaline sizing will continue to require excellent life potential and retention.

Product consistency and a generally improved product will continue to be searched. In general, fabrics with higher quality, longer life and lower cost will be developed. Improved consistency will result in repeatable and more dependable products.

Forming fabrics will need to offer greater levels of fiber support without sacrificing drainage or structural integrity. The extra support double-layer design family will continue to be actively used in the near future to meet the new demands. However, much more versatility with much less compromise is afforded to both the forming fabric design engineer and the papermaker through the use of lower caliper, three layer structures.

Forming fabrics will also need to be designed to run cleaner offering less affinity for contaminant attraction and fiber carryback. Anti-contaminant properties will be required even more as the use of recycled fiber is increased, especially in the brown grades. Because of recycled fiber and the changes it will make to wet-end chemistry, the forming fabric will be modified to help minimize contamination and plugging from these sources.

Increased demand from the printing and converting industries for smoother, stronger sheets with better printing surfaces will affect the forming fabric design and structure. In addition, the paper industry will have to produce these higher-quality sheets from lower-grade, less-expensive furnishes. Therefore, fabrics will have to be finer, giving more sheet support while improving their mechanical properties with regard to dimensional stability, wear resistance and ease of cleaning. These basically conflicting requirements can be met by multilayer designs and probably will be met by three-layer designs.

Paper machine speeds will continue to increase. The very short dewatering time then necessitates forming fabrics with good drainage and high fiber support. In addition, the increased speed will require an ever-increasing need for fabric stability and resistance to stretching due to higher tensions. Easy fabric installation for less downtime will require development of pin and thin line coil seamed forming fabrics.

Paper weights will continue to decrease especially for fine paper while maintaining or improving current standards for opacity, smoothness, strength, printability, etc. In addition, other grades, while maintaining sheet weight, will need improvements in printability. All of these demands will require finer forming fabrics to enable the papermaker to obtain smoother sheet properties and more uniform fine distribution while achieving excellent formation. Finer fabrics will require multilayer designs to obtain necessary stability on the wider, high speed machines. Kraft papers will continue to see expansion in the multi-ply sheets and will also see finer fabrics to accommodate lighter weight plies and improvements in formation to improve ply bonding and printability.

Lower basis weights in tissue will be developed. Low basis weight in tissue can be achieved in different ways. The fibers can be oriented in the thickness direction (Z direction) during forming to give higher bulk. Another way to achieve low basis weight is multilayer through air dryer (TAD) fabrics with small pits. TAD will require higher temperature resistant fabrics with better hydrolysis resistance for greater efficiency and productivity.

9.3 Pressing

It is well known that better dewatering requires less energy and produces a superior product. Paper machine builders will continue working with paper machine clothing manufacturers to make further improvements in this area.

In the 1950's, press felts, which were composed of 100% wool fibers, had a life of approximately one week. Use of synthetic fibers and new needling technologies increased the life drastically and revolutionized fabric designs. Double and triple layer base fabrics, twisted mono-filaments and stratified batt structures resulted in superior press fabric performance. More stable base structures and low batt-to-base ratio fabrics that will provide more uniform sheet profiles will be developed. Needling techniques will improve.

Usage of grooved belts and blind-drilled sleeves for shoe presses will increase on machines that are double-felted for kraft and other grades, as well as on single-felted operations.

The concept of no-open-draw is becoming a reality. No-open-draw will allow higher speeds by preventing sheet break. It is expected that shoe press type calender belts will replace soft nip calendering for board in the future.

Use of laminated press fabric structures will continue. Seamed fabrics utilizing a laminated top fabric are already being evaluated as are laminated types which are nonwovens with unique properties [9].

Alkaline papermaking for fine papers requires good water handling capacity which will require higher void volume structures. Couch consistency and pressability will be important. Felt filling due to carbonate and aluminum hydroxide will require coarser batt fibers with more open structures. Mechanical and chemical felt wear and shedding require high molecular weight batts and better chemical treatments.

Machine speeds and press loads will increase. Dwell times will be longer. Multiple shoe press sections and new Long Nip Press (LNP) sections will be utilized. To meet these challenges, single monofilament CD fabric constructions will be developed. Multilayer, multiple base fabric and pin seam constructions will increase. Chemical and wear resistant fiber content will increase. New and improved topical treatments will be developed for chemical and wear resistance.

To control the vibration and bounce in the fourth press, high caliper structures with improved mass uniformity are required. To improve sheet properties, pressure uniformity

will be improved. Press fabrics with densely woven base and fine denier fibers will be developed. New techniques to manufacture densely needled fabrics may prevent blowing.

Speeds will increase for bleached board pressing. There will be less double-felted pressing. High void volume designs will be used in first press positions. Smoother surface finish needling techniques will be developed.

For tissue and toweling, single-felted operations will increase. Softer pressure rolls will be used. Face-side high pressure showering and vacuum boxes will be utilized. Steam showers will be used in the nip.

Use of thermally stable nylon will increase dimensional stability in steam boxes/devronizers. High base-to-batt ratio designs will be developed for better compaction.

De-inking will become more important for newsprint. Higher machine speeds will require an extended nip press (ENP). Fourth presses will be common. Multi-nip or roll presses will likely be used for printing and writing grades. Third-position shoe presses are being used for various paper grades. Suction press rolls may be phased out and open draws between press and dryer sections may be eliminated [3].

Research and development work in impulse drying will continue. Combination of press loads and high temperatures provide better water removal with impulse drying. In this context, pressing and drying operations may be combined in the future. As a result, the paper machine length may be shorter.

9.4 Drying

Since the most difficult and expensive water removal takes place at the dryer section of the paper machine, there is room to make advances in the dryer section. Future dryer fabrics should maximize mass transfer and drying rate.

Use of high velocity, high temperature air impingement for more efficient drying will increase. A new concept in drying is Valmet's "Condebelt" dryer. In this system, in place of cylinders, two heated steel bands are used "between which the paper web is dried under high pressure." It is claimed that this system increases dryer speed [3].

With increasing use of recycled fibers, there is a need to develop still better contaminant resistant dryer fabrics. Residual adhesives, pitch and ink from the recycled fiber, coating and size press solids can fill the dryer fabrics prematurely and affect drying efficiency, dryer fabric performance and fabric life. Better contaminant resistant fabrics will increase steam savings and fabric life, reduce downtime for cleaning and improve sheet quality.

Higher temperature, wear resistant dryer fabrics with good hydrolysis resistant will be developed. Less expensive yarn materials that will provide better hydrolysis resistance are in demand.

Low caliper, smooth surface monofilament fabrics will be developed. Properly shaped monofilaments both in MD and CD directions will be used to achieve this. Improvements in surface contact area will improve dryer efficiency. Minimal seam marking is required. Fabrics with less air carrying properties need to be developed.

Use of single run and single tier dryer configurations has increased in recent years. These configurations support the sheet continuously without open draws. In these machines, speed variations may occur between the top and bottom cylinders which may cause sheet stretching, slippage and break. The dryer fabric plays an important roll to reduce the amount of speed difference. There is still room for better dryer fabric stability and resistance to distortion and damage. The current move to single tier dryer systems will continue. Steam-heated, rotating, cast-iron cylinders will likely continue to be used. Closed transfers will become standard and fabric-driven dryers, with no ropes, will be used [3].

9.5   References

1 "Forest, Wood and Paper Industry Grows Toward Future Health and Sustainability", *PaperAge,* Dec. 1996.

2 O'Brien, J., "Valmet Emphasizes Customer Focus and Technological Excellence", *PaperAge,* August 1996.

3 Meadows, D. G., "Panel Discusses Future Papermaking Technology", *TAPPI Journal,* Vol. 79, No. 6, 1996.

4 Kertula, R., "Design for the Future: What to Expect in the New Generation of Paper Machines", *TAPPI Journal,* Vol. 79, No. 6, June 1996.

5 O'Brien, J., "Some Big Challenges in Recycling", *PaperAge,* November 1 995.

6 O'Brien, M., "US Paper Industry Sets 50 Percent Paper Recovery Goal for 2000", PaperAge, February 1994.

7 O'Brien, J., "We are Approaching Limits for Recycling Percentages", *PaperAge,* November 1996.

8 Price, D., "Is the Case for Recycling Weakening?", *PaperAge,* November 1996.

9 Cunnane, F., "Laminated Press Fabric Structures: Their Usage, Their Past and Their Future", *PaperAge,* June 1996.

General References

Antos, D., "Paper Machine Clothing Developments Aimed at Increased Production and Quality", *PaperAge,* June 1995.

Edwards, J. C., "Machine Clothing Trends", *TAPPI Journal,* Vol. 77, No. 6, June 1994.

Fadum, O., "The Process Information Renaissance", *TAPPI Journal,* Vol. 79, No. 10, 1996.

Hawes, J. M., "Paper Machine Clothing and Recycled Fiber Usage", *PaperAge,* Nov. 1996.

Fekete, K., "Advances in Dryer Fabric Technology for Contamination-Prone Positions", *PaperAge,* Nov. 1996.

Luciano, B., and Fagerholm, L., "Technology of Dryer Fabric Design for High Speed Single Run—Single Tier Positions", *PaperAge,* January 1996.

Meadows, D. G., and Shearin, R. H., "Trends in Press Felt Technology", *TAPPI Journal,* Vol. 78, No. 7, July 1995.

Meadows, D. G., et al, "The Role of Machine Clothing in Paper Machine Development and Performance", *TAPPI Journal,* Vol. 79, No. 4, April 1996.

O'Brien, J., "What's New in Forming Technology?", *PaperAge,* February 1996.

O'Brien, J., "Some Thoughts on the Future of Machine Clothing", *PaperAge,* June 1996.

Ow Yang, G. B., "On-line Sensors are Growing Smarter and Smarter", *TAPPI Journal,* Vol. 79, No. 10, 1996.

"Papermachine Clothing: Emphasis is on Better Dewatering and Low Marking", *PPI,* March 1991.

Patrick, K. L., "Clothing Companies Brace for Change This Decade, See Winding Road Ahead", *Pulp & Paper,* October 1990.

Pauksta, P. M., "A Current Perspective on the Chlorine Debate", *TAPPI Journal,* Vol. 78, No. 7, July 1995.

Shearin, R. H., "The Current State of Recycled Paper Supply", *TAPPI Journal,* Vol. 78, No. 2, February 1995.

Size-denier Relationships for Polyester and Nylon Monofilaments

Courtesy of Shakespeare Monofilament.

POLYESTER MONOFILAMENT SIZE EQUIVALENCY CHART

Diameter Thousandths of an Inch (mil.)	Diameter Millimeters	Denier	Decitex	Yards/lb.	Meters/Kg.
.0039	.10	95	106	45,591	93,881
.004	.1016	100	112	44,289	89,243
.0043	.11	116	129	38,327	77,230
.0047	.12	139	154	32,078	64,638
.005	.1270	157	175	28,345	57,115
.0051	.13	163	182	27,245	54,899
.0055	.14	190	211	23,425	47,201
.0059	.15	219	243	20,357	41,020
.006	.1524	226	252	19,684	39,663
.0063	.16	250	277	17,854	35,977
.0067	.17	282	314	15,786	31,809
.007	.1788	308	343	14,461	29,140
.0071	.18	317	352	14,057	28,325
.0075	.19	354	393	12,598	25,385
.0079	.20	393	436	11,354	22,879
.008	.2032	403	448	11,072	22,310
.0083	.21	434	482	10,286	20,727
.0087	.22	476	529	9,362	18,865
.009	.2286	510	567	8,748	17,628
.0091	.23	521	579	8,557	17,243
.0094	.24	556	618	8,020	16,160
.0098	.25	605	672	7,378	14,867
.010	.2540	630	700	7,086	14,278
.0102	.26	655	728	6,811	13,724
.0106	.27	707	786	6,306	12,708
.011	.28	765	850	5,832	11,751
.0114	.29	818	909	5,452	10,987
.0118	.30	877	974	5,089	10,254
.012	.3048	907	1,008	4,921	9,915
.0122	.31	937	1,041	4,761	9,593
.0126	.32	1,000	1,111	4,463	8,994
.013	.33	1,064	1,183	4,193	8,449
.0134	.34	1,131	1,256	3,946	7,952
.0138	.35	1,199	1,333	3,721	7,497
.014	.3556	1,234	1,372	3,615	7,285
.0142	.36	1,270	1,411	3,514	7,081
.0146	.37	1,342	1,492	3,324	6,698
.015	.38	1,417	1,575	3,149	6,346
.0154	.39	1,494	1,660	2,988	6,020
.0157	.40	1,552	1,725	2,874	5,792
.016	.4064	1,612	1,792	2,768	5,577
.0161	.41	1,633	1,814	2,733	5,508
.0165	.42	1,715	1,905	2,602	5,244
.0169	.43	1,799	1,999	2,481	4,999
.017	.4318	1,820	2,023	2,452	4,940
.0173	.44	1,885	2,095	2,367	4,770
.0177	.45	1,973	2,193	2,261	4,557
.018	.4572	2,041	2,268	2,187	4,407
.0181	.46	2,063	2,293	2,163	4,358
.0185	.47	2,156	2,395	2,070	4,172
.0189	.48	2,250	2,500	1,983	3,997
.019	.4826	2,274	2,527	1,962	3,955
.0193	.49	2,346	2,607	1,902	3,833
.0197	.50	2,444	2,716	1,825	3,676

POLYESTER MONOFILAMENT SIZE EQUIVALENCY CHART

Diameter Thousandths of an Inch (mil.)	Diameter Millimeters	Denier	Decitex	Yards/lb.	Meters/Kg.
.020	.5080	2,520	2,800	1,771	3,567
.0201	.51	2,545	2,828	1,754	3,534
.0205	.52	2,647	2,941	1,686	3,397
.0209	.53	2,751	3,057	1,622	3,268
.021	.5334	2,778	3,087	1,606	3,237
.0213	.54	2,858	3,175	1,561	3,147
.0217	.55	2,966	3,296	1,504	3,032
.022	.56	3,060	3,400	1,458	2,939
.0224	.57	3,161	3,512	1,412	2,845
.0228	.58	3,274	3,638	1,363	2,746
.023	.5842	3,332	3,703	1,339	2,699
.0232	.59	3,390	3,767	1,316	2,652
.0236	.60	3,508	3,898	1,272	2,563
.024	.61	3,628	4,032	1,230	2,478
.0244	.62	3,750	4,167	1,190	2,398
.0248	.63	3,874	4,305	1,152	2,321
.025	.6350	3,937	4,375	1,133	2,284
.0252	.64	4,000	4,445	1,115	2,248
.0256	.65	4,128	4,587	1,081	2,178
.026	.66	4,258	4,732	1,048	2,112
.0264	.67	4,390	4,878	1,016	2,048
.0268	.68	4,524	5,027	986	1,988
.027	.6858	4,592	5,103	972	1,958
.0272	.69	4,660	5,178	957	1,929
.0276	.70	4,799	5,332	930	1,874
.028	.71	4,939	5,488	903	1,821
.0283	.72	5,045	5,606	844	1,782
.0287	.73	5,189	5,765	860	1,733
.029	.7366	5,298	5,887	842	1,697
.0291	.74	5,334	5,927	836	1,686
.0295	.75	5,482	6,091	814	1,640
.0299	.76	5,632	6,258	792	1,597
.030	.7620	5,670	6,300	787	1,586
.0303	.77	5,783	6,426	771	1,555
.0307	.78	5,937	6,596	751	1,515
.031	.7874	6,054	6,727	737	1,485
.0311	.79	6,093	6,770	732	1,476
.0315	.80	6,251	6,945	714	1,439
.0319	.81	6,410	7,123	696	1,403
.032	.8128	6,451	7,168	692	1,394
.0323	.82	6,572	7,303	679	1,368
.0326	.83	6,695	7,439	666	1,343
.033	.84	6,860	7,623	650	1,311
.0334	.85	7,028	7,808	635	1,279
.0338	.86	7,197	7,997	620	1,249
.034	.8636	7,282	8,092	613	1,235
.0342	.87	7,368	8,187	605	1,220
.0346	.88	7,542	8,380	591	1,192
.035	.89	7,717	8,575	578	1,165
.0354	.90	7,894	8,772	565	1,139
.0358	.91	8,074	8,971	552	1,114
.036	.9144	8,164	9,072	546	1,101
.0362	.92	8,255	9,173	540	1,089
.0366	.93	8,439	9,376	529	1,065
.037	.94	8,624	9,583	517	1,043
.0374	.95	8,812	9,791	506	1,020
.0377	.96	8,954	9,949	498	1,004
.038	.9652	9,097	10,108	490	988
.0381	.97	9,145	10,161	488	983
.0385	.98	9,338	10,375	478	963
.0389	.99	9,533	10,592	468	943
.039	.9906	9,582	10,646	465	938
.0393	1.00	9,730	10,811	458	924
.0397	1.01	9,929	11,032	449	905
.040	1.0160	10,080	11,200	442	892

NYLON MONOFILAMENT SIZE EQUIVALENCY CHART

Diameter Thousandths of an Inch (mil.)	Diameter Millimeters	Denier	Decitex	Yards/lb.	Meters/Kg.
.004	.1016	83	92	53,658	108,122
.0043	.11	96	106	46,436	93,569
.0047	.12	114	127	38,868	78,319
.005	.1270	130	144	34,341	69,198
.0051	.13	135	150	33,008	66,512
.0055	.14	157	174	28,381	57,188
.0059	.15	181	201	24,663	49,697
.006	.1524	187	208	23,848	48,054
.0063	.16	206	229	21,632	43,588
.0067	.17	233	259	19,126	38,538
.007	.1778	254	283	17,521	35,305
.0071	.18	262	291	17,031	34,317
.0075	.19	292	325	15,262	30,754
.0079	.20	324	360	13,756	27,719
.008	.2032	332	369	13,414	27,030
.0083	.21	358	398	12,462	25,112
.0087	.22	393	437	11,343	22,856
.009	.2286	421	468	10,599	21,357
.0091	.23	430	478	10,367	20,890
.0094	.24	459	510	9,716	19,578
.0098	.25	499	554	8,939	18,013
.010	.2540	520	577	8,585	17,299
.0102	.26	541	601	8,252	16,628
.0106	.27	584	649	7,641	15,396
.011	.28	629	699	7,095	14,297
.0114	.29	675	750	6,606	13,311
.0118	.30	724	804	6,165	12,424
.012	.3048	748	832	5,962	12,013
.0122	.31	773	859	5,768	11,623
.0126	.32	825	917	5,407	10,896
.013	.33	878	976	5,080	10,236
.0134	.34	933	1,037	4,781	9,634
.0138	.35	990	1,100	4,508	9,084
.014	.3556	1,019	1,132	4,380	8,826
.0142	.36	1,048	1,165	4,257	8,579
.0146	.37	1,108	1,231	4,027	8,115
.015	.38	1,170	1,300	3,815	7,688
.0154	.39	1,233	1,370	3,620	7,294
.0157	.40	1,281	1,424	3,483	7,018
.016	.4064	1,331	1,479	3,353	6,757
.0161	.41	1,347	1,497	3,312	6,673
.0165	.42	1,415	1,573	3,153	6,354
.0169	.43	1,485	1,650	3,005	6,057
.017	.4318	1,502	1,669	2,970	5,985
.0173	.44	1,556	1,729	2,868	5,780
.0177	.45	1,629	1,810	2,740	5,521
.018	.4572	1,684	1,872	2,649	5,339
.0181	.46	1,703	1,892	2,620	5,280
.0185	.47	1,779	1,977	2,508	5,054
.0189	.48	1,857	2,063	2,403	4,842
.019	.4826	1,877	2,085	2,378	4,792
.0193	.49	1,936	2,152	2,304	4,644
.0197	.50	2,018	2,242	2,212	4,457
.020	.5080	2,080	2,311	2,146	4,324
.0201	.51	2,100	2,334	2,125	4,281
.0205	.52	2,185	2,428	2,042	4,116
.0209	.53	2,271	2,523	1,965	3,960

NYLON MONOFILAMENT SIZE EQUIVALENCY CHART

Diameter Thousandths of an Inch (mil.)	Diameter Millimeters	Denier	Decitex	Yards/lb.	Meters/Kg.
.021	.5334	2,293	2,548	1,946	3,922
.0213	.54	2,359	2,621	1,892	3,813
.0217	.55	2,448	2,720	1,823	3,673
.022	.56	2,516	2,796	1,773	3,574
.0224	.57	2,609	2,899	1,711	3,447
.0228	.58	2,703	3,003	1,651	3,327
.023	.5842	2,750	3,056	1,622	3,270
.0232	.59	2,798	3,109	1,595	3,214
.0236	.60	2,896	3,217	1,541	3,106
.024	.61	2,995	3,328	1,490	3,003
.0244	.62	3,095	3,439	1,442	2,905
.0248	.63	3,198	3,553	1,395	2,812
.025	.6350	3,250	3,611	1,373	2,767
.0252	.64	3,302	3,669	1,351	2,724
.0256	.65	3,407	3,786	1,310	2,639
.026	.66	3,515	3,905	1,270	2,559
.0264	.67	3,624	4,026	1,231	2,482
.0268	.68	3,734	4,149	1,195	2,408
.027	.6858	3,790	4,212	1,177	2,373
.0272	.69	3,847	4,274	1,160	2,338
.0276	.70	3,961	4,401	1,127	2,270
.028	.71	4,076	4,529	1,095	2,206
.0283	.72	4,164	4,627	1,071	2,106
.0287	.73	4,283	4,759	1,042	2,100
.029	.7366	4,373	4,859	1,020	2,057
.0291	.74	4,403	4,892	1,013	2,042
.0295	.75	4,525	5,028	986	1,987
.0299	.76	4,648	5,165	960	1,935
.030	.7620	4,680	5,200	953	1,922
.0303	.77	4,774	5,304	935	1,884
.0307	.78	4,900	5,445	910	1,835
.031	.7874	4,997	5,552	893	1,800
.0311	.79	5 029	5,588	887	1,788
.0315	.80	5,159	5,733	865	1,743
.0319	.81	5,291	5,879	843	1,700
.032	.8128	5,324	5,916	838	1,689
.0323	.82	5,425	6,027	822	1,658
.0326	.83	5,526	6,140	807	1,627
.033	.84	5,662	6,292	788	1,588
.0334	.85	5,800	6,445	769	1,550
.0338	.86	5,940	6,600	751	1,514
.034	.8636	6,011	6,679	742	1,496
.0342	.87	6,082	6,757	734	1,479
.0346	.88	6,225	6,916	717	1,445
.035	.89	6,370	7,077	700	1,412
.0354	.90	6,516	7,240	685	1,380
.0358	.91	6,664	7,405	669	1,349
.036	.9144	6,739	7,488	662	1,334
.0362	.92	6,814	7,571	655	1,320
.0366	.93	6,965	7,739	640	1,291
.037	.94	7,118	7,909	627	1,263
.0374	.95	7,273	8,081	613	1,236
.0377	.96	7,390	8,211	604	1,217
.038	.9652	7,508	8,343	594	1,198
.0381	.97	7,548	8,387	591	1,191
.0385	.98	7,707	8,564	579	1,167
.0389	.99	7,868	8,742	567	1,143
.039	.9906	7,909	8,788	564	1,137
.0393	1,00	8,031	8,923	555	1,120
.0397	1.01	8,195	9,106	544	1,097
.040	1.016	8,320	9,244	536	1,081
.045	1.143	10,530	11,700	423	854
.050	1.270	13,000	14,444	343	691
.055	1.397	15,730	17,477	283	571
.060	1.524	18,720	20,800	328	480
.065	1.651	21,970	24,411	203	409
.070	1.778	25,480	28,311	175	353
.075	1.905	29,250	32,500	152	307
.080	3.032	33,280	36,977	134	270

Forming Section Troubleshooting Guide

Symptom	Possible Causes/Check Points/Possible Solutions

Barring	—Check conditions of rolls (vibration)
	—Check seals of the couch roll
	—Check for variation on all machine elements
	—Check for loose bearing houses
	—Compare mark to felt pattern
	—Check for pressure pulses
	   • headbox
	   • pressure screen
	   • stock delivery system

Cockling	—Poor formation —Water jumps—stock jump —Water circulation from suction couch roll —Reduce refining —Pulsating on the stock system —Rewetting of the fabric from showers —Water spots from dandy roll or couch press
Couch Shadow Marking	—Check drainage—may be flooding couch —Check vacuums—increase drainage on table —Check drag to jet ratio —Check fabric tension—may be too low —Check for excessive fabric wear —Review fabric drainage capacity—may need to increase fabric drainage —Freeness too low
Cracks	—Check machine deckles —Check trim —Rewetting on edges

Symptom	Possible Causes/Check Points/Possible Solutions
Crushing	—Sheet running wet into pressing elements —Check draws (speed differences) —Check couch press —Check dandy roll —Dandy out of balance —Sheet too wet at dandy —Dandy set too hard —Consistency too high —Fabric plugged —Suction couch roll showers not set correctly
Curl	—Fiber alignment —Two sidedness —Refining —Draw —Edge dewatering —Increase long fiber —Reduce freeness of stock —Retention aid uneven —Fabric dewatering uneven —Check headbox slice opening
Dandy Roll Marks	—Plugged dandy shower —Damaged dandy roll —Dandy not level —Check alignment of dandy —Sheet coming into dandy, too wet —Sheet too dry—picking sheet —Check vacuum levels of suction boxes
Edge Breaks	—Trim squirt plugged—not cutting clean —Recouching of sheet and edge trim in pick nip —Foil, couch roll—trim too small —Dirt build up on bottom lip —Dirt build up on fabric edge —Stock build up on deckles —Poor dewatering of edge area —Rewetting at couch —Fabric damaged —Stock build up on foils or suction boxes —Fabric edge damage from suction box sealing —Deckles touching fabric —Stock build up on return rolls
Guiding	—High vacuum on suction boxes —Improper guide palm sensitivity —Non-uniform fabric tension—front to back —Insufficient wrap on guide roll

Symptom	Possible Causes/Check Points/Possible Solutions
	—Diagonal of drilled boxes
	—Check drive sync
	—Check for fabric slippage
	—Check doctor blades
	—Check roll bearings
	—Check fabric tension
	—Check draw
	—Check suction box vacuum and surface
	—Check for unbalance rolls
	—Check guide movement
Moisture Profile Uneven	—Uneven head box slice caliper
	—Stock jet speed difference in tubes
	—Stock speed difference
	—Uneven plates in headbox
	—Check rectifier roll
	—Deflection of breast roll
	—Check foils for unevenness—level side to side
	—Check foils for wear
	—Uneven fabric tension
	—Uneven shower water supply
	—Check for plugged suction rolls
Pin Holes	—Check for air in the system
	—Check for excessive drainage—stapling
	—Change table elements—too slow drainage
	—May need to close up fabric
	—Check vacuum of suction boxes—too high
	—Check for thin spots in sheet
	—Poor sheet release
Ply Bonding/Delamination	—Increase refining
	—Improve formation
	—Decrease internal sizing
	—Increase long fiber content
	—Increase strength additives
Poor Drainage	—Check if freeness is too high
	—Check if stock temperature is too low
	—Check pH
	—Check retention aids used
	—Check if refining is excessive
	—Check for air in stock
	—Check fabric for excessive wear
	—Check machine elements for wear
	—Check table alignment
	—Check fabric tension
	—Check for dirty fabrics

Symptom	Possible Causes/Check Points/Possible Solutions
	—Check for air leaks —Sheet sealing

Poor Formation	—Too high fabric oscillation—air
	—Vibration of table elements
	—Fabric slippage
	—Check for excessive fabric wear
	—High friction at suction boxes
	   • vacuum too high
	   • dirt
	   • sharp edges
	—Increase fabric tension
	—Check for fabric deflection
	—Check for vacuum pulses
	—Check rush/drag—impingement—elements

Release	—Check for sheet stapling —Trim squirts not cutting edge —Check fabric for excessive drainage —Check fabric for excessive sheet side wear —Couch vacuum too high —Pickup vacuum too low —Pickup felt plugged—add water to pickup felt —Check fabric tension —Check draws —Check for wrinkles/ridges
Ridges/Wrinkles of Forming Fabrics	—Check machine alignment —Excessive table shake —Check high pressure shower nozzles —Check for deposits on table elements —Check for damaged foils—table elements —Check operating tension —Check pressure of high pressure showers —Check rolls and doctor blades for excessive wear —Check rolls and doctor blades for alignment, deflection and stability —Check shower flows and pressures —Check drive ratios —Check guide system, pressure and air supply —Check for wrong geometry of fabric oscillation
Sheet Wrinkles	—Too high stock speed in head box tube system —Base plates uneven —Plugged tubes —Slice uneven —Check for damaged slice —Check breast roll shower —Check for worn shower nozzles

Symptom	*Possible Causes/Check Points/Possible Solutions*
	—Check for breast roll alignment
	—Check for build up on breast roll
	—Deckle setting incorrect
	—Dirty fabric
	—Worn fabric
	—Waves in fabrics
Speed Variation on Fabrics	—Check for fabric slippage
	—Check for excessive friction
	—Check table elements
	—Check bearings
	—Check machine alignment
	—Check for drive overload—drag load
	—Check suction box vacuum—too high
	—Check for roll surfaces—too smooth
	—Check for slot width being too wide
	—Check for vacuum variation
	—Check for couch stability
	—Check for frame instability
	—Check for couch roll cover hardness variation
	—Check for roll balance
Streaky Formation	—Check headbox slice opening
	—Check for stock build up on slice
	—Check for slice distortion
	—Check headbox
	—Check rectifier roll
	—Check for wear streaks in fabric
	—Worn out slice
	—Check alignment of table
	—Check if fabric is dirty
	—Dirty dandy roll
Wire Mark	—Check refining
	—Check vacuum—reduce if possible
	—Check fabric tension—increase if possible
	—Check drag to jet ratio
	—Check for excessive fabric drainage
	—Reduce loading on dandy-couch press pickup felt
	—Check fabric for excessive sheet side wear
	—Different fabric may be needed

Press Section Troubleshooting Guide

Symptom	Possible Causes/Check Points/Possible Solutions
Poor Pickup	—Increase felt water just ahead of the pickup zone —Check weight profile, especially if the pickup problem is not confined to the edges —Speed differential: Are loads and amperages normal? —"Wire" tension—try to increase—below 20 is too low (25 pli on suction breast roll, SBR) —Felt tension: Is it normal? Try slack, then tight. —Holes in the sheet—Is forming fabric contaminated? Check for fibers in spray as felt and forming fabric separate. Lots of fibers indicate poor pickup, not contamination. —Sheet stapled to forming fabric: Lower vacuum at SBR; check suction box position. —Lower felt pickup point (into forming fabric)
Crushing	—Decrease felt water content —Increase vacuum (check vacuum) —Clean felt —Inspect for plugged uhle box, suction roll holes, blind-drilled hole, etc. —Lower machine speed while high pressure showering —Check for felt surface damage —Check for roll cover slippage
Blistering	—Check for uniform Yankee coating (check spray boom coverage) —Check for uneven loading/wear on Yankee doctor blade —Hot edges (add water to the edges) —Lower steam pressure in Yankee, increase hood temperature —Check Yankee differential pressure (high differential creates hot spots resulting in blistering. Good number is 12–16 psi). —Check thermal topography of the Yankee

369

Symptom	Possible Causes/Check Points/Possible Solutions
Sheet Following	—Check felt moisture ratio —Check for uniform Yankee coating. If uneven, check Yankee doctor blade and spray bottom —Check for filled, compacted, smooth areas in felt and clean accordingly —Check suction pressure roll box position (throw off should be 8°–10° from the exiting felt tangent) and seal strips —Lower felt water—has pressure roll crown been changed —Check pressure roll loading and increase
Felt Bagginess	—Check felt tension (fabric slippage) —Check for worn uhle box strips, pickup shoes/bars and their proper lubrication —Check for temperature differentials across the face of the fabric —Check for Yankee hood leaks on the surface of the fabric
Sheet Moisture Profile Problems	—Determine where problem was seen first —Check for uneven basis weight profile —Check for damaged forming fabric —Measure felt moisture—does problem correlate? —Is the felt skewed? —Clean felt —Uneven shower application—uneven cleaning —Check air pressure on the seal strips —Are seal strips loaded evenly? —Plugged/worn uhle box —Check hood profiling dampers —Check for roll plugging —Check for roll cover slippage —Is the sheet profiler operating correctly? —Doctor loading on the pressure rolls or main rolls —Doctor loading on the Yankee —Check spray boom for nozzle plugging
Sheet Stealing (to bottom felt)	—Reduce loading at the main press —Reduce bottom felt tension —Vary water flow (usually lower) —Check uhle box vacuum —Rotate the main press suction box toward the wet end —Check air pressure on the seal strips —Are seal strips loaded evenly? —Is bottom felt filled? If so, clean. —Lightly brush the bottom felt surface —Turn bottom felt inside out

Tension Force Calculations of Dryer Fabrics

INTRODUCTION

To enable a dryer fabric to perfectly fulfil its role, it is necessary that it is properly tightened.

This tension is generally expressed in kilograms per centimeter (kg/cm) of width of fabric and ideally it should be between 1.5 and 1.8 kg/cm.

Below 1.5 kg/cm, the contact between the paper and the cylinder is negatively affected and consequently, the drying capacity is diminished.

Above 1.8 kg/cm, excessive mechanical tension could lead to excessive deflection of the rolls and can cause premature rupture of fabric and/or seam.

1. GENERAL SITUATION

On most paper machines, the tension regulation systems of dryer fabrics (if they exist) are schematized as follows:

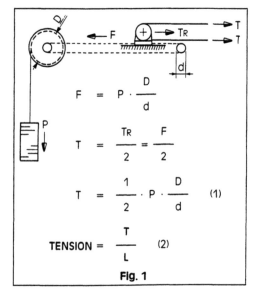

$$F = P \cdot \frac{D}{d}$$

$$T = \frac{T_R}{2} = \frac{F}{2}$$

$$T = \frac{1}{2} \cdot P \cdot \frac{D}{d} \qquad (1)$$

$$\text{TENSION} = \frac{T}{L} \qquad (2)$$

Fig. 1

Force "F" which acts on the stretch roll depends on force "P", shown here by weights but can equally be a hydraulic system or compressed air or be exerted by a motor.

Dividing force "T" by the width of the fabric "L" presents the effective tension.

2. PARTICULAR CASES

The preceding formulas can be used with good approximation when the wrap on the stretch roll is approximately 180°C and the force "F" acting on it is parallel to the reaction force "TR" of the stretch roll.

When these two conditions are not simultaneously met, it is important to know that in establishing the formula (1) above, the cosine values of the intervening angles with the vectorial breakdown should be used.

2.1 The movement of the stretch roll follows the bisector of the angle formed by the fabric

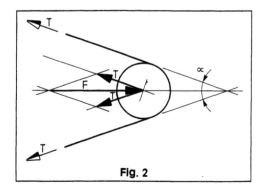

Fig. 2

F = force acting on the stretch roll in equilibrium with TR

TR = total reaction force of the fabric

After breaking down the vectorial component, we obtain:

$$T = \frac{F}{2} \cdot \frac{1}{\cos \frac{\infty}{2}}$$

2.2 The movement of the stretch roll occurs parallel to one of the edges of the fabric

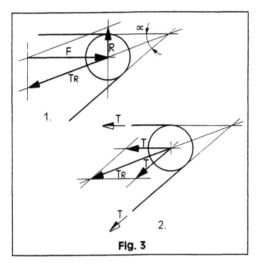

Fig. 3

The total reaction force "TR" of the fabric only balances the acting force "F" with the aid of the counteracting force "R" which represents the reaction of the frame of the machine.

By breaking down force "F" vectorially in Pos. 1, we obtain:

$$T_R = F \cdot \frac{1}{\cos \frac{\alpha}{2}}$$

The decomposition of "TR" on Pos. 2 gives:

$$T = \frac{T_R}{2} \cdot \frac{1}{\cos \frac{\alpha}{2}}$$

Finally:

$$T = \frac{F}{2} \cdot \frac{1}{\cos^2 \frac{\alpha}{2}}$$

2.3 The movement of the stretch roll can occur in any direction in relation to the direction of the acting forces

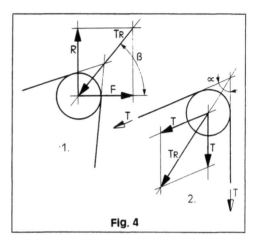

Fig. 4

F = Force acting on the stretch roll
R = Counteracting force of the machine frame
TR = Total counteracting force on the dryer fabric

The vectorial breakdown of "F" on Pos. 1 produces:

$$T_R = F \cdot \frac{1}{\cos \beta}$$

The breakdown of "TR" on Pos. 2 shows:

$$T = \frac{T_R}{2} \cdot \frac{1}{\cos \frac{\alpha}{2}}$$

Finally:

$$T = \frac{F}{2} \cdot \frac{1}{\cos \beta \cdot \cos \frac{\alpha}{2}}$$

2.4 Diagram of the values of

$$\frac{1}{\cos\dfrac{\propto}{2}} \quad \text{and of} \quad \frac{1}{\cos^2\dfrac{\propto}{2}} \quad \text{as}$$

a function of the angle "∝"

Depending on the machines, the angle "∝" can be strongly variable and can reach sometimes relatively high values when, for example, a dryer fabric has been shortened as a result of a new seam.

Another frequent example: a change in the return run of the fabric.

The diagram shows us that for "∝" = 60°, the over-tension on the fabric compared to the classical case surpasses 30% in case 2.2 (relatively frequent) and even more than 15% in case 2 1.

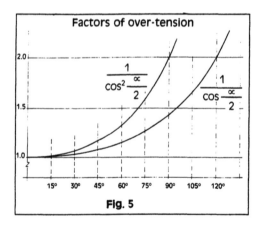

Fig. 5

3. SOME EXAMPLES OF THE CALCULATIONS OF "F" AND "T"

3.1 Bottom position of the section Effect of the weight of the stretch roll

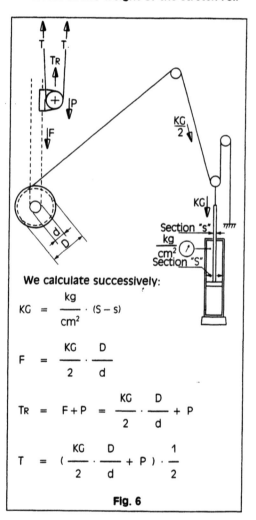

We calculate successively:

$$KG = \frac{kg}{cm^2} \cdot (S - s)$$

$$F = \frac{KG}{2} \cdot \frac{D}{d}$$

$$T_R = F + P = \frac{KG}{2} \cdot \frac{D}{d} + P$$

$$T = \left(\frac{KG}{2} \cdot \frac{D}{d} + P \right) \cdot \frac{1}{2}$$

Fig. 6

374

3.2 Influence of a second system of pulleys

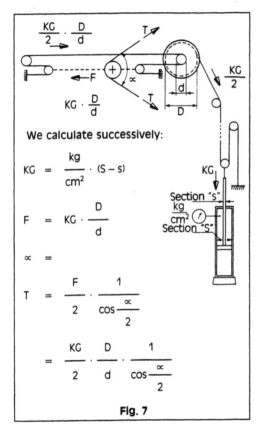

$$\frac{KG}{2} \cdot \frac{D}{d} \longrightarrow$$

$$\longleftarrow F$$

$$KG \cdot \frac{D}{d}$$

We calculate successively:

$$KG = \frac{kg}{cm^2} \cdot (S-s)$$

$$F = KG \cdot \frac{D}{d}$$

$$\propto =$$

$$T = \frac{F}{2} \cdot \frac{1}{\cos \dfrac{\propto}{2}}$$

$$= \frac{KG}{2} \cdot \frac{D}{d} \cdot \frac{1}{\cos \dfrac{\propto}{2}}$$

Fig. 7

3.3 Influence of a pneumatic returning stretch roll

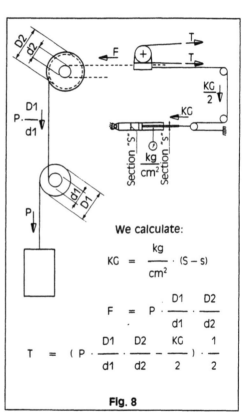

$$P \cdot \frac{D1}{d1}$$

We calculate:

$$KG = \frac{kg}{cm^2} \cdot (S-s)$$

$$F = P \cdot \frac{D1}{d1} \cdot \frac{D2}{d2}$$

$$T = \left(P \cdot \frac{D1}{d1} \cdot \frac{D2}{d2} - \frac{KG}{2} \right) \cdot \frac{1}{2}$$

Fig. 8

3.4 Influence of the angle "∝"
The movement of the stretch roll occurs according to the bisector of the angle " ∝ " formed by the fabric

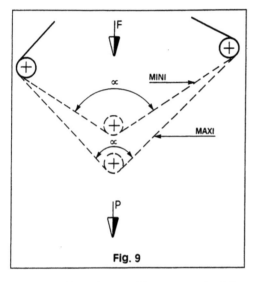

Fig. 9

In this machine, the possible movement of the stretch roll is not very important.

The angle "∝" varies only slightly between the maximum and minimum positions of the stretch roll.

$$90° \leqslant \propto \leqslant 115°$$

As per 2.1 and 3.1 above, the force on the fabric is calculated as follows:

$$T = (F + P) \cdot \frac{1}{2} \cdot \frac{1}{\cos\dfrac{\propto}{2}}$$

The factor of over-tension shall is therefore situated between the values:

$$1.41 \leqslant \frac{1}{\cos\dfrac{\propto}{2}} < 1.86$$

3.5 Influence of the angle "∝"
The movement of the stretch roll occurs parallel to one side of the fabric

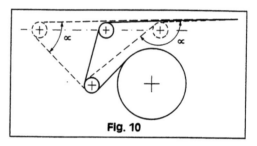

Fig. 10

On this machine, it was calculated that the angle "∝" for the maximum and minimum positions of the stretch roll varies between the extremities:

$$46° \leqslant \propto \leqslant 143°$$

As per 2.2 above, the tension of the fabric is calculated as follows:

$$T = \frac{F}{2} \cdot \frac{1}{\cos^2\dfrac{\propto}{2}}$$

The over-tension factor is therefore between the values:

$$1.18 \leqslant \frac{1}{\cos^2\dfrac{\propto}{2}} \leqslant 9.93$$

To regulate correctly the manometric pression required, it is vital to take into consideration the position of the stretch roll on the frame.

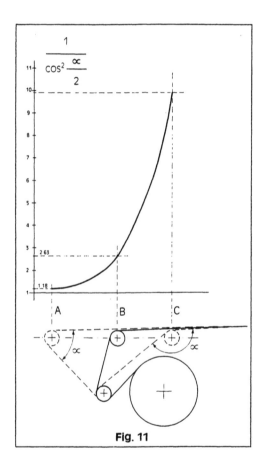

Fig. 11

3.5.1

In this machine, not considering the over-tension factor, a manometric pressure of 3.5 kg/cm² corresponds to a fabric tension of 1.6 kg/cm.

3.5.2

In fact, the extreme position "A" of the stretch roll would require a manometric pressure of:

$$\frac{3.5}{1.18} = 2.97\ kg/cm^2$$

in order to generate the fabric tension of 1.6 kg/cm.

3.5.3

Similarly, in position "B", the manometric pressure should be:

$$\frac{3.5}{2.63} = 1.33\ kg/cm^2$$

in order to generate a same tension of 1.6 kg/cm.

Ing. Marcel SIQUET
Applications Engineer
January 1990

List of Testing Laboratories/ Research Facilities

TAPPI does not sanction the companies cited in this list. TAPPI provides this information as a service to its members and the general public. This information was obtained from a survey of industry test laboratories and research facilities. To determine the services they offer, please contact individual laboratories. To be included in the current test laboratories list, please call (770) 209-7208.

APPLIED TECHNICAL SERVICES, INC.
1190 Atlanta Industrial Dr.
Marietta, GA 30066
Contact Person: Bob Dunning or Phil Rogers
Tel: (404)423-1400

1218 Donaldson Rd.
Greenville, SC 29605
Contact Person: Paul West, Manager
Tel: (803)299-0525

108-A Castle Dr.
Madison, AL 35758
Contact Person: John Cooley, Manager

ARTHUR D. LITTLE, INC.
25 Acorn Park
Cambridge, MA 02140
Contact Person: Dr. Judith Harris
Tel: (617)864-5770

AUBURN UNIVERSITY
Pulp and Paper Research and Education Center (PPREC)
242 Ross Hall
Auburn, AL 36849-5128
Director: Dr. Ronald D. Neuman
Tel: (205)826-5223

BADGER LABORATORIES & ENGINEERING COMPANY, INC.
1110 South Oneida Street
Appleton, WI 54915
President: Arthur B. Kaplan
Tel: (414)739-9213
      (800)242-3556

BATTELLE MEMORIAL INSTITUTE/COLUMBUS LABORATORIES
505 King Avenue

Columbus, OH 43201
Tel: (614)424-6424

CHICAGO TESTING LABORATORY, INC.
3360 Commerical Avenue
Northbrook, IL 60062
President: Conway C. Burton
Tel: (312)498-6400

CONTAINER TESTING LABORATORY, INC.
607 Fayette Ave.
Mamaroneck, NY 10543
Tel: (914) 381-2600
Fax: (914) 381-0143

CÉGEP DE TROIS-RIVIÈRES PULP & PAPER SPECIALIZED CENTER
2250, rue St.-Oliver
Troise-Rivières, QC G9A 5E6
Canada
Directeur: Mr. Pierre Lavoie, ing.
Tel: (819)372-0202
Fax: (819)372-9938

DSET LABORATORIES, INC.
Box 1850
Black Canyon Stage One
Phoenix, AZ 85029
Tel: (602)465-7356

ECONOTECH SERVICES LIMITED
852 Derwent Way
Annacis Island
New Westminister, BC V3M 5R1
Canada
Pulping & Bleaching: Randy Lowe and Tadas Macas
Fibre Testing: Wendy McAlpine
Analytical: Terry Peel
Tel: (604)526-4221
Fax: (604)526-1898

U.S. Mailing Address:
P.O. Box 952
Point Roberts, WA 98281

FOREST SERVICE
Forest Products Laboratory
1 Gifford Pinchot Dr.
Madison, WI 53705-2398
Director: John R. Erickson
Tel: (608)231-9200
Fax: (608)231-9592

GALBRAITH LABORATORIES, INC.
2323 Sycamore Drive
Knoxville, TN 37921
Tel: (615)546-1335

GORHAM INTERNATIONAL, INC.
P. O. Box 8
Gorham, ME 04038
Contact Person: Alfred Belisle
Tel: (207)892-2216

HERTY FOUNDATION
P. O. Box 7798
Garden City, GA 31418
Director: Dr. Michael J. Kocurek
Tel: (912)964-5541
Fax: (912)964-5614

INDUSTRIAL TESTING AND RESEARCH LABORATORY
P. O. Box 43643
650 Great S.W. Pkwy.
Atlanta, GA 30336
Contact Person: Roberta Louise Johnson
Tel: (404)344-7883

INSTITUTE OF PAPER SCIENCE AND TECHNOLOGY
575 14th Street, N.W.
Atlanta, GA 30318
President: James Ferris

Tel: (404)853-9500
Fax: (404)853-9510

INTEGRATED PAPER SERVICES, INC.
Suite 250
101 West Edison Avenue
P. O. Box 446
Appleton. WI 54912-0446

Director of Technology: Salman Aziz
Analytical Chemistry: Sally A. Berben or Betty J. Stevens
Aquatic Biology: David L. Rades or David F. Sanders
Fiber Sciences: Walter J. Rantanen or Sara J. Spielvogel
Information Systems & Services: Craig S. Booher
Paper & Paperboard Testing: Catherine J. DeLain
Pulping, Bleaching & Papermaking: Thomas W. Paulson
Tel: (414)749-3040
Fax: (414)749-3046

MCGILL UNIVERSITY
Department of Chemical Engineering
3480 University Street
Montreal. PQ H3A 2A7
Canada
Coordinator: M. R. Kamal or L. St. Pierre
Tel: (514)398-4262

MICHIGAN TECHNOLOGICAL UNIVERSITY
Institute of Wood Research
Houghton, MI 49931
Director: Dr. W. E. Frayer
Tel: (906)487-2464

NORTH CAROLINA STATE UNIVERSITY
Pulp and Paper Laboratory
Box 8005
Raleigh, NC 27695-8005
Department Head: Dr. Richard J. Thomas
Tel: (919)515-2888
Fax: (919)515-7231
Wood Products Laboratory
Box 8005
Raleigh, NC 27695-8005
Department Head: Dr. Richard J. Thomas
Tel: (919)515-2881
Fax: (919)515-7231

PRO-PACK TESTING LABORATORY, INC.
15 North Florida
Belleville, IL 62221
Contact Person: Manuel Rosa
Tel: (618)277-1160
Fax: (618)277-1163

PULP AND PAPER RESEARCH INSTITUTE OF CANADA
570 St. John's Boulevard
Pointe, PQ H9R 3J9
Canada
Contact Person: Wilfried Johasch
Tel: (514)630-4100
Fax: (514 6304134

SUNY COLLEGE OF ENVIRONMENTAL SCIENCE AND FORESTRY
Empire State Paper Research Institute
Syracuse, NY 13210
Director: Dr. Leland R. Schroeder
Tel: (315)470-6502

Paper Science & Engineering Pilot Plant
Syracuse, NY 13210
Director: Dr. William Holtzman
Tel: (315)470-6506
Fax: (315)470-6779

TECHNIDYNE CORPORATION
100 Quality Ave.
New Albany, IN 47150-2272
Contact Persons: Thomas Crawford and Pat Robertson
Tel: (812)948-2884
Fax: (812)945-6847

UNIVERSITY OF CALIFORNIA, BERKELEY
Forest Products Laboratory
1301 South 46th Street
Richmond, CA 94804
Director: Dr. Frank C. Beall
Tel: (415)231-9452

UNIVERISTY OF LOWELL RESEARCH FOUNDATION
450 Aiken Street
Lowell, MA 01854
Executive Director: Edward F. Miller, Jr.
Tel: (508)458-2508
Fax: (508)453-6586

UNIVERSITY OF MAINE
Environmental Studies Center
Coburn Hall #1

Orono, ME 04469
Director: Gregory K. White
Tel: (207)581-1490
Fax: (207)581-1426

Department of Chemical Engineering
Jenness Hall
Orono, ME 04469
Laboratory Manager: Proserfina Bennett
Tel: (207)581-2321
Fax: (207)581-2223

UNIVERSITY OF OREGON
Forest Industries Management Center
Eugene, OR 97403
Director: Dr. Stuart U. Rich
Tel: (503)686-3335

UNIVERSITY OF QUEBEC AT TROIS-RIVIERES
Pulp and Paper Research Center
3351 Boulevard des Forges
Trois-Rivieres, PQ G9A 5H7
Canada
Director: Dr. H. O. Lavalle
Tel: (819)376-5075
Fax: (819)376-5012

UNIVERSITY OF TORONTO
Pulp and Paper Centre
Dept. of Chemical Engineering & Applied Chemistry
200 College Street
Toronto, ON M5S 1A4
Canada
Director: Prof. Douglas Reeve
Tel: (416)978-3062

UNIVERSITY OF WASHINGTON
Center for International Trade in Forest Products
College of Forest Resources AR-10
Seattle, WA 98195
Acting Director: Dena Daivid B. Thorud
Tel: (206)543-8684
Fax: (206)543-3254

WESTERN MICHIGAN UNIVERSITY
Paper and Printing Pilot Plants
Department of Paper and Printing Science and Engineering
Kalamazoo, MI 49008
Interim Chairperson: Aryon D. Byle
Tel: (616)387-2770

Units and Conversion Factors

1. U.S. CUSTOMARY UNITS

Length	1 foot = 12 inches
	1 yard = 3 feet
	1 mile = 1,760 yards = 5,280 feet
	1 inch = 1,000 mils

Area	1 square foot = 144 square inches
	1 square yard = 9 square feet
	1 acre = 43,560 square feet
	1 square mile = 640 acres

Volume	1 cubic yard = 27 cubic feet
	1 cubic foot = 1,728 cubic inches
	1 gallon = 231 cubic inches

liquid or fluid measures

1 pint = 4 gills = 16 ounces
1 quart = 2 pints
1 gallon = 4 quarts
1 cubic foot = 7.4805 gallon

dry measures

1 quart = 2 pints
1 peck = 8 quarts
1 bushel = 4 pecks

Weights	1 ounce = 16 drams = 437.5 grams
	1 pound = 16 ounces = 7,000 grains
	1 ton (US) = 2,000 lb

Pressure	1 atmosphere = 14.7 psi

2. SI PREFIXES

multiplication factors	prefix	SI symbol
1 000 000 000 000 000 000 = 10^{18}	exa	E
1 000 000 000 000 000 = 10^{15}	pecta	P
1 000 000 000 000 = 10^{12}	tera	T
1 000 000 000 = 10^{9}	giga	G
1 000 000 = 10^{6}	mega	M
1 000 = 10^{3}	kilo	k
100 = 10^{2}	hecto	h
10 = 10^{1}	deka	da
0.1 = 10^{-1}	deci	d
0.01 = 10^{-2}	centi	c
0.001 = 10^{-3}	milli	m
0.000 001 = 10^{-6}	micro	μ
0.000 000 001 = 10^{-9}	nano	n
0.000 000 000 001 = 10^{-12}	pico	p
0.000 000 000 000 001 = 10^{-15}	femto	f
0.000 000 000 000 000 001 = 10^{-18}	atto	a

3. SI UNITS

quantity	unit	SI symbol	Formula
Base Units			
length	meter	m	
mass	kilogram	kg	
time	second	s	
electric current	ampere	A	
thermodynamic temperature	Kelvin	K	
amount of substance	mole	mol	
luminous intensity	candela	cd	
Derived Units			
acceleration	meter per second square		m/s^2
area	square meter		m^2
density	kilogram per cubic meter		kg/m^3
energy	joule	J	N·m
force	Newton	N	$kg·m/s^2$
frequency	hertz	Hz	1/s
power	Watt	W	J/s
pressure	Pascal	Pa	N/m^2
velocity	meter per second		m/s
voltage	volt	V	W/A
volume	cubic meter		m^3
work	joule	J	N·m

1 meter = 100 cm = 1,000 millimeter
1 angstrom = 10^{-8} centimeter

1 ton = 1,000 kg
1 kilogram = 1,000 gram
1 kilogram = 9.807 Newton
1 Newton = 100,000 dyn

1 liter = 100 centiliter = 1,000 milliliter = 1,000 cubic centimeter
1 cubic meter = 1,000,000 cubic centimeter

1 atmosphere = 1.01×10^3 Pascal
1 kg/cm^2 = 0.9678 atmosphere
1 bar = 10^6 dynes/cm^2 = 10^5 Newton/m^2
1 Pascal = 1 Newton/m^2
1 cm of mercury (0° C) = 13.60 cm of water (4° C)

1 kWh = 3.6×10^6 Joules
1 Watts = 1 Joules/second = 1.34×10^{-3} HP

4. CONVERSION BETWEEN SI AND U.S. CUSTOMARY UNITS

1 miles = 1.609 kilometer
1 yard = 0.9144 m
1 inch = 25.4 mm
1 m^2 = 1.196 square yard
1 acre = 4,047 m^2
1 cubic inch = 16.39 cubic centimeter
1 lb = 0.453 59 kg = 453.59 g
1 quart = 0.9464 liters
1 ounce = 28.35 grams
1 ounce (liquid) = 0.02957 liters
1 kg/m^2 = 9.807 Pascal = 0.001422 pounds/square inch
1 lb/square foot = 4.882 kg/m^2
1 ton (metric) = 2,205 lbs
1 BTU = 1,054 Joules
1 Joule = 0.7376 ft-lbs
1 HP = 550 ft-lb/second
1 Newton = 0.2248 lbs
1 m/s^2 = 3.281 feet/s^2
1 cfm = 18.29 m^3/m^2/h

Metric to US Customary Conversion Factors

to get	multiply	by
inches	centimeters	0.3937007874
feet	meters	3.280839895
yards	meters	1.093613298
miles	kilometers	0.6213711922
ounces	grams	0.03527396195
pounds	kilograms	2.204622622
gallons (liquid)	liters	0.2641720524
fluid ounces	mililiters (cc)	0.03381402270

to get	multiply	by
square inches	centimeter squares	0.1550003100
square feet	meter squares	10.76391042
square yards	meter squares	1.195990046
cubic inches	milliliters (cc)	0.06102374409
cubic feet	cubic meters	35.31466672
cubic yards	cubic meters	1.307950619

US Customary to Metric

to get	multiply	by
microns	mils	25.4
centimeters	inches	2.54
meters	feet	0.3048
meters	yards	0.9144
kilometers	miles	1.609344
grams	ounces	28.34952313
kilograms	pounds	0.45359237
liters	gallons (liquid)	3.785411784
milliliters (cc)	fluid ounces	29.57352956
square centimeters	square inches	6.4516
square meters	square feet	0.09290304
square meters	square yards	0.83612736
milliliters (cc)	cubic inches	16.387064
cubic meters	cubic feet	0.02831684659
cubic meters	cubic yards	0.764554858

5. TEMPERATURE CONVERSION

$^\circ F = 9/5\ (^\circ C) + 32$

$^\circ C = 5/9\ [(^\circ F) - 32]$

$^\circ K\ (\text{Kelvin}) = {}^\circ C + 273$

SOURCES:

—ASTM

—Mark's Standard Handbook for Mechanical Engineers, McGraw-Hill Book Company, 1978.

—Standard Mathematical Tables, CRC Press, 1982.

—Handbook of Mathematical, Scientific and Engineering Formulas, Tables, Functions, Graphs, Transforms, Research and Education Association, 1994.

Index

Abrasion, 104, 205
Abrasion resistance, 63, 228
Absorbency, 301
Acid papermaking, 136
Acid resistance, 228
Acidic sulfite process, 3
Acrylic, 198
Additives, 4
Adhesion force, 247
AF&PA, 11
Air carrying, 194
Air carrying capacity, 228
Air laying, 138
Air permeability, 89, 171, 228, 301
Air volume, 286
Algor®, 348
Alkaline kraft process, 3
Alkaline papermaking, 136
American Association of Textile Chemists and
    Colorists (AATCC), 69, 337
American National Standards Institute (ANSI),
    291, 337
American Paperboard Packaging Environmental
    Council, 31
American Society for Quality Control (ASQC),
    310, 337
American Society of Testing and Materials (ASTM),
    64, 291, 337
Andritz system, 248
Animal fiber, 281
Anticontamination impregnation, 225
Aragonite, 136
Aramid, 196
Assembly language, 341
AutoCAD®, 348
Autoclave, 228

Baby dryer, 245, 246
Bar diagram, 318, 319
Bark, 2

Barker, 10
Base fabric, 159
Base pressure uniformity, 154
Baseless felt, 161
Basic language, 343
Basis weight, 298
Bast fiber, 282
Batt, 160
Batt laying 166
Batt on base, 216
Batt stratification, 154
Batt-on-base felt, 161
Batt-on-mesh felt, 162
Batt-to-base ratio, 171
Beating, 4
Beating up, 72
Bel Champ, 252
Bel Run, 251
Beloit ENP, 190
Bending stiffness, 296
Binder yarn, 54
BiNip™ press, 189
Bitnet, 343
BiVent™ press, 189
Black liquor, 4
Bleach chemicals, 10
Bleached paperboard, 25, 27
Bleacher, 10
Bleaching, 4
Bleed-through, 107, 179
Bending, 166
Blow box, 247
Blow tank, 10
Blowing, 179, 182
Bonding, 226
Bottleneck, 89
Bottom layer, 54
Bottom roll signal, 269, 271
Bounce, 179
Bow, 104

Milton Keynes UK
Ingram Content Group UK Ltd.
UKHW052021071024
449327UK00027B/2374